# HANDBOOK OF

# *Vapor* PRESSURE

## VOLUME 4

### INORGANIC COMPOUNDS AND ELEMENTS

# LIBRARY OF PHYSICO-CHEMICAL PROPERTY DATA

## Handbook of Vapor Pressure

Volume 1: $C_1$ to $C_4$ Compounds (Product #5189)
Volume 2: $C_5$ to $C_7$ Compounds (Product #5190)
Volume 3: $C_8$ to $C_{28}$ Compounds (Product #5191)
Volume 4: Inorganic Compounds and Elements (Product #5394)
*Carl L. Yaws*

## Handbook of Viscosity

Volume 1: $C_1$ to $C_4$ Compounds (Product #5362)
Volume 2: $C_5$ to $C_7$ Compounds (Product #5364)
Volume 3: $C_8$ to $C_{28}$ Compounds (Product #5368)
*Carl L. Yaws*

## Handbook of Thermal Conductivity

Volume 1: $C_1$ to $C_4$ Compounds (Product #5382)
Volume 2: $C_5$ to $C_7$ Compounds (Product #5383)
Volume 3: $C_8$ to $C_{28}$ Compounds (Product #5384)
*Carl L. Yaws*

Each of the above series contains data for more than 1,000 organic compounds, including hydrocarbons, oxygenates, halogenates, nitrogenates, sulfur compounds, and silicon compounds. The data are presented in graphs for *vapor pressure, viscosity,* or *thermal conductivity* as a function of temperature and are arranged by carbon number and chemical formula to enable the engineer to quickly determine values at the desired temperatures.

---

## Handbook of Transport Property Data (Product #5392)

*Carl L. Yaws*

Comprehensive data on viscosity, thermal conductivity, and diffusion coefficients of gases and liquids are presented in convenient tabular format.

---

## Physical Properties of Hydrocarbons

Volume 1, Second Edition (Product #5067)
Volume 2, Third Edition (Product #5175)
Volume 3 (Product #5176)
Volume 4 (Product #5272)
*R. W. Gallant and Carl L. Yaws*

The four-volume series provides chemical, environmental, and safety engineers with quick and easy access to vital physical property data needed for production and process design calculations.

---

## Thermodynamic and Physical Property Data (Product #5031)

*Carl L. Yaws*

Property data for 700 major hydrocarbons and organic chemicals, including oxygen, nitrogen, fluorine, chlorine, bromine, iodine, and sulfur compounds, are provided.

# HANDBOOK OF

# *Vapor* PRESSURE

## VOLUME 4
### INORGANIC COMPOUNDS AND ELEMENTS

# Carl L. Yaws

Gulf Publishing Company
Houston, London, Paris, Zurich, Tokyo

# Handbook of Vapor Pressure, Volume 4

ISBN 0-88415-394-0

Gulf Publishing Company
Book Division
P.O. Box 2608, Houston, Texas 77252-2608

10  9  8  7  6  5  4  3  2  1

Printed on Acid-Free Paper (∞)

### Library of Congress Cataloging-in-Publication Data

Yaws, Carl L.
    Handbook of vapor pressure / Carl L. Yaws.
        p.   cm.
    Includes bibliographical references and index.
    Contents: v. 1. $C_1$ to $C_4$ compounds, — v. 2. $C_5$ to $C_7$ compounds, — v. 3. $C_8$ to $C_{28}$ compounds — v. 4. Inorganic Compounds and Elements.
    ISBN 0-88415-189-1 (v.  1: acid-free). — ISBN 0-88415-190-5 (v.  2: acid-free). — ISBN 0-88415-191-3 (v.  3: acid-free) — ISBN 0-88415-394-0 (v. 4: acid-free)
    1. Organic compounds—Tables.    2. Vapor pressure—Tables. I. Title.
QD257.7.Y39    1994
661'.8—dc20                                                    93-34106
                                                                   CIP

British Library Cataloguing in Publication Data. A catalogue record for this book is available from the British Library.

# CONTENTS

# CONTRIBUTORS

Li Bu — Graduate student, Chemical Engineering Department, Lamar University, Beaumont, Texas 77710, USA

Sachin Nijhawan — Graduate student, Chemical Engineering Department, Lamar University, Beaumont, Texas 77710, USA

Carl L. Yaws — Professor, Chemical Engineering Department, Lamar University, Beaumont, Texas 77710, USA

# ACKNOWLEDGMENTS

Many colleagues and students have made contributions and helpful comments over the years. The author is grateful to each: Jack R. Hopper, Joe W. Miller, Jr., C.S. Fang, K.Y. Li, Keith C. Hansen, Daniel H. Chen, P.Y. Chiang, H.C. Yang, Xiang Pan, Xiaoyan Lin, Li Bu, and Sachin Nijhawan.

The author wishes to extend special appreciation to his wife (Annette) and family (Kent, Michele, Chelsea, and Brandon; Lindsay and Rebecca; and Matthew and Sarah).

The author wishes to acknowledge that the Gulf Coast Hazardous Substance Research Center provided partial support to this work.

## DISCLAIMER

# PREFACE

Vapor Pressure data are important in many engineering applications in the chemical processing and petroleum refining industries. The objective of this book is to provide the engineer with such vapor pressure data. The data are presented in graphs covering a wide temperature range to enable the engineer to quickly determine values at the temperatures of interest.

The contents of the book are arranged in the following order:

- Graphs
- References
- List of compounds
- Coefficients for vapor pressure equation
- Critical properties and acentric factor

The graphs for vapor pressure as a function of temperature are arranged by chemical formula to provide ease of use. Many of the graphs cover the full liquid range from melting point to boiling point to critical point. English units are used, but for those involved in metric and SI usage, each graph displays a conversion factor to provide SI units.

The coverage of inorganics is comprehensive and encompasses 343 compounds, including carbon oxides such as carbon monoxide and carbon dioxide; nitrogen oxides such as nitric oxide and nitrous oxide; sulfur oxides such as sulfur dioxide and sulfur trioxide; hydrogen oxides such as water and hydrogen peroxide; ammonias such as ammonia and ammonium hydroxide; hydrogen halides such as hydrogen chloride and hydrogen fluoride; sulfur acids such as sulfuric acid and hydrogen sulfide; hydroxides such as sodium hydroxide and potassium hydroxide; silicon halides such as trichlorosilane and silicon tetrachloride; ureas such as urea and thiourea; cyanides such as hydrogen cyanide and cyanogen chloride; hydrides such as silane and diborane; sodium derivatives such as sodium chloride and sodium fluoride; aluminum derivatives such as aluminum borohydride and aluminum fluoride; and many other compound types. A total of 82 elements are covered, including hydrogen, nitrogen, oxygen, helium, argon, neon, chlorine, bromine, iodine, fluorine, sulfur, phosphorous, aluminum, lead, tin, mercury, sodium, magnesium, silicon, antimony, boron, iron, chromium, cobalt, titanium, tantalum, silver, gold, platinum, radon, uranium, and many others.

The literature has been carefully searched. The following primary references were used extensively in construction of the graphs:

- Daubert, T. E. and R. P. Danner, *Data Compilation of Properties of Pure Compounds,* Parts 1, 2, 3, and 4, Supplements 1 and 2, DIPPR Project, AIChE, New York, NY (1985–1992).

- Nesmeyanov, A. N., *Vapor Pressure of the Chemical Elements,* Elsevier, New York, NY (1963).

- Yaws, C. L., *Physical Properties,* McGraw-Hill, New York, NY (1977).

- Yaws, C. L., *Thermodynamic and Physical Property Data,* Gulf Publishing Co., Houston, TX (1992).

- Yaws, C. L. and R. W. Gallant, *Physical Properties of Hydrocarbons,* Vols. 1 (2nd ed.), 2 (3rd ed.), and 3 (1st ed.), Gulf Publishing Co., Houston, TX (1992, 1993, 1993).

Additional references are given in the section near the end of this book. These primary and additional references provide full documentation for the original sources used in regression of the data.

A list of compounds is given near the end of the book to aid the user in quickly locating the compound of interest from the chemical formula or name.

Coefficients for the vapor pressure equation are provided near the end of the book. Critical properties and acentric factor are given in the last section of the book. The tabulated values are especially arranged for quick usage with hand calculator or computer. Computer programs, containing data for all compounds, are available for a nominal fee. The programs are in ASCII which can be accessed by other software. Contact Carl L. Yaws, P.O. Box 10053, Lamar University, Beaumont, TX 77710, phone/FAX 409-880-8787.

# HANDBOOK OF

# *Vapor*
# PRESSURE

## VOLUME 4
### INORGANIC COMPOUNDS
### AND ELEMENTS

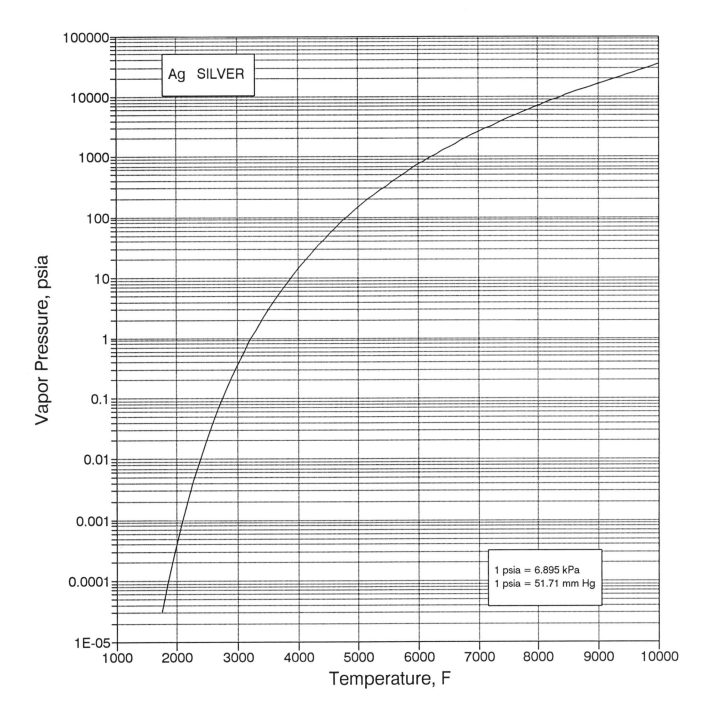

Ag SILVER

Vapor Pressure, psia vs Temperature, F

1 psia = 6.895 kPa
1 psia = 51.71 mm Hg

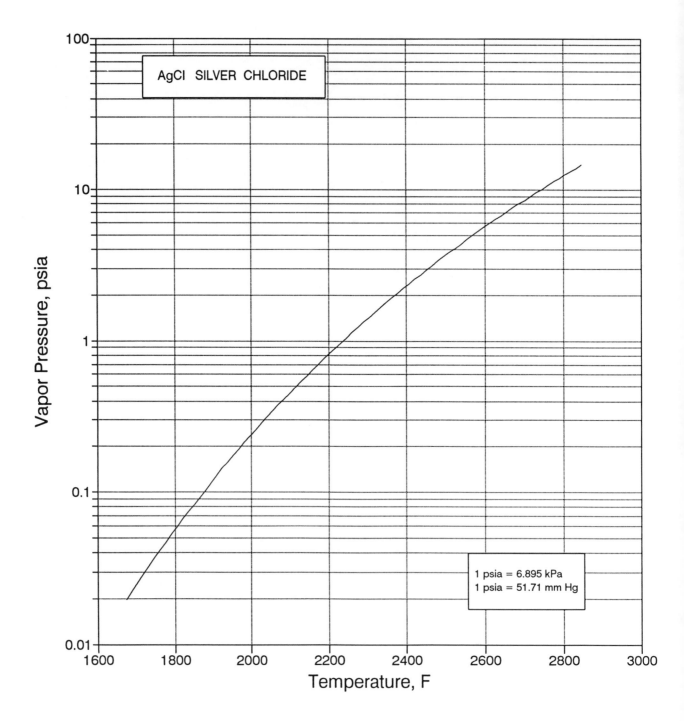

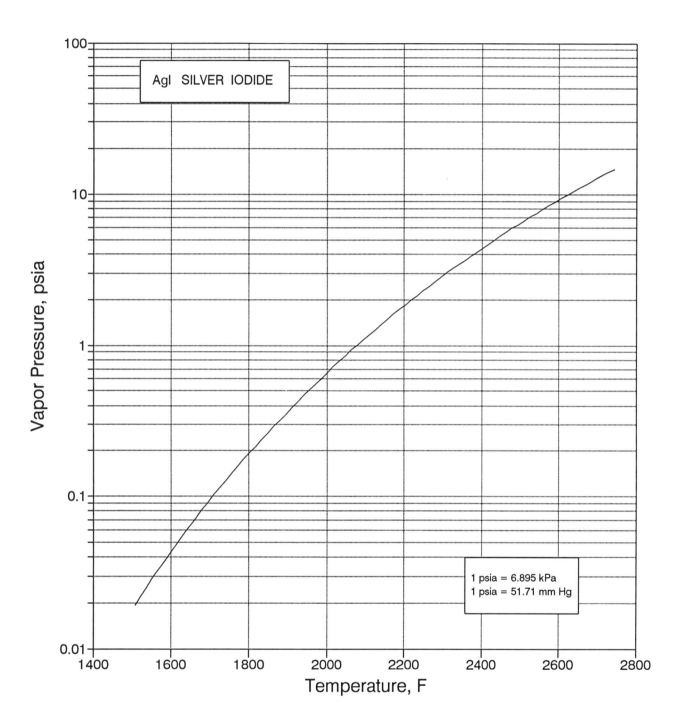

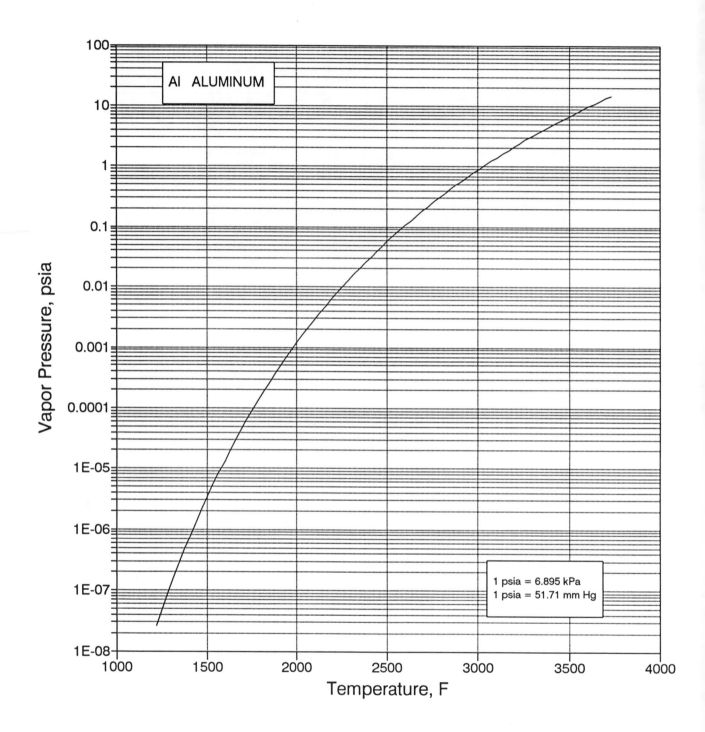

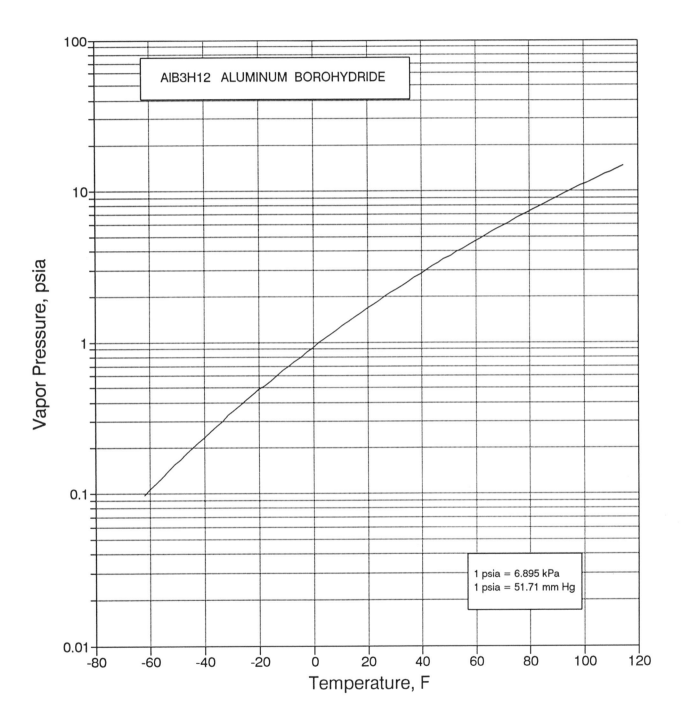

AlB3H12   ALUMINUM BOROHYDRIDE

Vapor Pressure, psia

Temperature, F

1 psia = 6.895 kPa
1 psia = 51.71 mm Hg

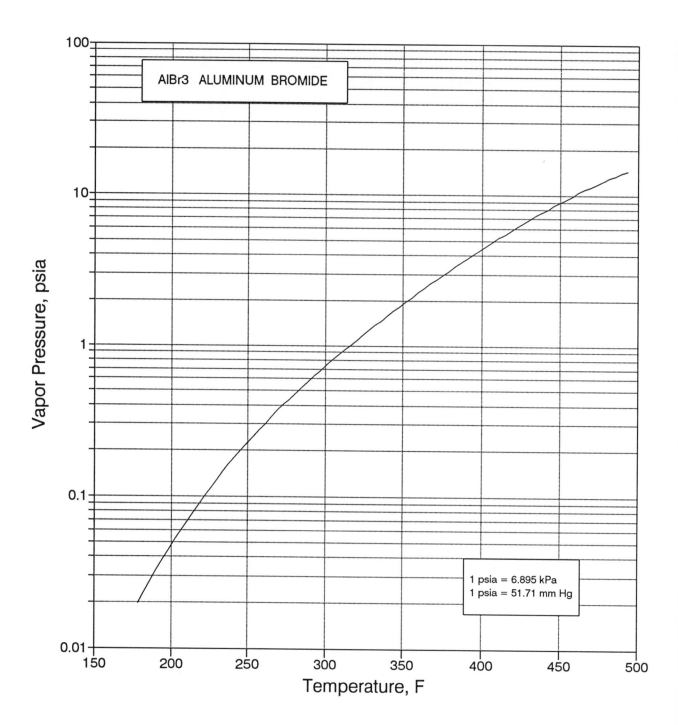

AlBr3  ALUMINUM BROMIDE

Vapor Pressure, psia

Temperature, F

1 psia = 6.895 kPa
1 psia = 51.71 mm Hg

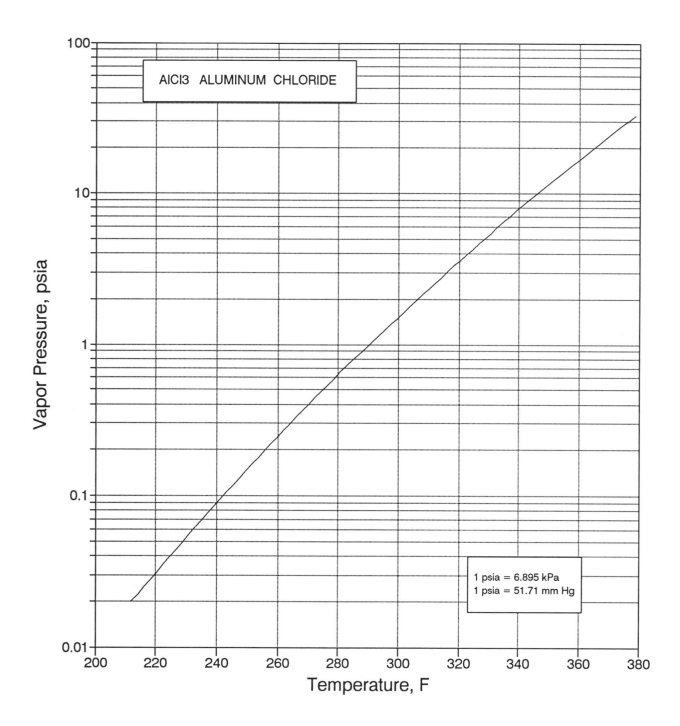

AlCl3  ALUMINUM CHLORIDE

Vapor Pressure, psia

Temperature, F

1 psia = 6.895 kPa
1 psia = 51.71 mm Hg

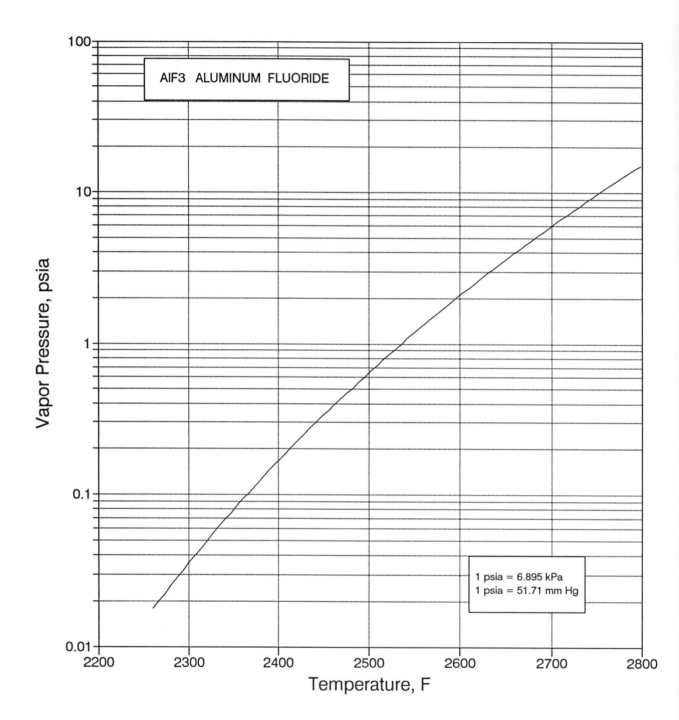

8

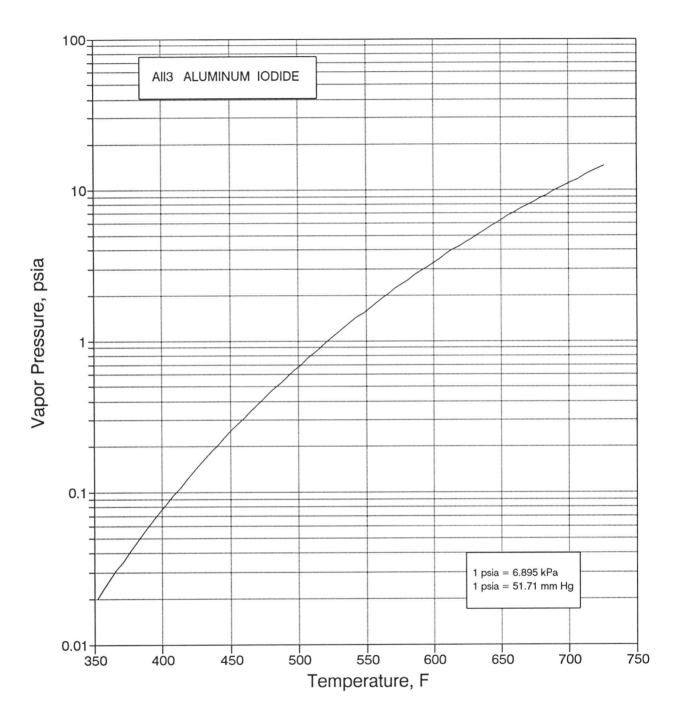

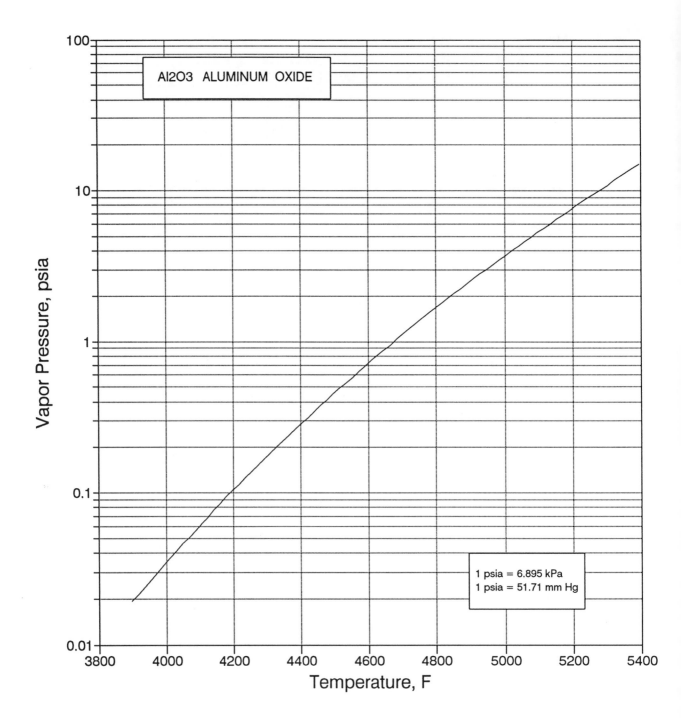

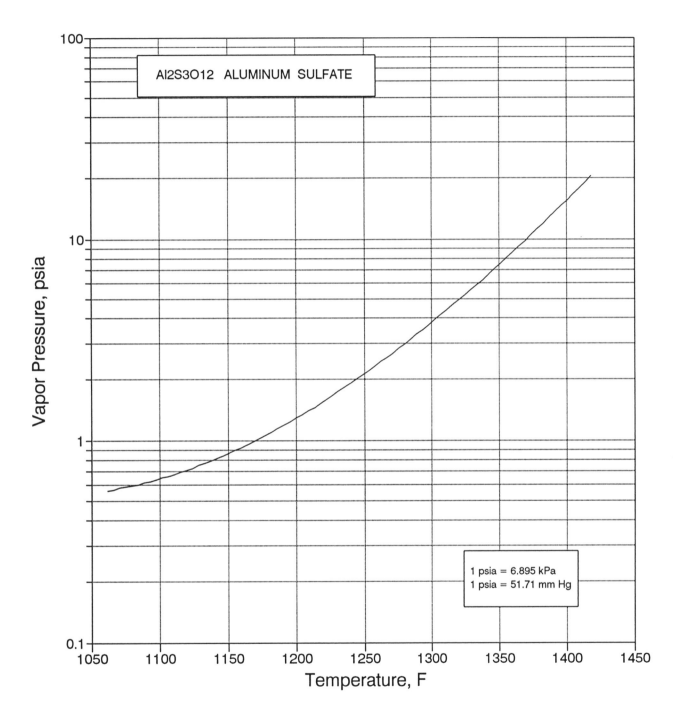

Al2S3O12 ALUMINUM SULFATE

1 psia = 6.895 kPa
1 psia = 51.71 mm Hg

Vapor Pressure, psia

Temperature, F

11

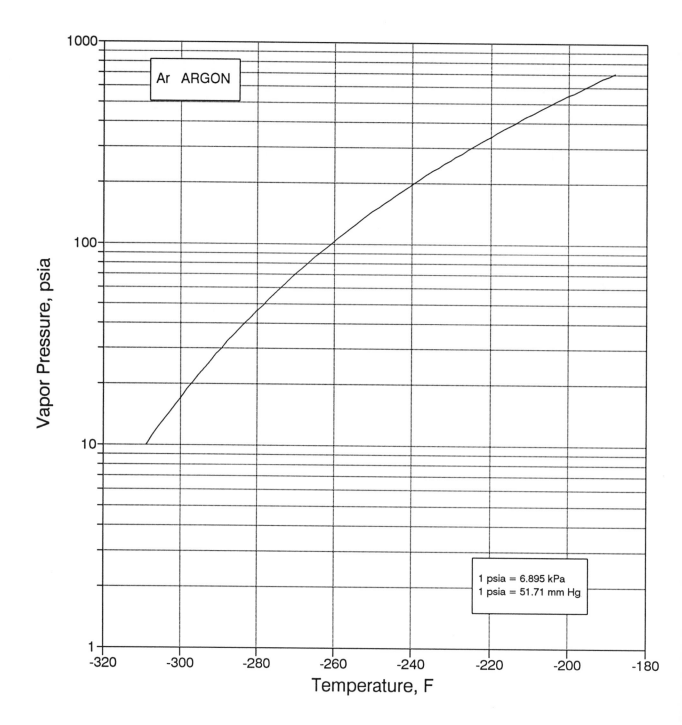

1000

Ar ARGON

Vapor Pressure, psia

100

10

1 psia = 6.895 kPa
1 psia = 51.71 mm Hg

1
-320    -300    -280    -260    -240    -220    -200    -180
Temperature, F

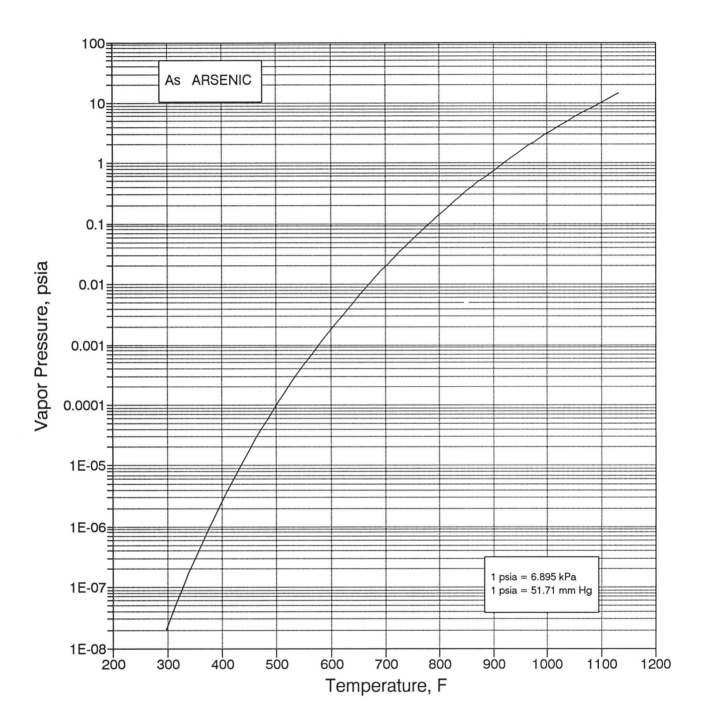

As  ARSENIC

Vapor Pressure, psia

Temperature, F

1 psia = 6.895 kPa
1 psia = 51.71 mm Hg

13

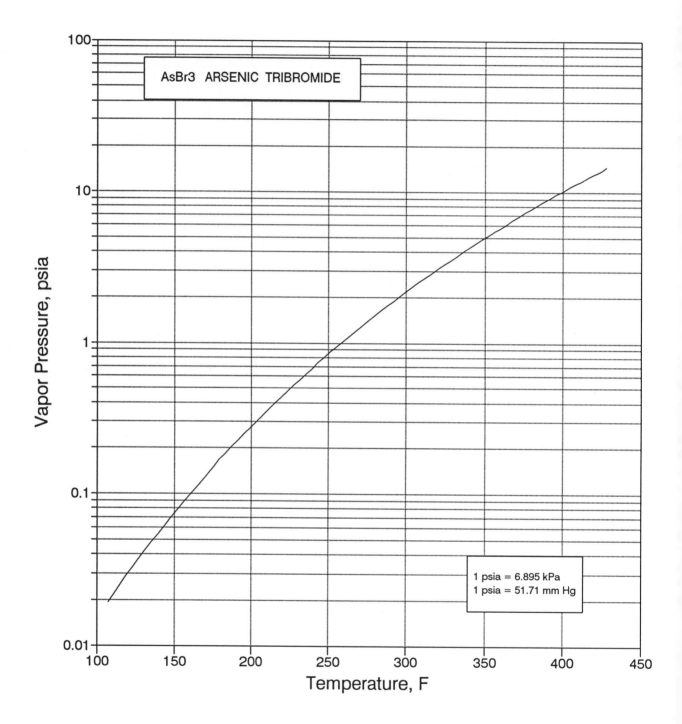

AsBr3  ARSENIC TRIBROMIDE

1 psia = 6.895 kPa
1 psia = 51.71 mm Hg

Vapor Pressure, psia

Temperature, F

14

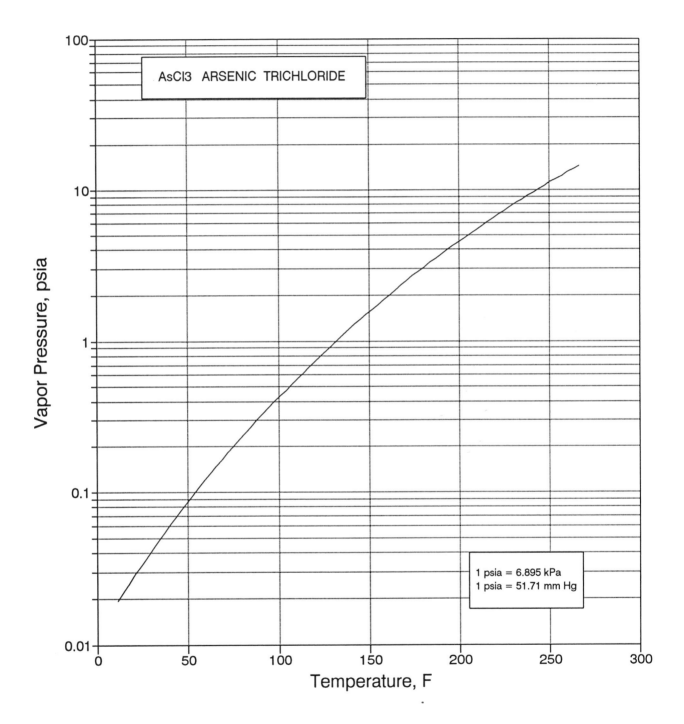

AsCl3  ARSENIC TRICHLORIDE

1 psia = 6.895 kPa
1 psia = 51.71 mm Hg

Vapor Pressure, psia

Temperature, F

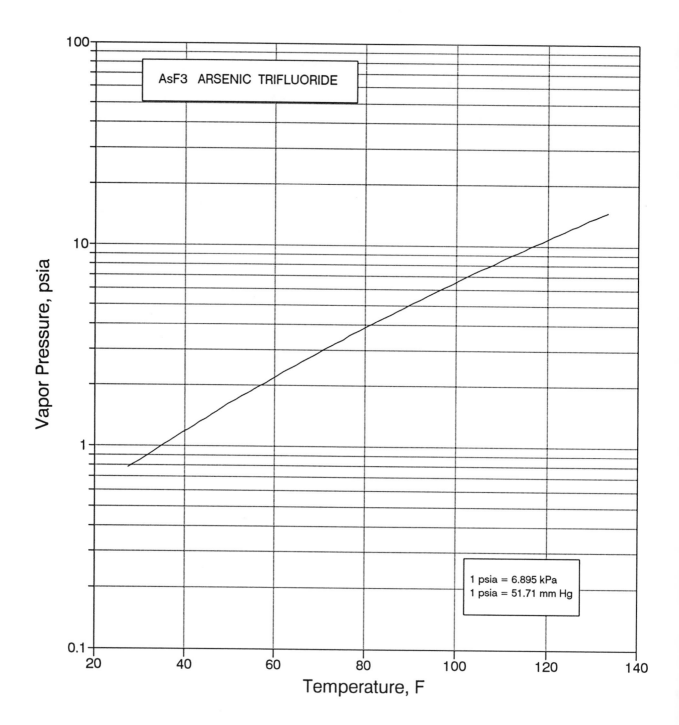

AsF3   ARSENIC   TRIFLUORIDE

Vapor Pressure, psia

1 psia = 6.895 kPa
1 psia = 51.71 mm Hg

Temperature, F

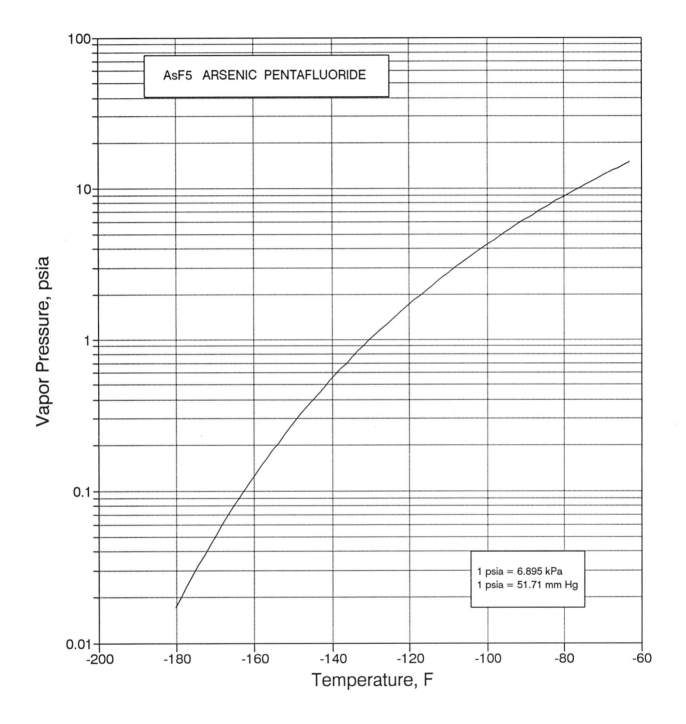

AsF5  ARSENIC PENTAFLUORIDE

Vapor Pressure, psia

Temperature, F

1 psia = 6.895 kPa
1 psia = 51.71 mm Hg

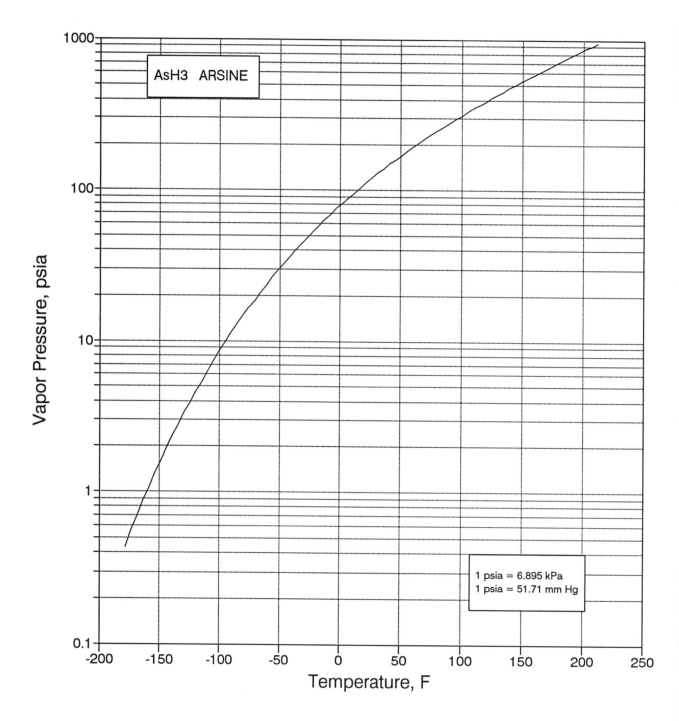

1 psia = 6.895 kPa
1 psia = 51.71 mm Hg

AsH3  ARSINE

Vapor Pressure, psia

Temperature, F

18

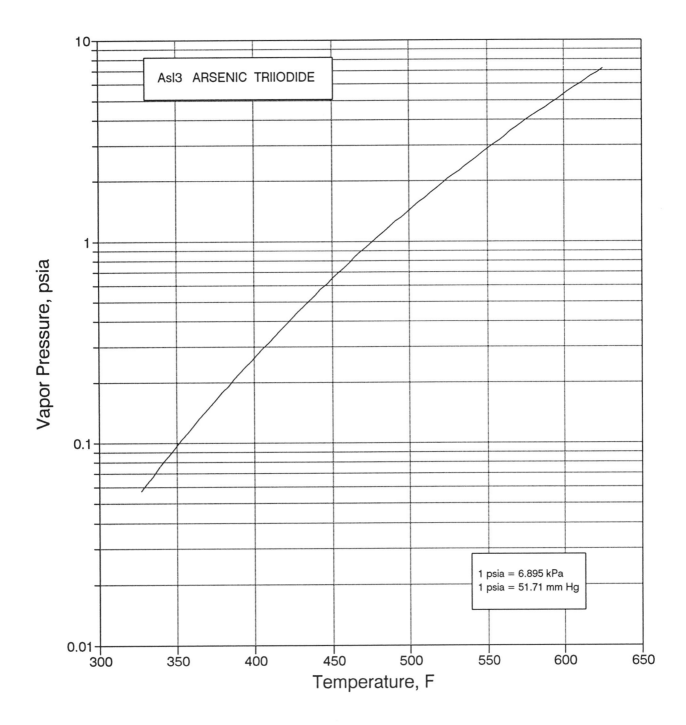

AsI3  ARSENIC TRIIODIDE

Vapor Pressure, psia

Temperature, F

1 psia = 6.895 kPa
1 psia = 51.71 mm Hg

19

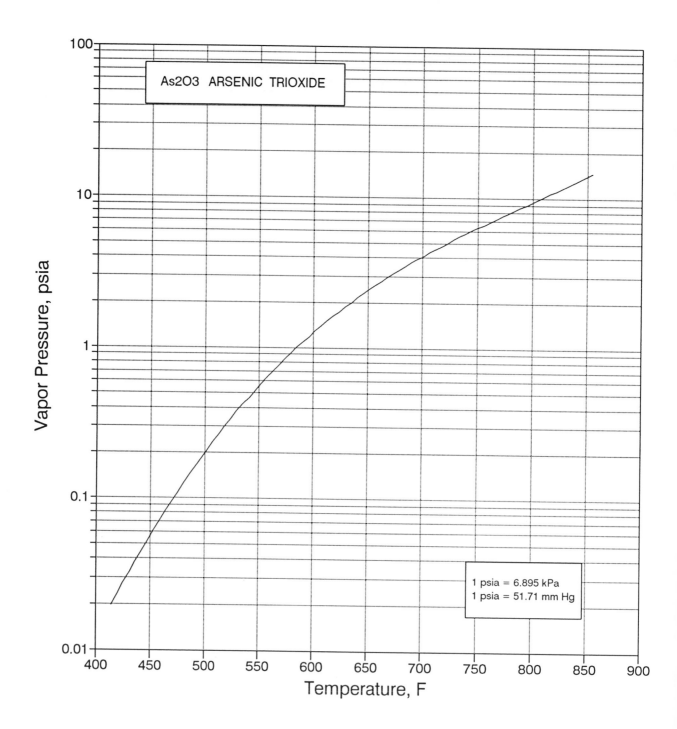

As2O3  ARSENIC TRIOXIDE

Vapor Pressure, psia

Temperature, F

1 psia = 6.895 kPa
1 psia = 51.71 mm Hg

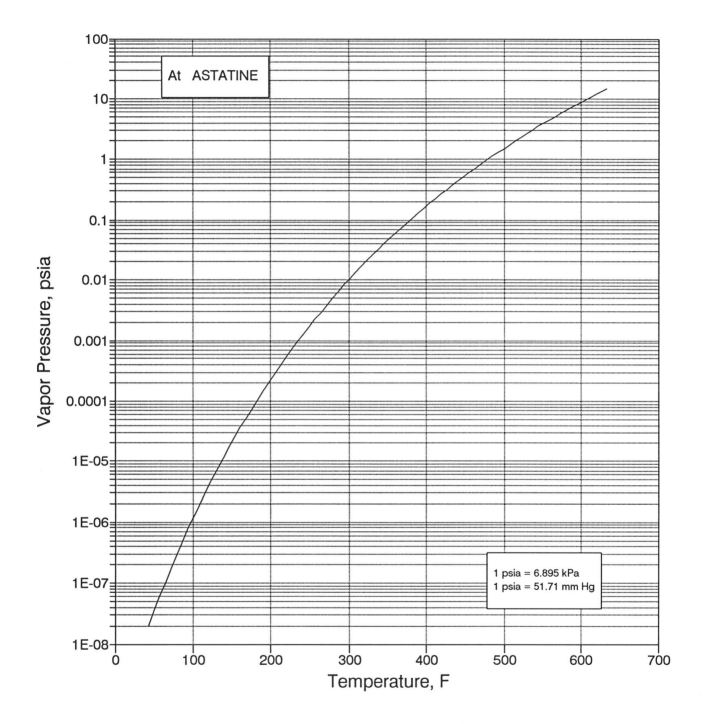

At ASTATINE

1 psia = 6.895 kPa
1 psia = 51.71 mm Hg

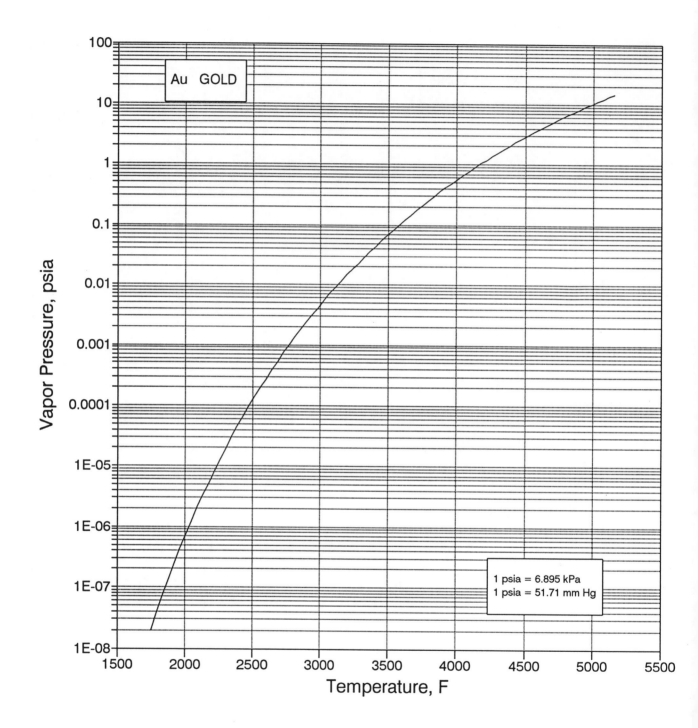

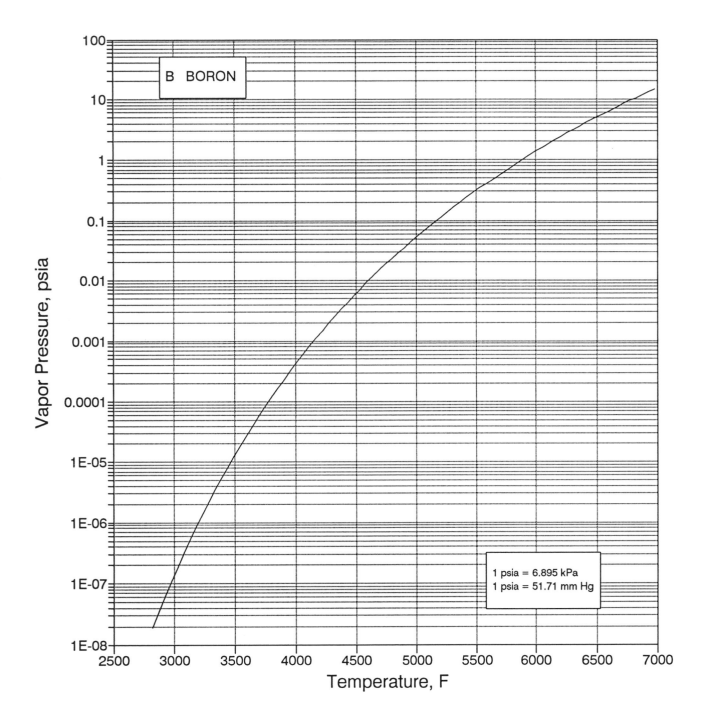

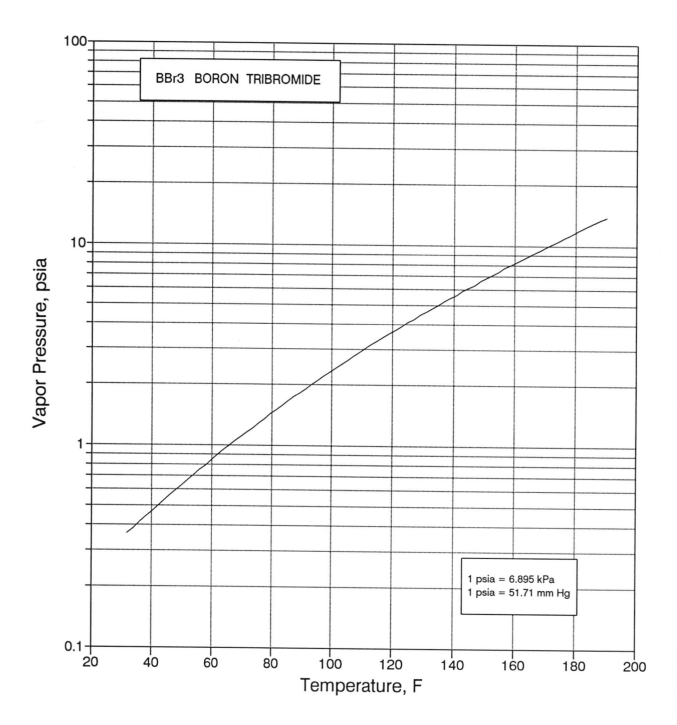

BBr3  BORON TRIBROMIDE

1 psia = 6.895 kPa
1 psia = 51.71 mm Hg

Vapor Pressure, psia

Temperature, F

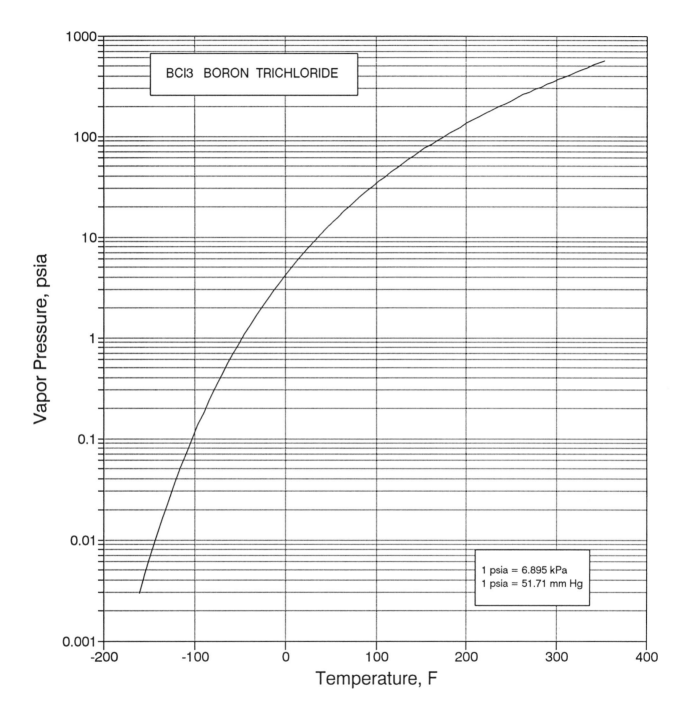

BCl3  BORON TRICHLORIDE

Vapor Pressure, psia

Temperature, F

1 psia = 6.895 kPa
1 psia = 51.71 mm Hg

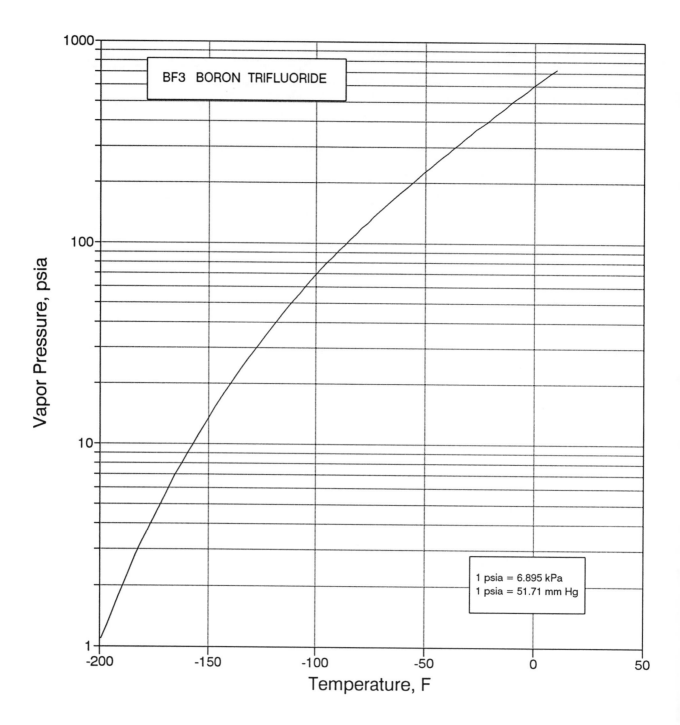

BF3 BORON TRIFLUORIDE

Vapor Pressure, psia

Temperature, F

1 psia = 6.895 kPa
1 psia = 51.71 mm Hg

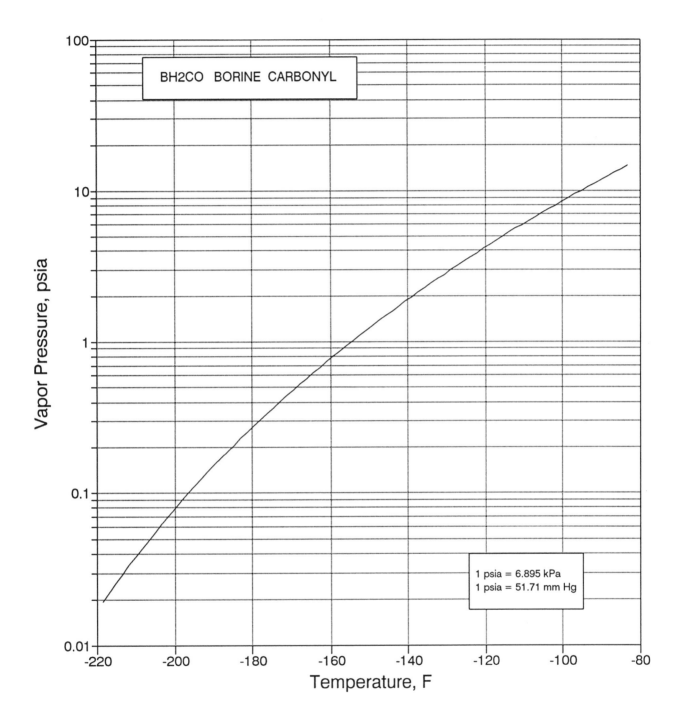

BH2CO   BORINE CARBONYL

Vapor Pressure, psia

Temperature, F

1 psia = 6.895 kPa
1 psia = 51.71 mm Hg

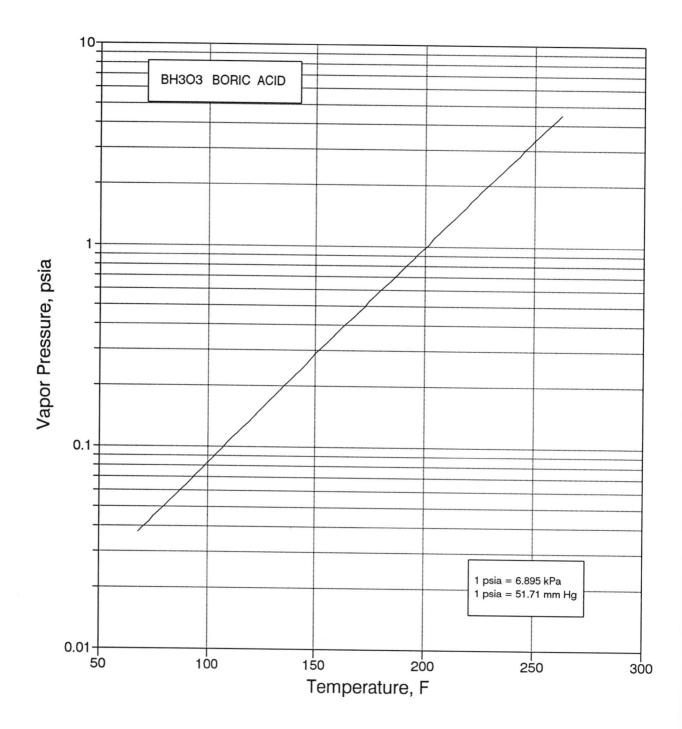

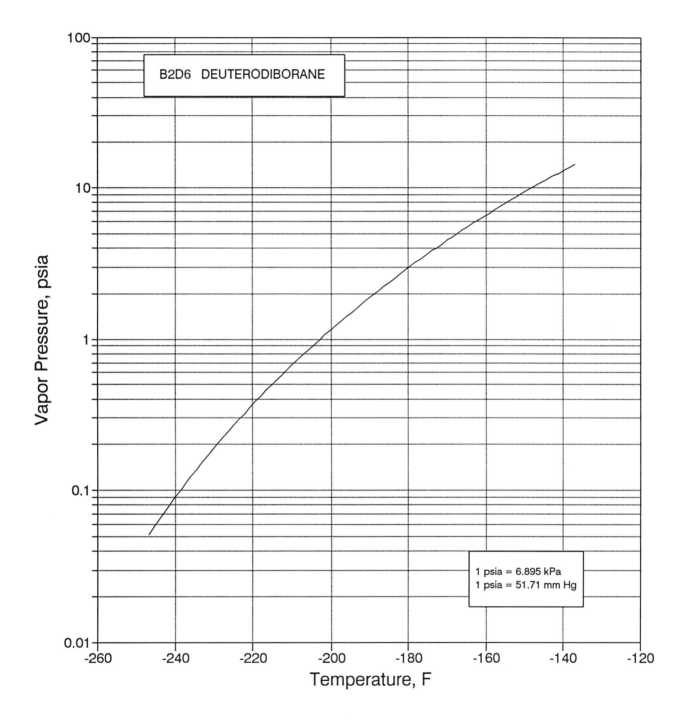

Vapor Pressure, psia

Temperature, F

B2D6   DEUTERODIBORANE

1 psia = 6.895 kPa
1 psia = 51.71 mm Hg

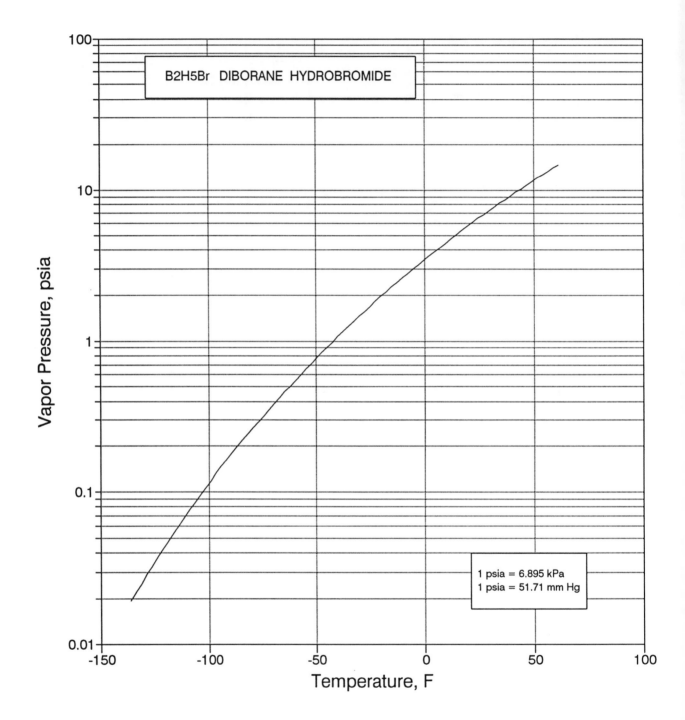

B2H5Br  DIBORANE HYDROBROMIDE

1 psia = 6.895 kPa
1 psia = 51.71 mm Hg

Vapor Pressure, psia

Temperature, F

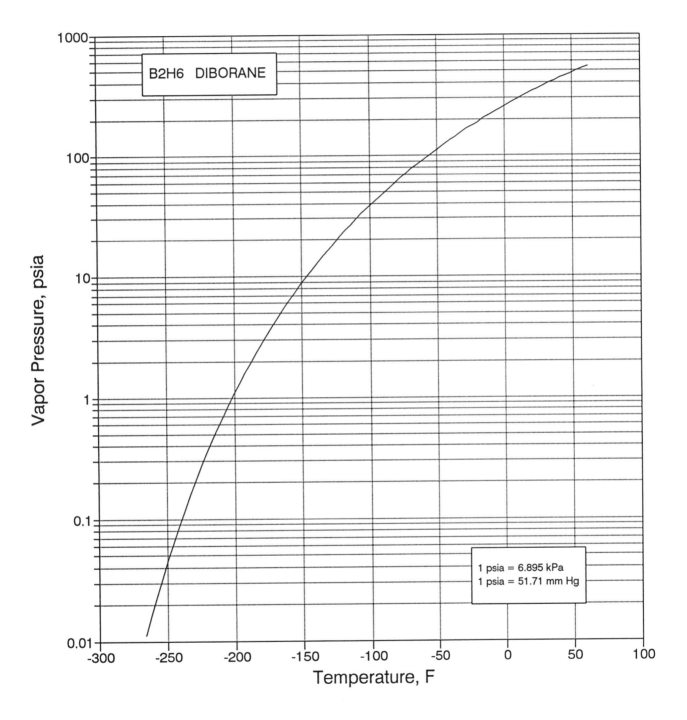

B2H6  DIBORANE

1 psia = 6.895 kPa
1 psia = 51.71 mm Hg

Vapor Pressure, psia

Temperature, F

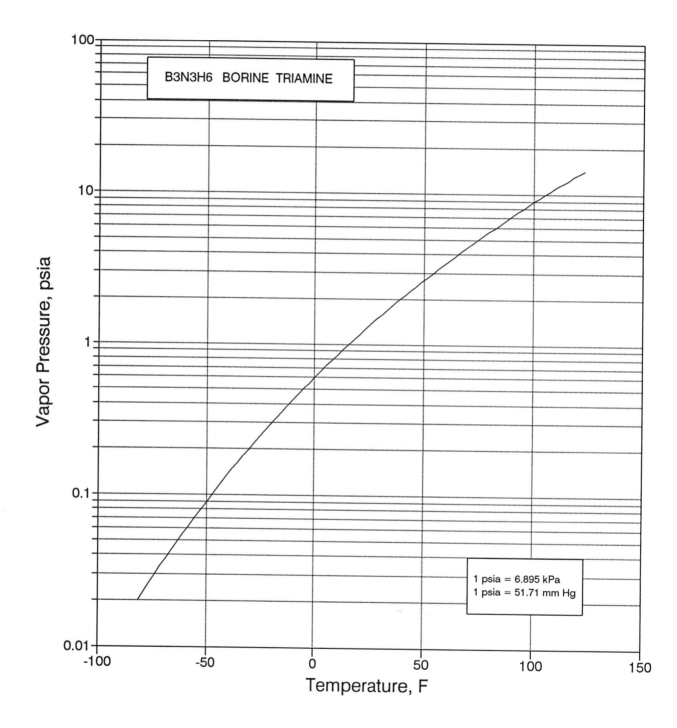

B3N3H6  BORINE  TRIAMINE

Vapor Pressure, psia

Temperature, F

1 psia = 6.895 kPa
1 psia = 51.71 mm Hg

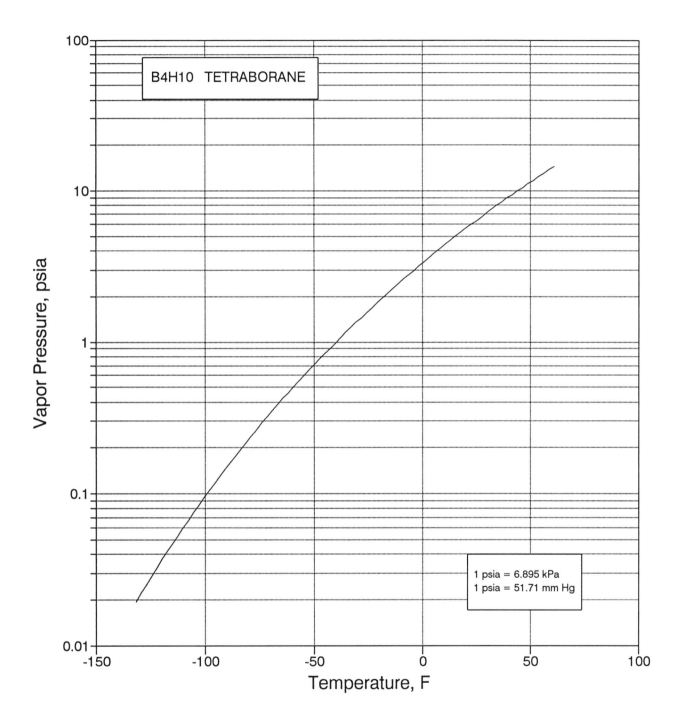

B4H10  TETRABORANE

Vapor Pressure, psia

Temperature, F

1 psia = 6.895 kPa
1 psia = 51.71 mm Hg

33

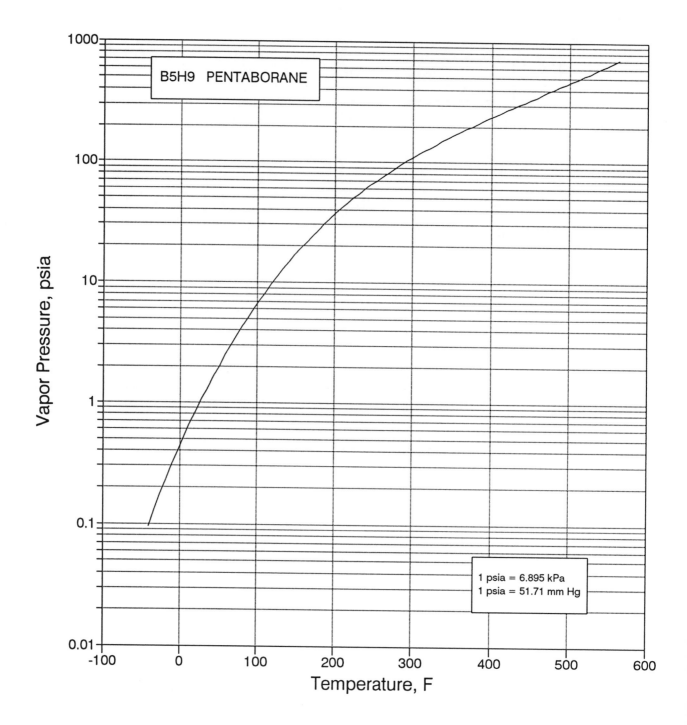

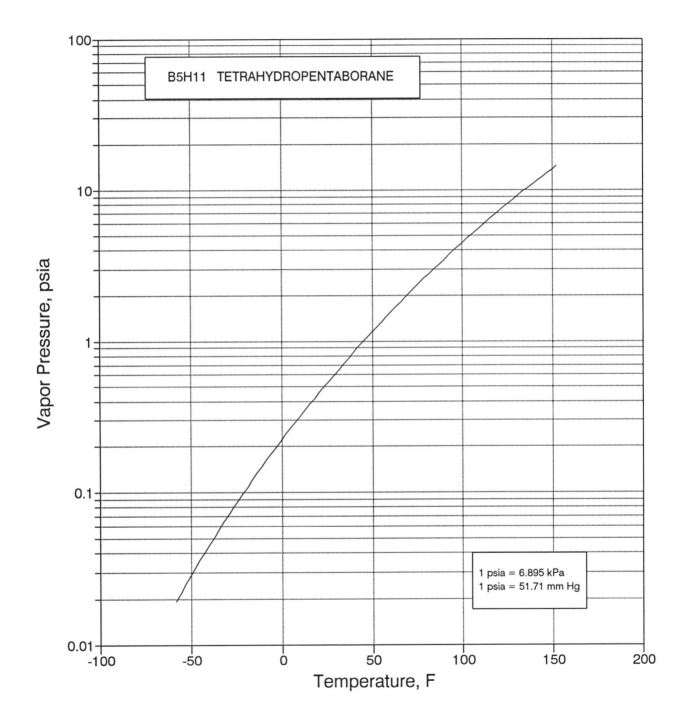

B5H11   TETRAHYDROPENTABORANE

1 psia = 6.895 kPa
1 psia = 51.71 mm Hg

Vapor Pressure, psia

Temperature, F

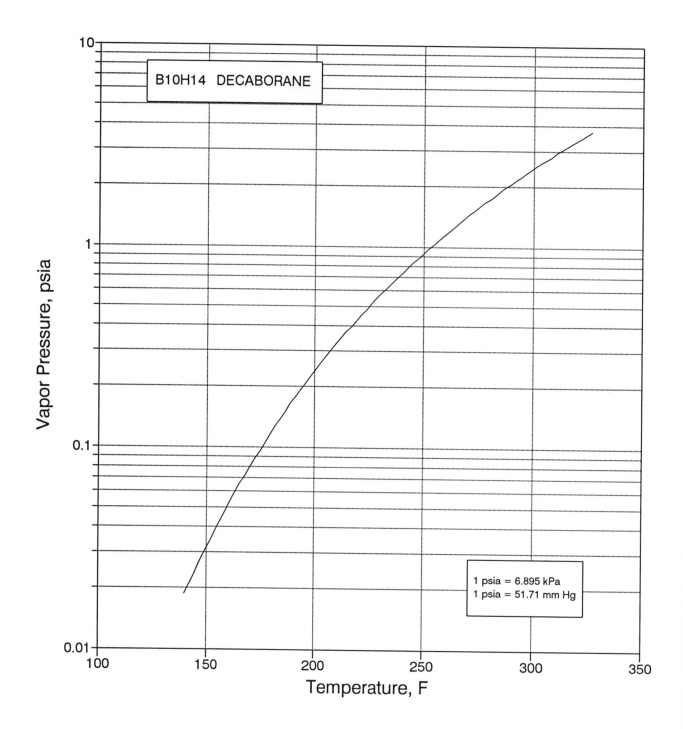

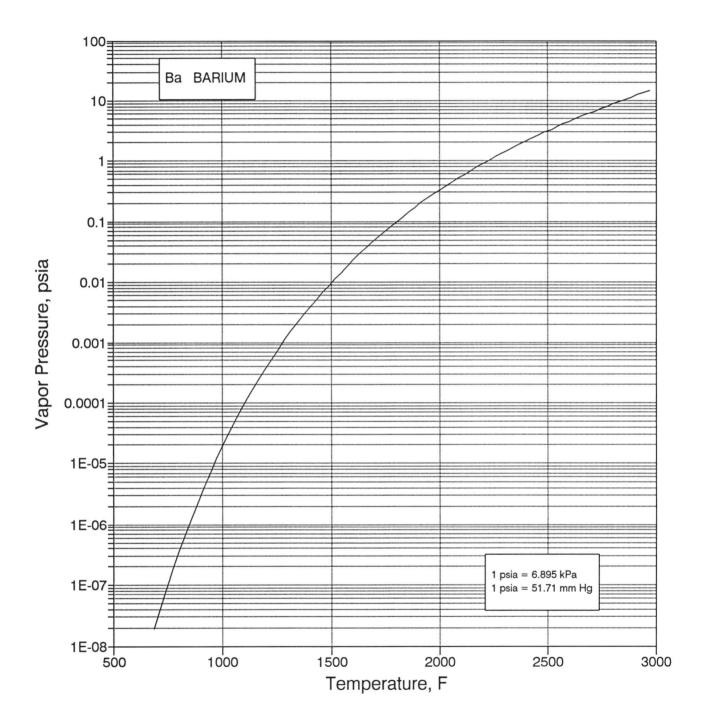

Ba BARIUM

Vapor Pressure, psia

Temperature, F

1 psia = 6.895 kPa
1 psia = 51.71 mm Hg

37

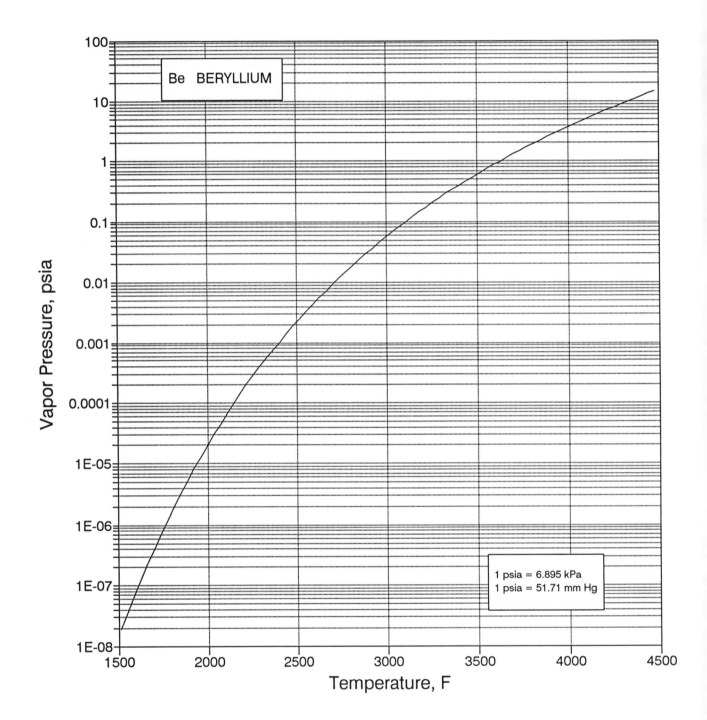

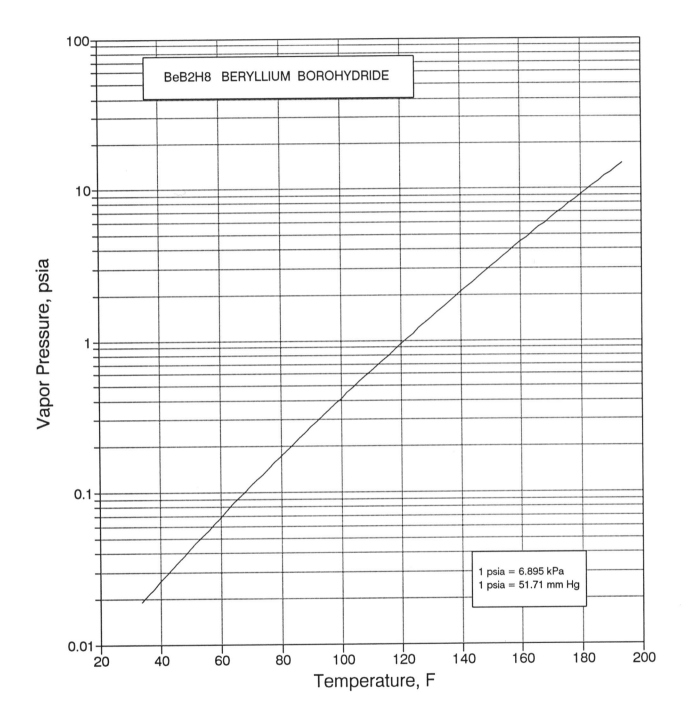

BeB2H8  BERYLLIUM  BOROHYDRIDE

Vapor Pressure, psia

Temperature, F

1 psia = 6.895 kPa
1 psia = 51.71 mm Hg

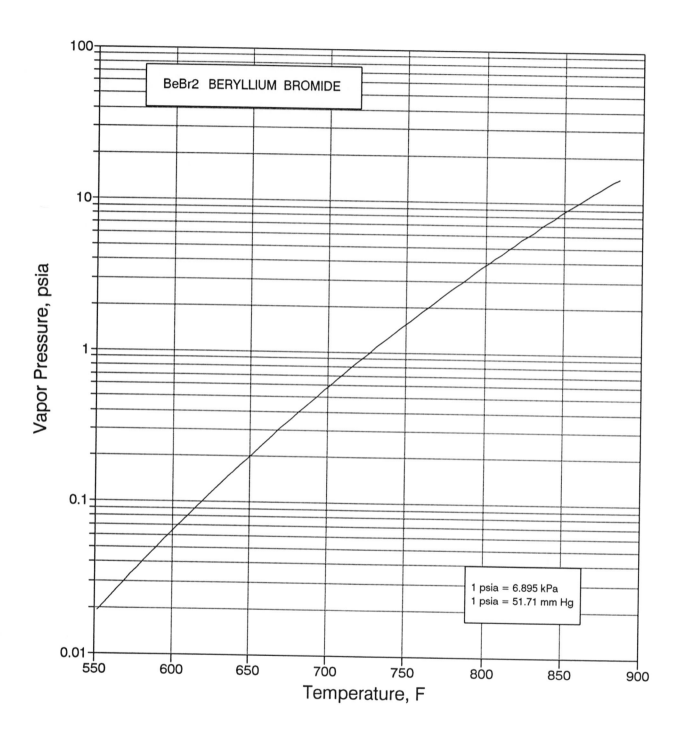

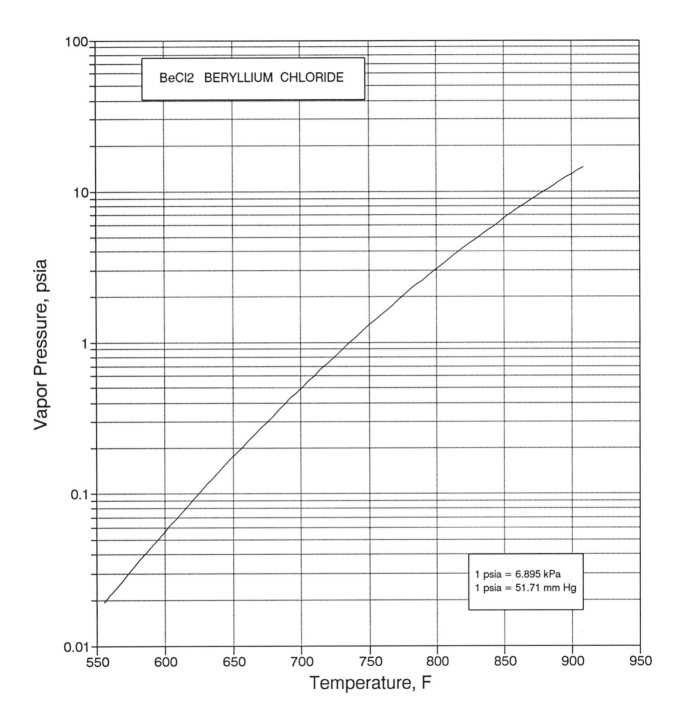

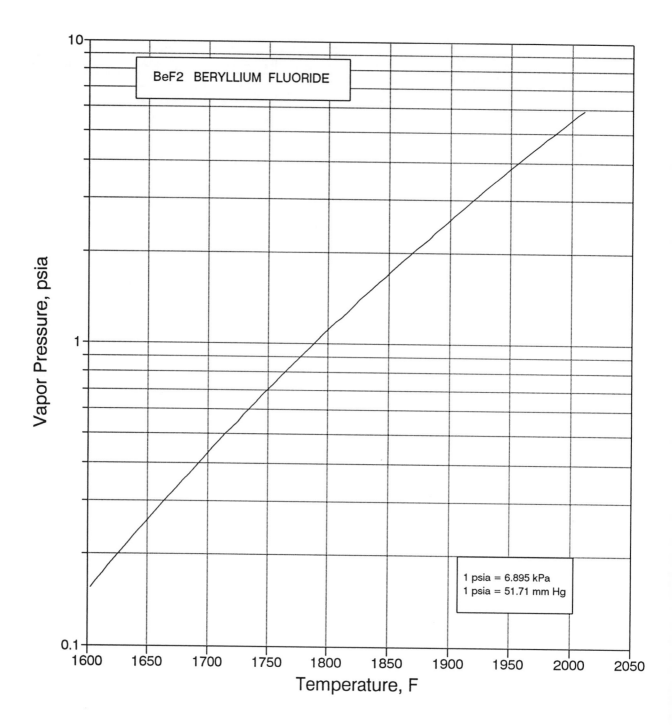

Vapor Pressure, psia

BeF2  BERYLLIUM FLUORIDE

1 psia = 6.895 kPa
1 psia = 51.71 mm Hg

Temperature, F

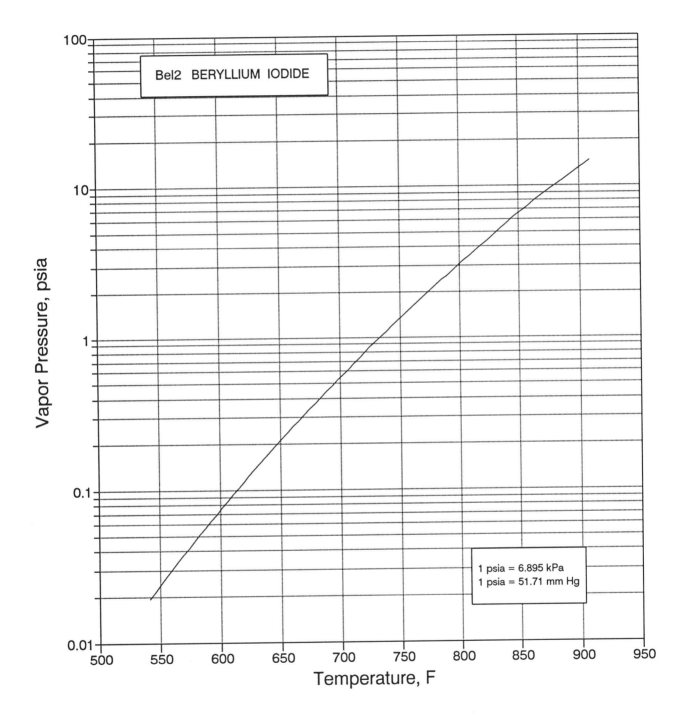

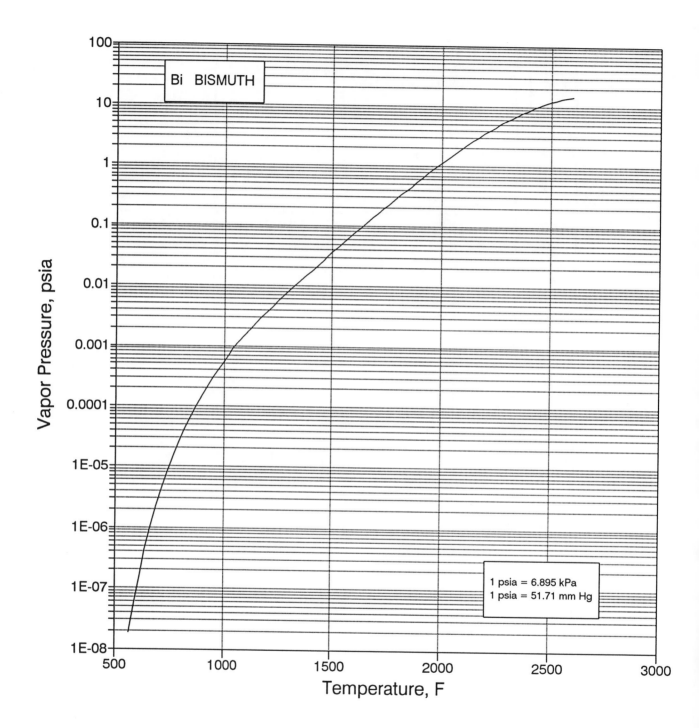

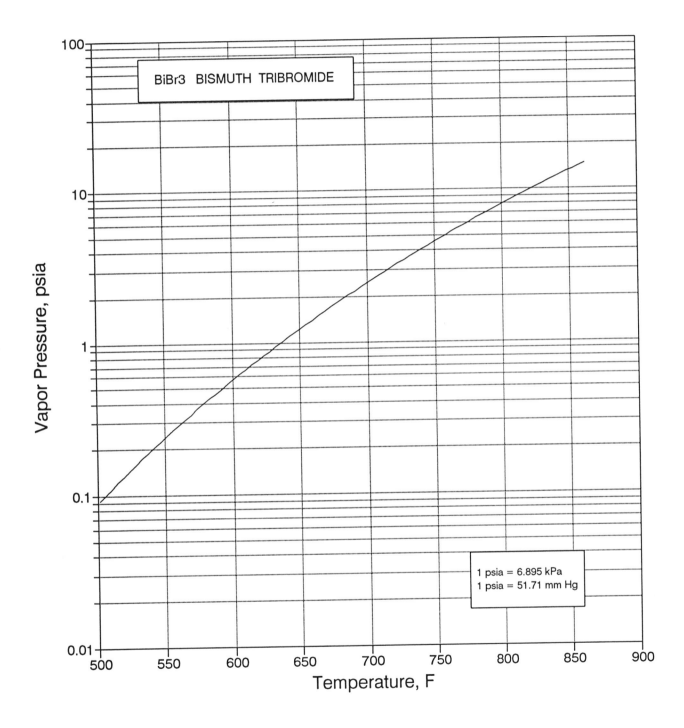

BiBr3  BISMUTH  TRIBROMIDE

1 psia = 6.895 kPa
1 psia = 51.71 mm Hg

Vapor Pressure, psia

Temperature, F

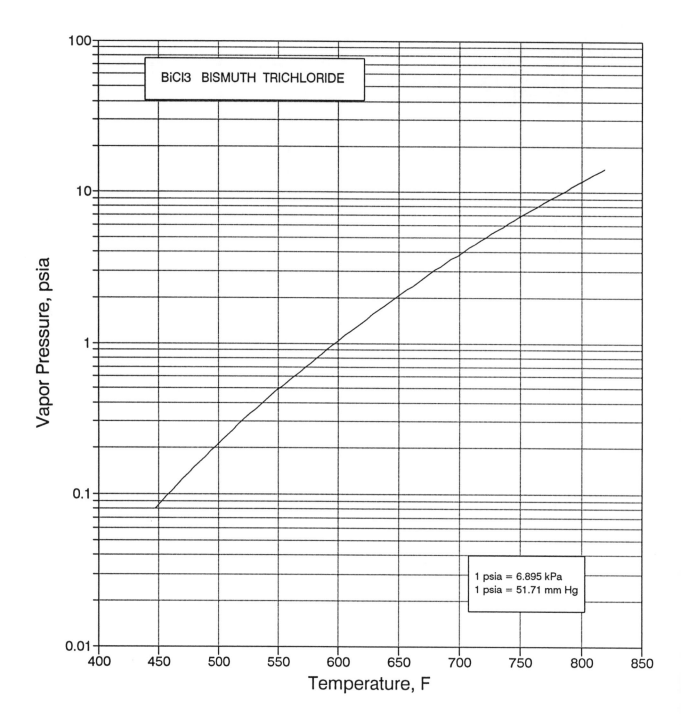

BiCl3  BISMUTH TRICHLORIDE

Vapor Pressure, psia

Temperature, F

1 psia = 6.895 kPa
1 psia = 51.71 mm Hg

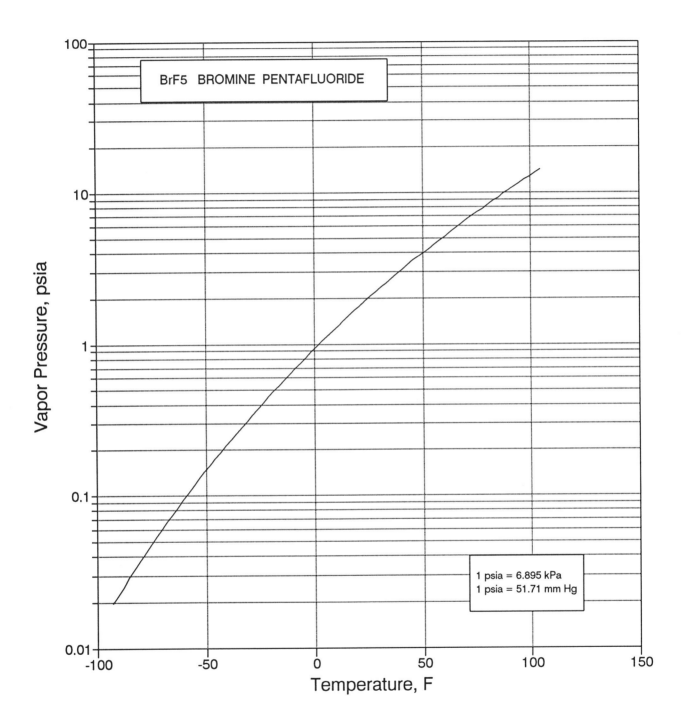

BrF5 BROMINE PENTAFLUORIDE

Vapor Pressure, psia

Temperature, F

1 psia = 6.895 kPa
1 psia = 51.71 mm Hg

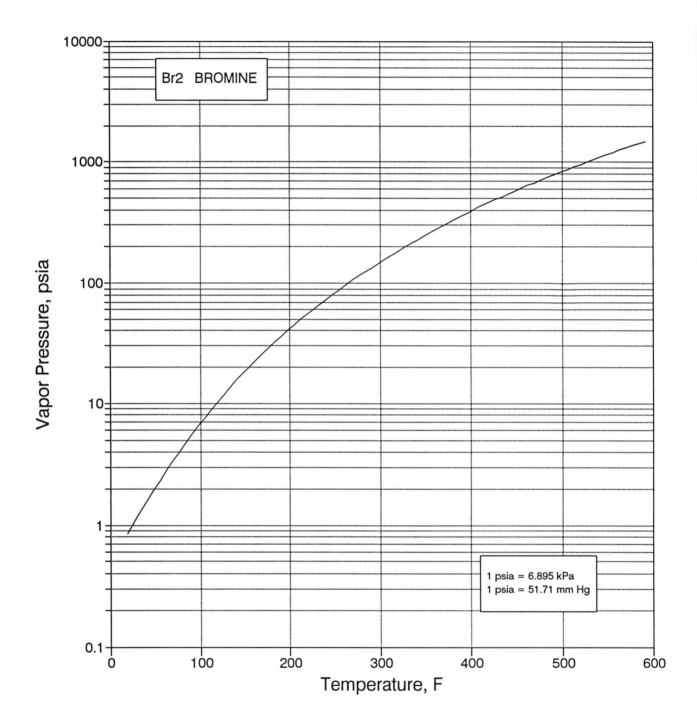

Br2  BROMINE

Vapor Pressure, psia

Temperature, F

1 psia = 6.895 kPa
1 psia = 51.71 mm Hg

48

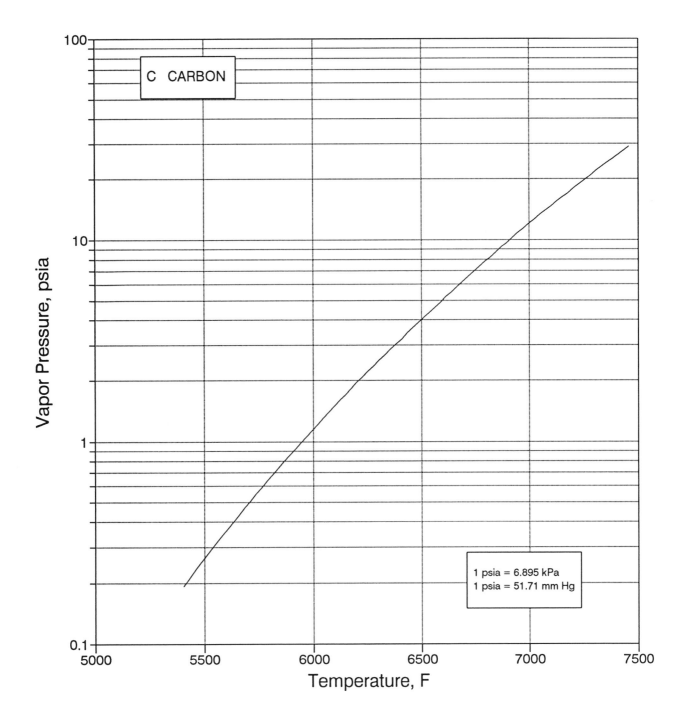

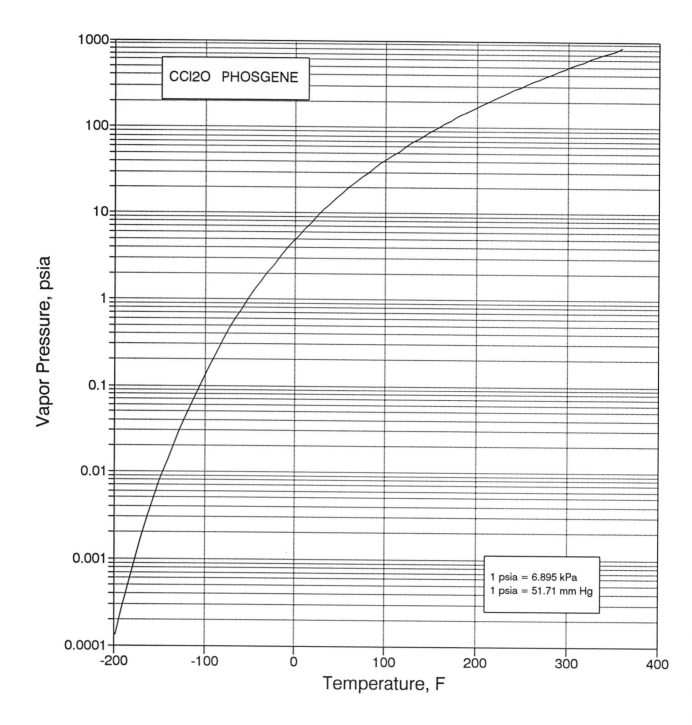

CCl2O  PHOSGENE

Vapor Pressure, psia

Temperature, F

1 psia = 6.895 kPa
1 psia = 51.71 mm Hg

50

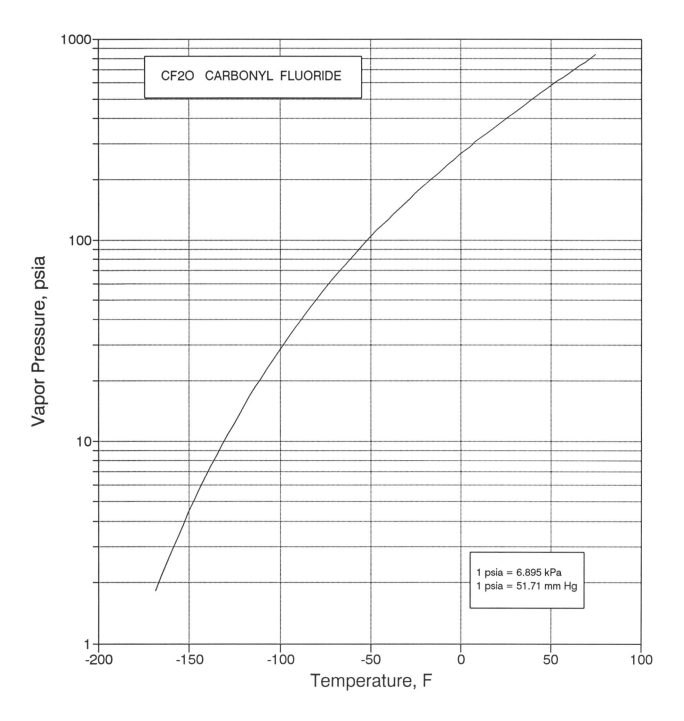

CF2O  CARBONYL FLUORIDE

Vapor Pressure, psia

Temperature, F

1 psia = 6.895 kPa
1 psia = 51.71 mm Hg

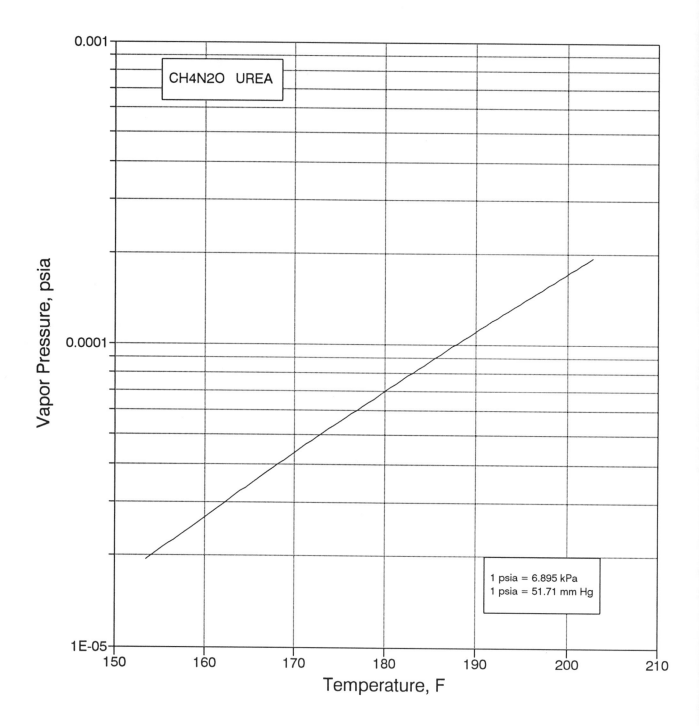

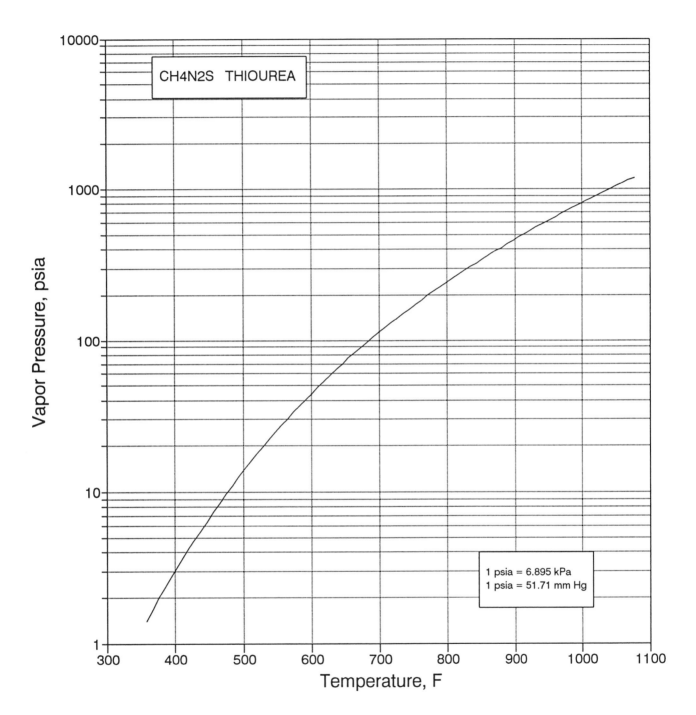

1 psia = 6.895 kPa
1 psia = 51.71 mm Hg

CH4N2S  THIOUREA

Vapor Pressure, psia

Temperature, F

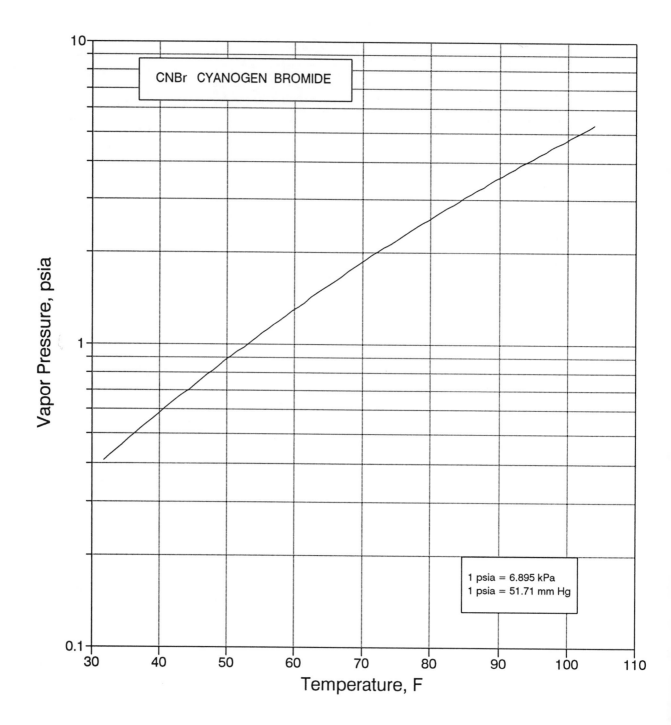

CNBr CYANOGEN BROMIDE

Vapor Pressure, psia

Temperature, F

1 psia = 6.895 kPa
1 psia = 51.71 mm Hg

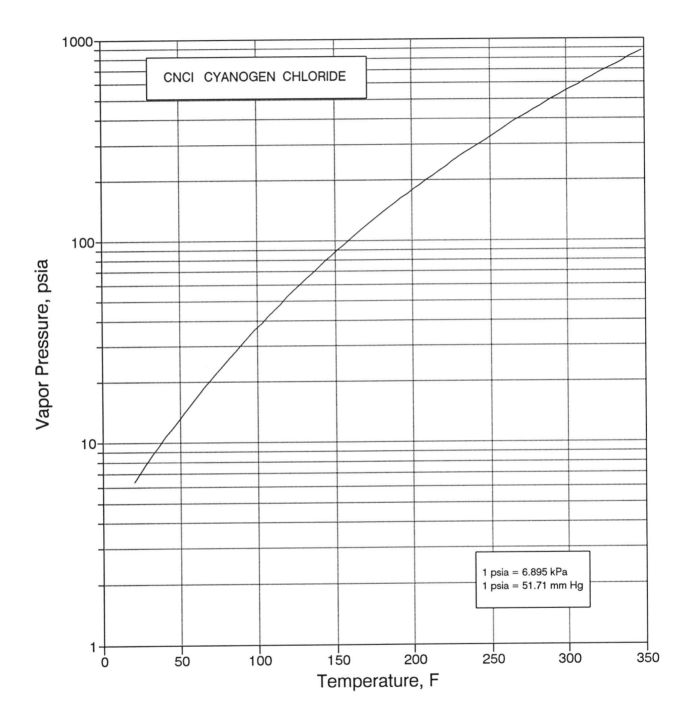

Vapor Pressure, psia

Temperature, F

CNCl  CYANOGEN CHLORIDE

1 psia = 6.895 kPa
1 psia = 51.71 mm Hg

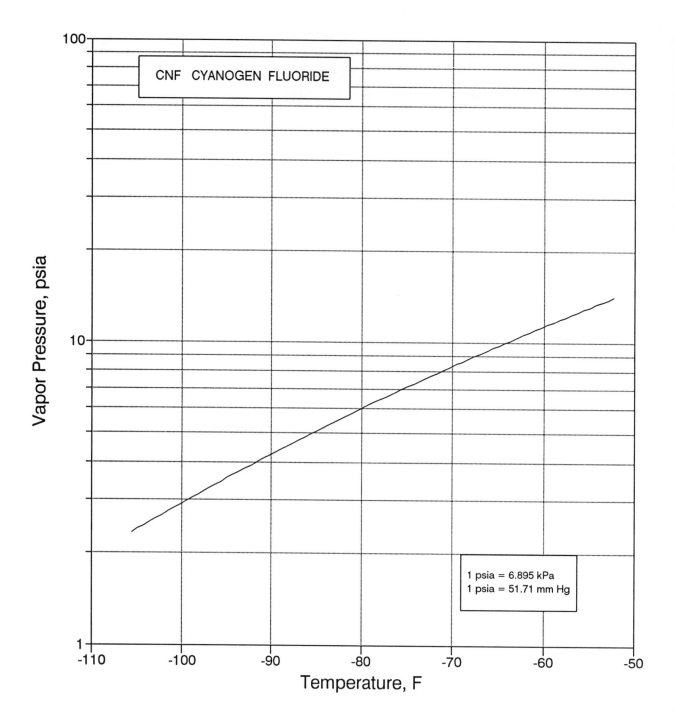

CNF  CYANOGEN FLUORIDE

Vapor Pressure, psia

Temperature, F

1 psia = 6.895 kPa
1 psia = 51.71 mm Hg

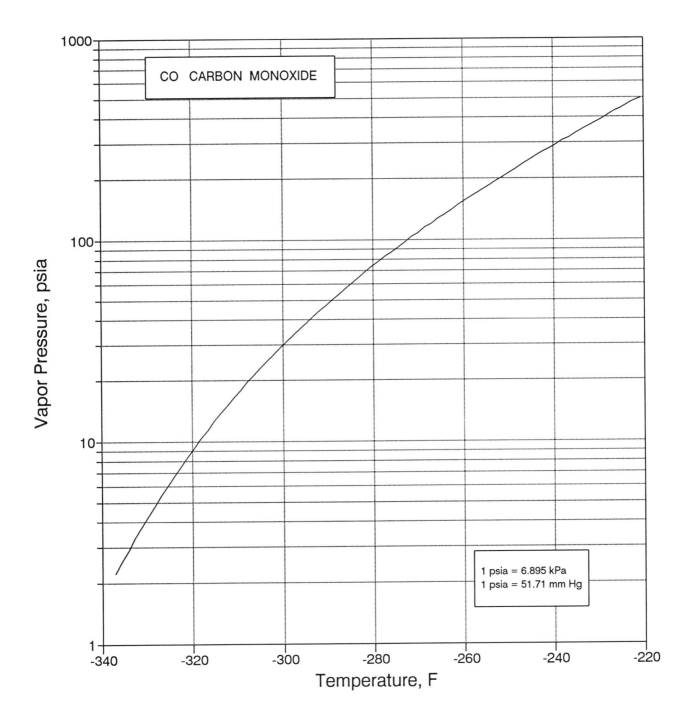

CO  CARBON MONOXIDE

Vapor Pressure, psia

Temperature, F

1 psia = 6.895 kPa
1 psia = 51.71 mm Hg

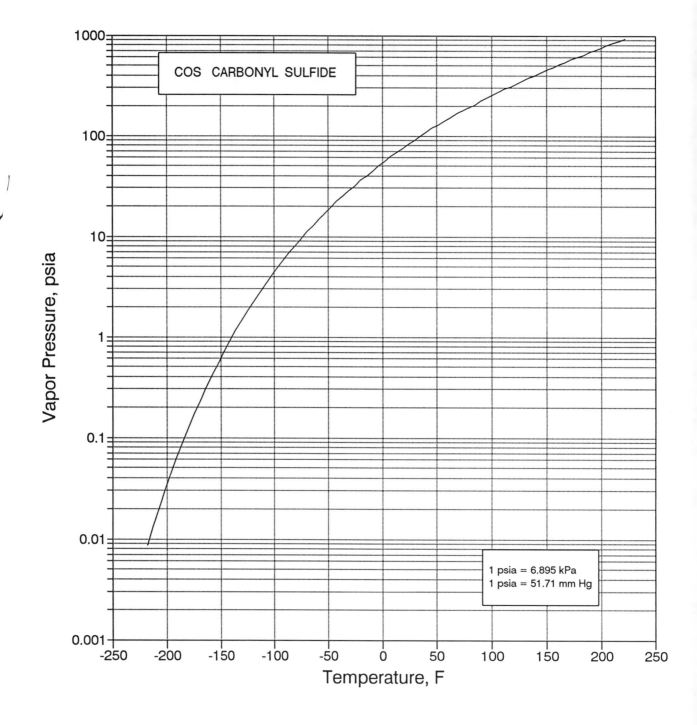

COS   CARBONYL SULFIDE

1 psia = 6.895 kPa
1 psia = 51.71 mm Hg

Vapor Pressure, psia

Temperature, F

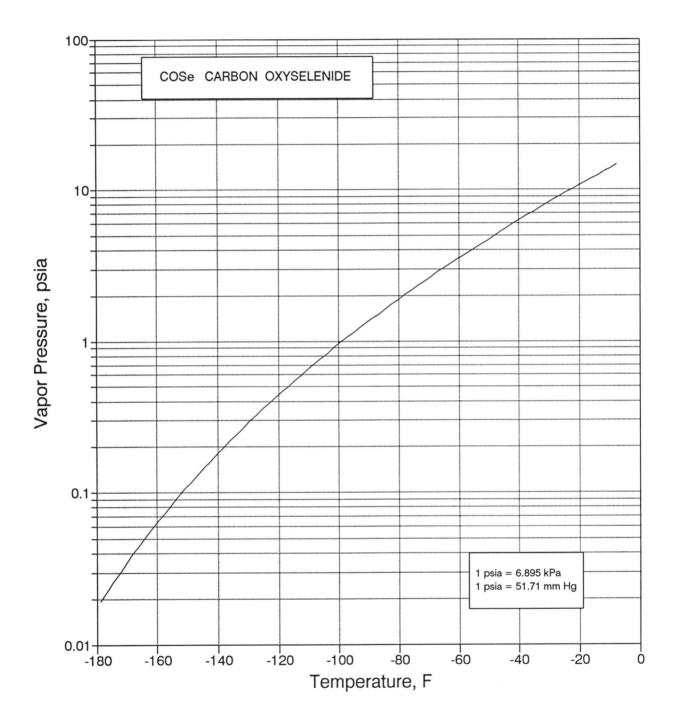

COSe CARBON OXYSELENIDE

Vapor Pressure, psia

Temperature, F

1 psia = 6.895 kPa
1 psia = 51.71 mm Hg

59

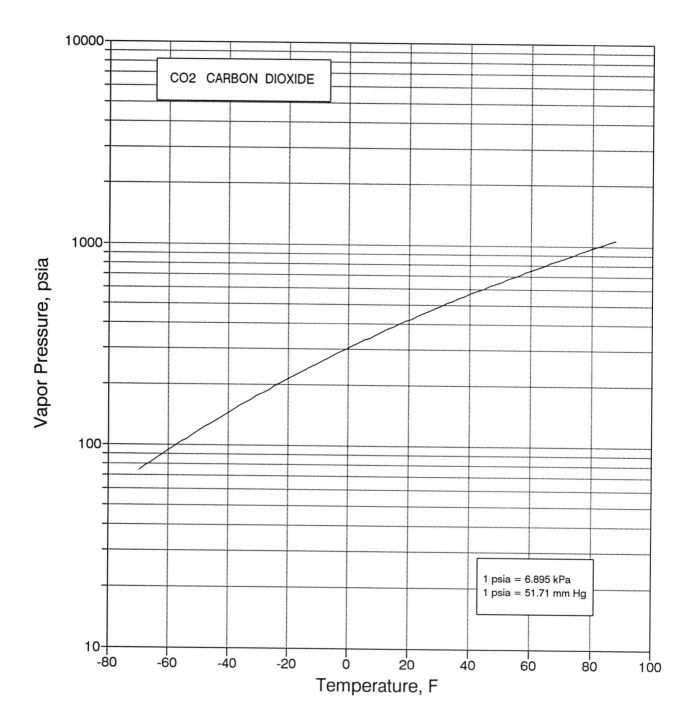

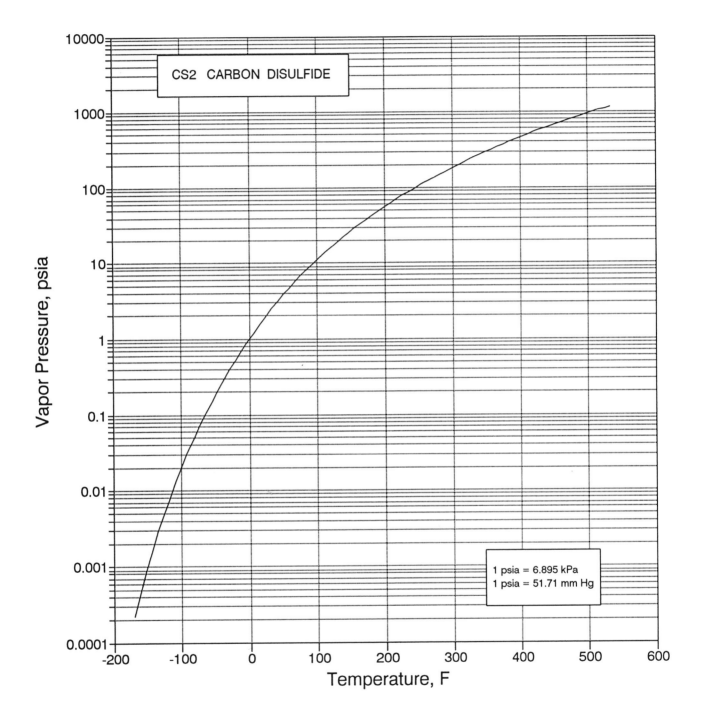

CS2  CARBON DISULFIDE

1 psia = 6.895 kPa
1 psia = 51.71 mm Hg

61

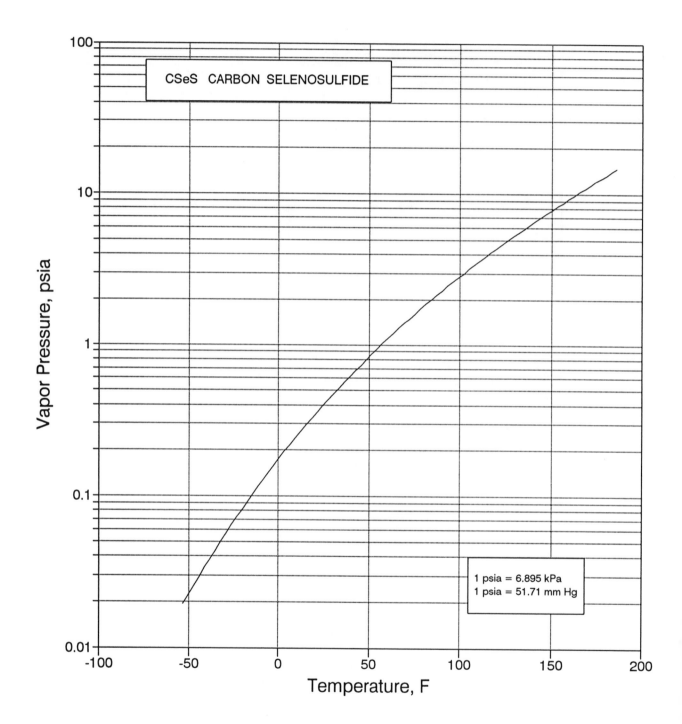

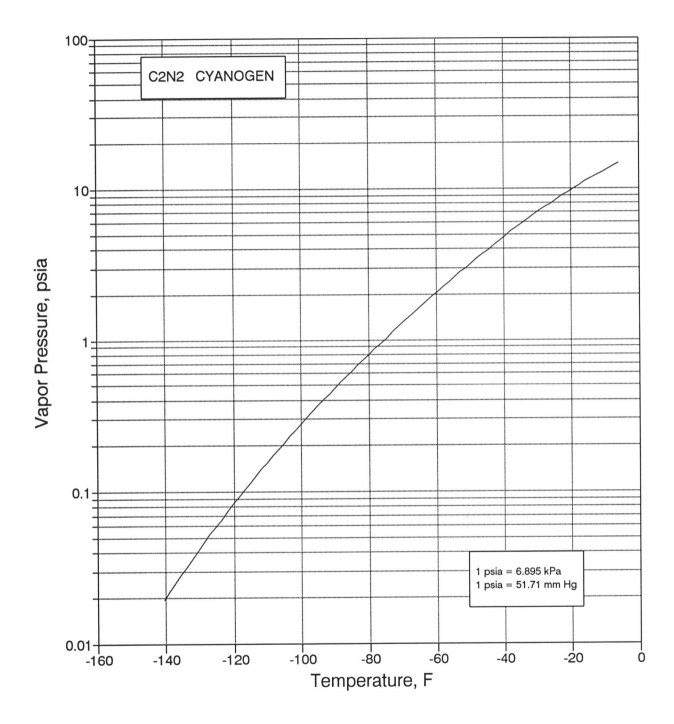

C2N2  CYANOGEN

Vapor Pressure, psia

Temperature, F

1 psia = 6.895 kPa
1 psia = 51.71 mm Hg

63

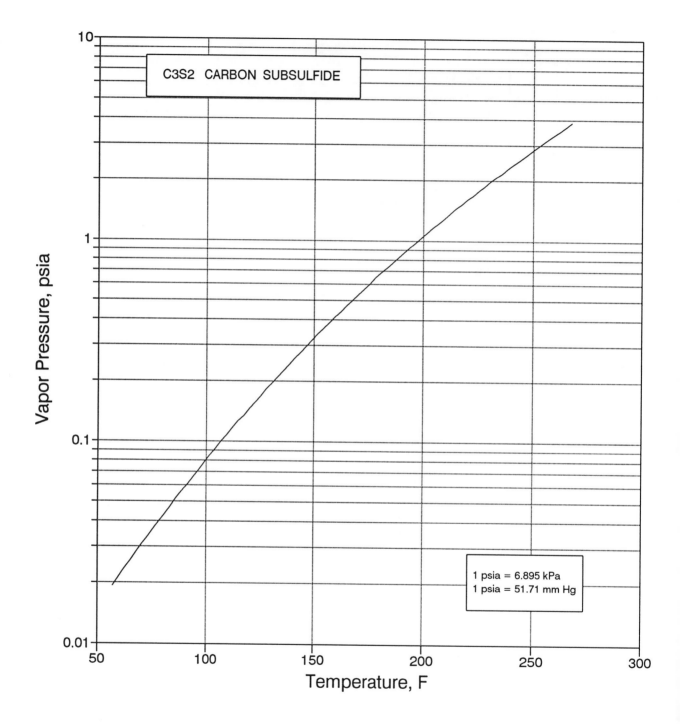

64

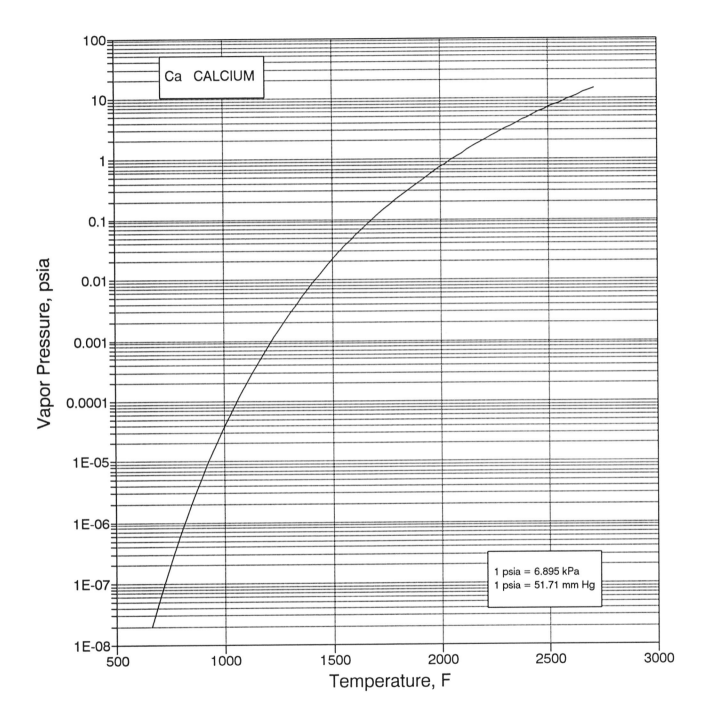

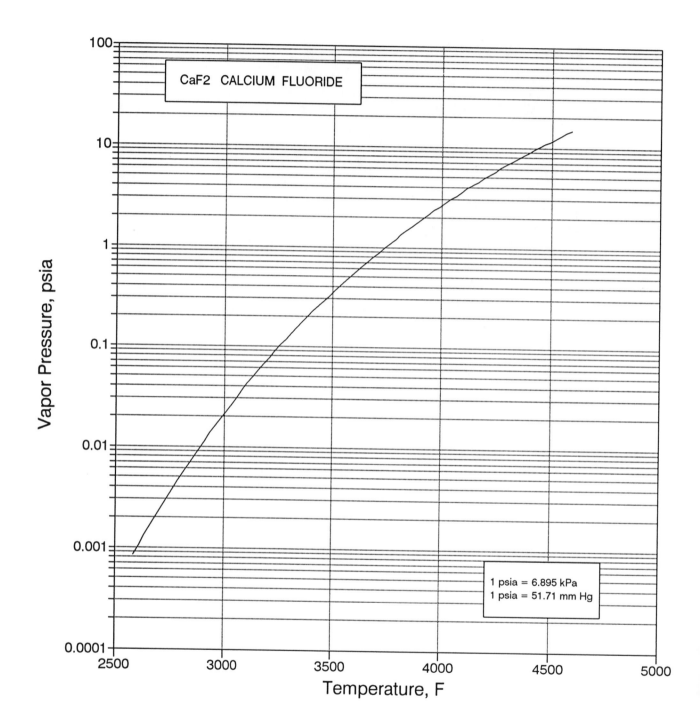

CaF2 CALCIUM FLUORIDE

Vapor Pressure, psia

Temperature, F

1 psia = 6.895 kPa
1 psia = 51.71 mm Hg

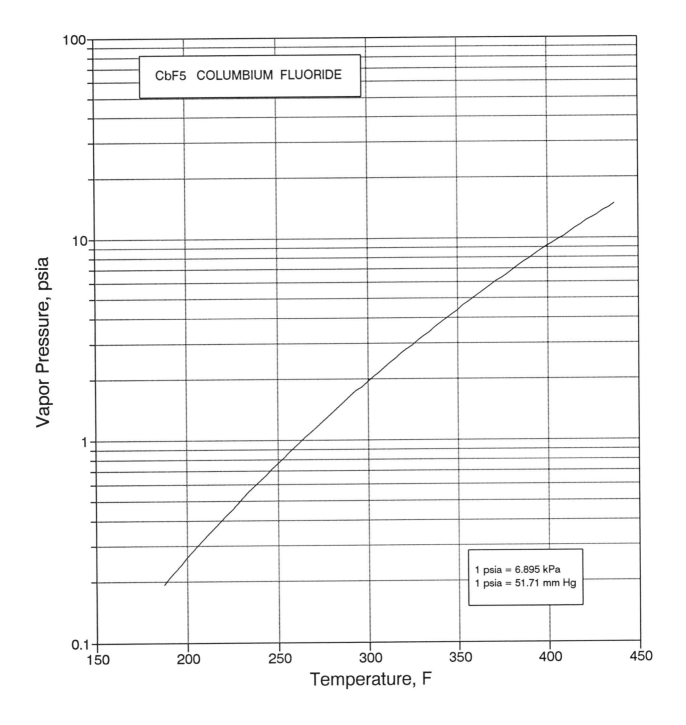

CbF5  COLUMBIUM FLUORIDE

1 psia = 6.895 kPa
1 psia = 51.71 mm Hg

Vapor Pressure, psia

Temperature, F

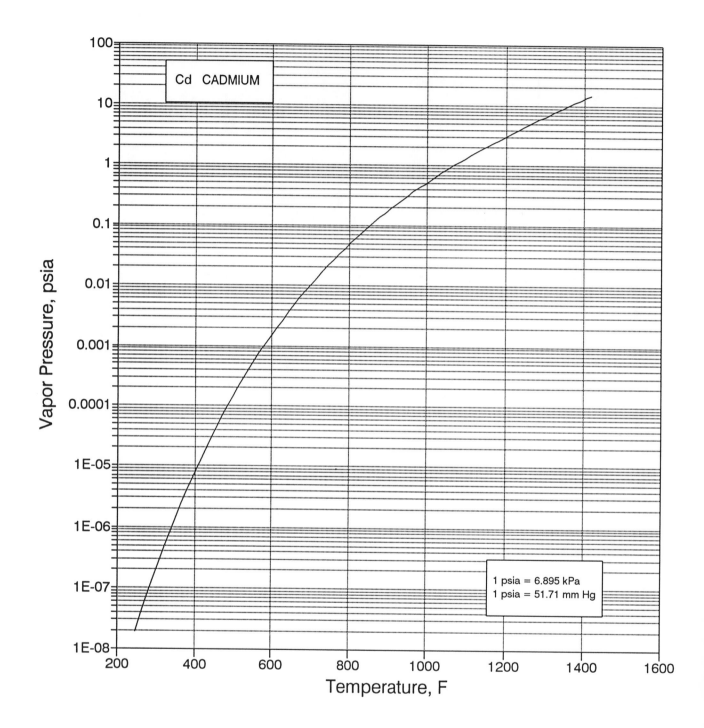

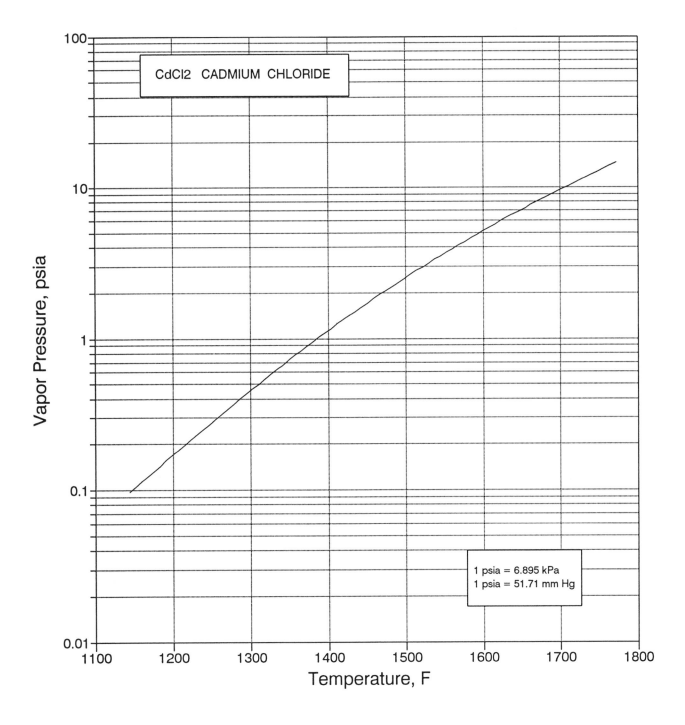

CdCl2  CADMIUM CHLORIDE

1 psia = 6.895 kPa
1 psia = 51.71 mm Hg

Vapor Pressure, psia

Temperature, F

69

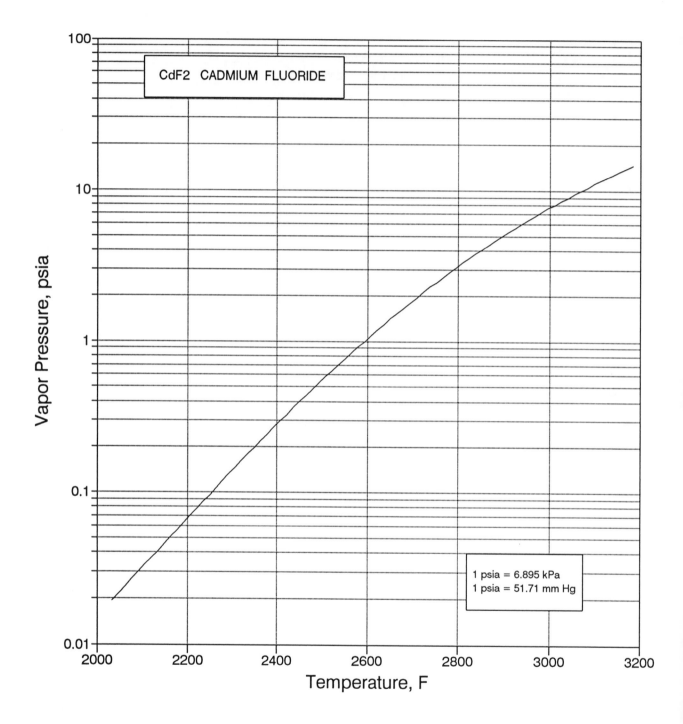

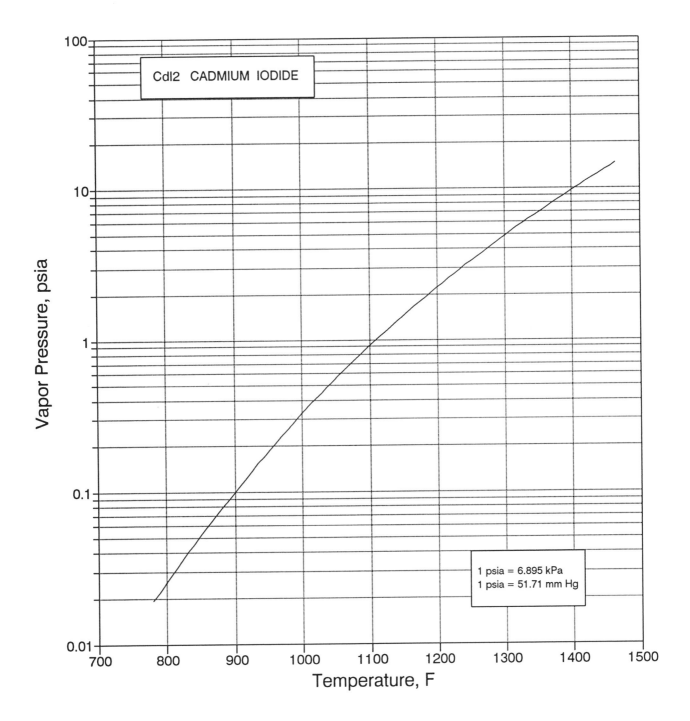

Vapor Pressure, psia

Temperature, F

CdI2  CADMIUM IODIDE

1 psia = 6.895 kPa
1 psia = 51.71 mm Hg

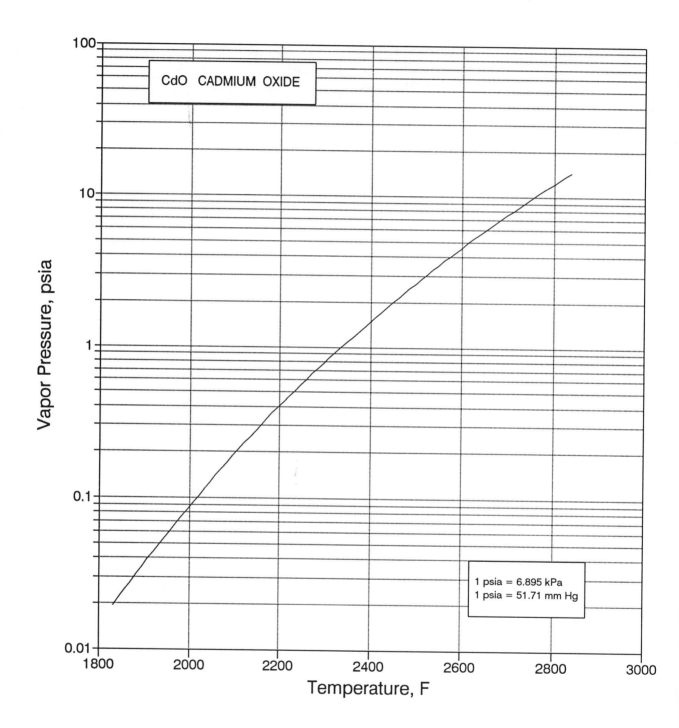

CdO  CADMIUM OXIDE

Vapor Pressure, psia

Temperature, F

1 psia = 6.895 kPa
1 psia = 51.71 mm Hg

72

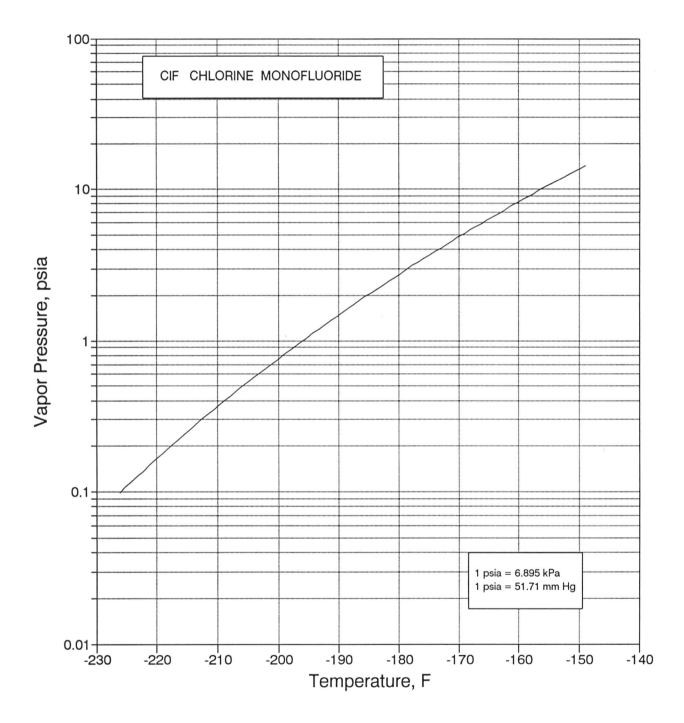

CIF CHLORINE MONOFLUORIDE

1 psia = 6.895 kPa
1 psia = 51.71 mm Hg

Vapor Pressure, psia

Temperature, F

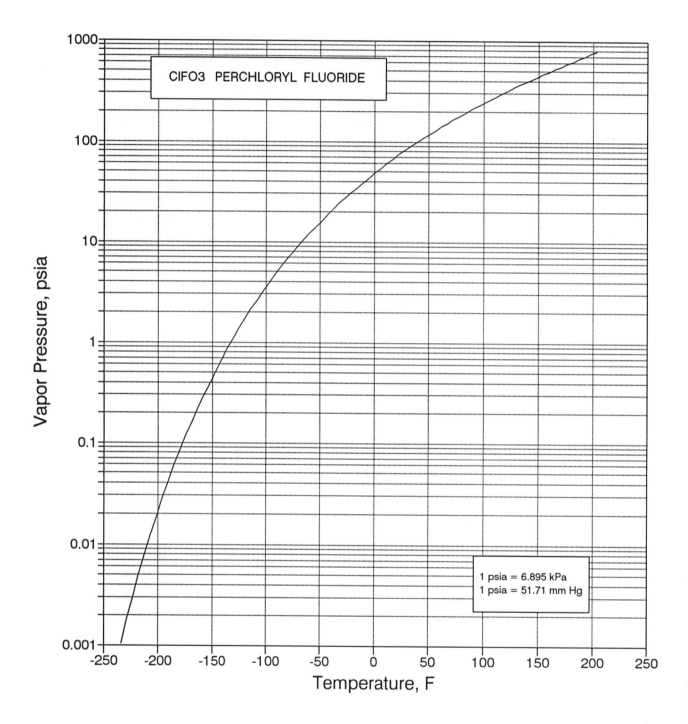

CIFO3 PERCHLORYL FLUORIDE

1 psia = 6.895 kPa
1 psia = 51.71 mm Hg

Vapor Pressure, psia

Temperature, F

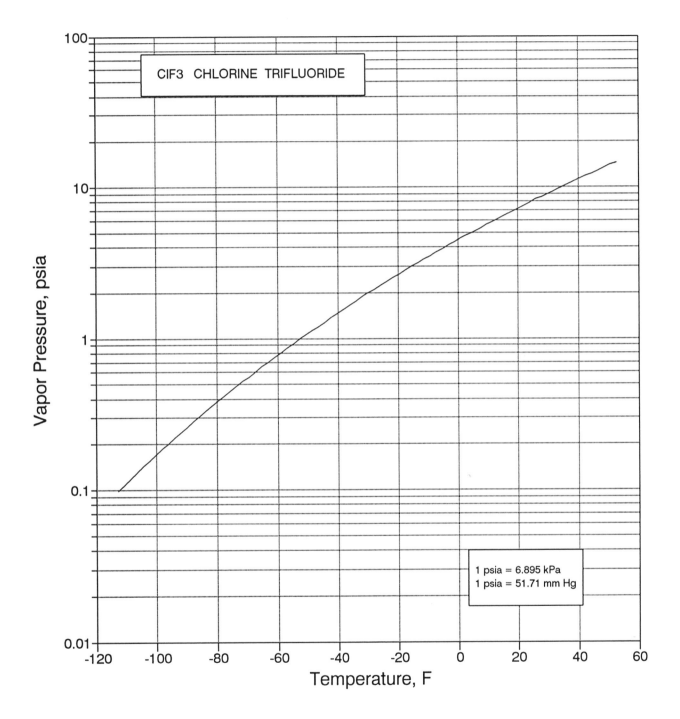

CIF3  CHLORINE  TRIFLUORIDE

Vapor Pressure, psia

Temperature, F

1 psia = 6.895 kPa
1 psia = 51.71 mm Hg

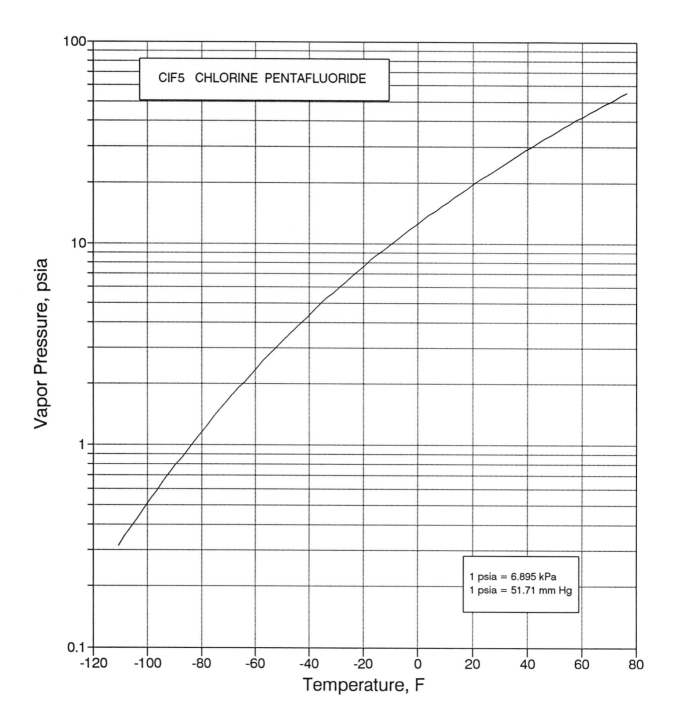

ClF5 CHLORINE PENTAFLUORIDE

1 psia = 6.895 kPa
1 psia = 51.71 mm Hg

Vapor Pressure, psia

Temperature, F

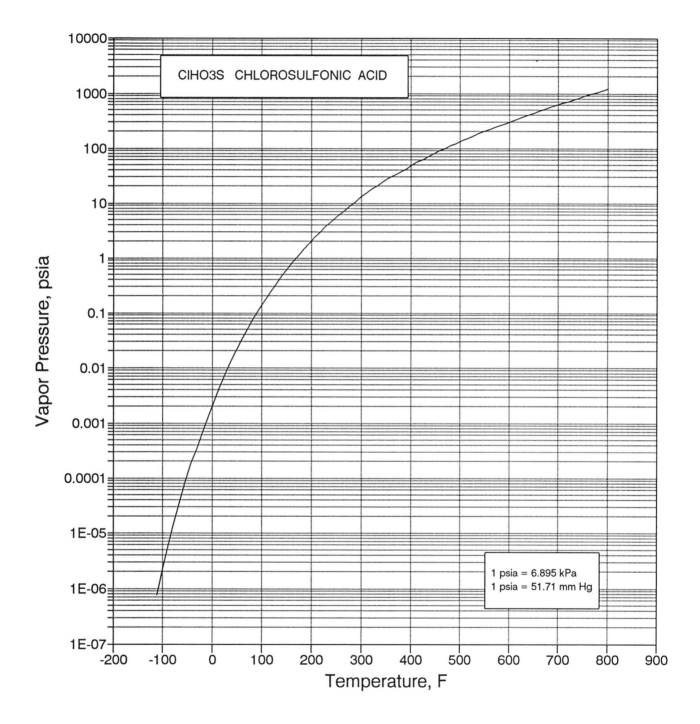

ClHO3S   CHLOROSULFONIC ACID

1 psia = 6.895 kPa
1 psia = 51.71 mm Hg

Vapor Pressure, psia

Temperature, F

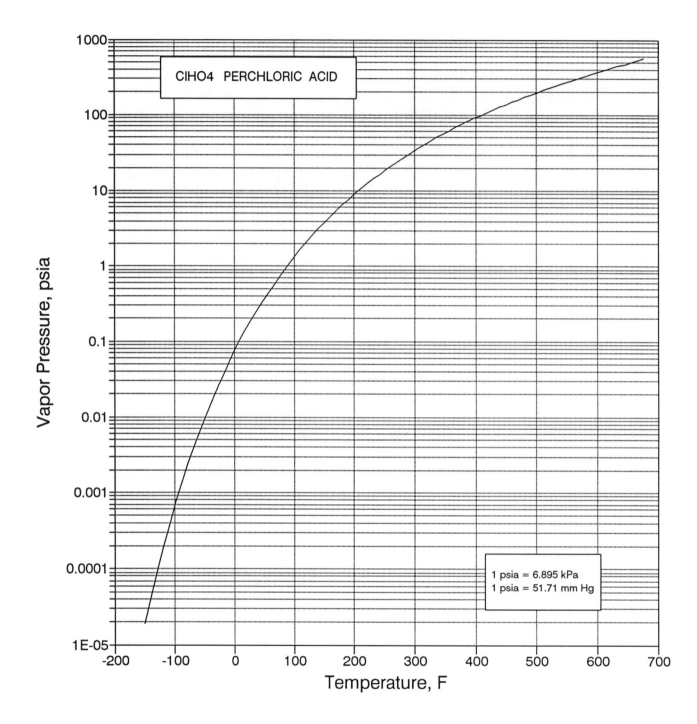

CIHO4  PERCHLORIC ACID

1 psia = 6.895 kPa
1 psia = 51.71 mm Hg

Vapor Pressure, psia

Temperature, F

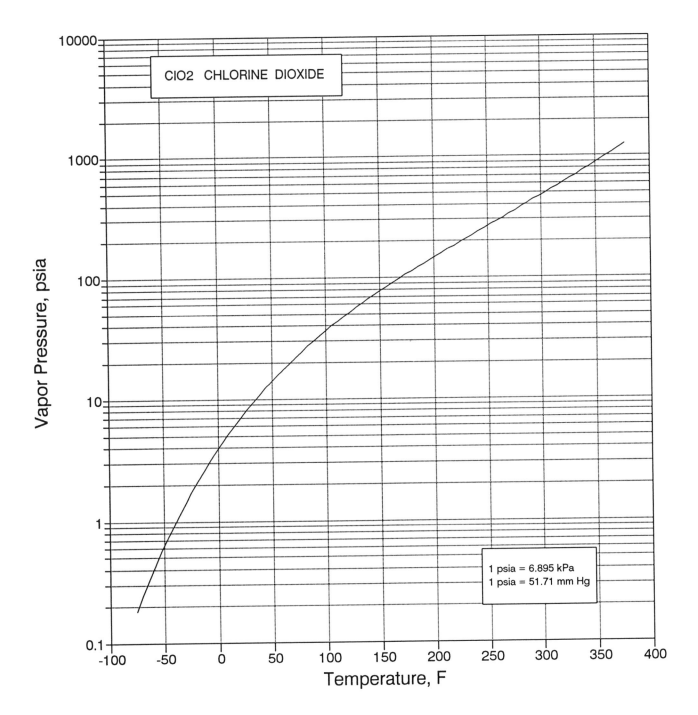

CIO2  CHLORINE DIOXIDE

1 psia = 6.895 kPa
1 psia = 51.71 mm Hg

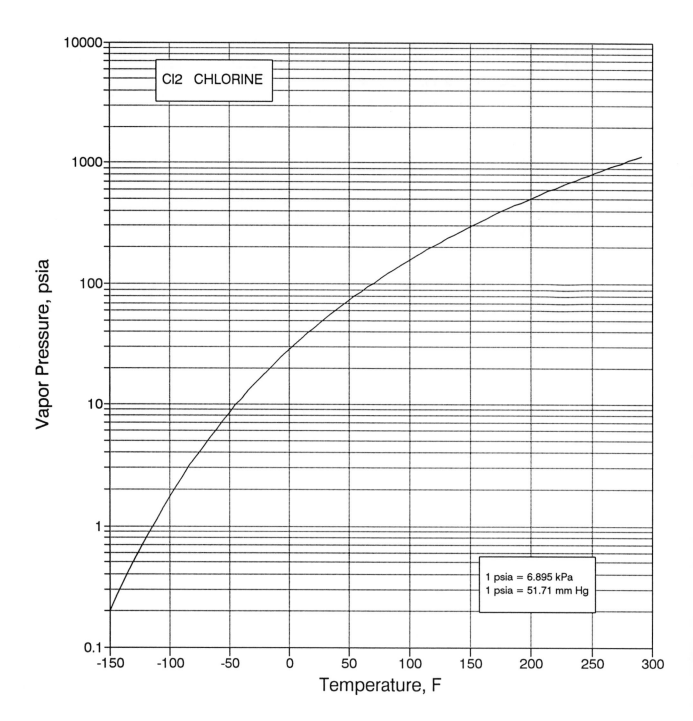

Cl2 CHLORINE

1 psia = 6.895 kPa
1 psia = 51.71 mm Hg

Vapor Pressure, psia

Temperature, F

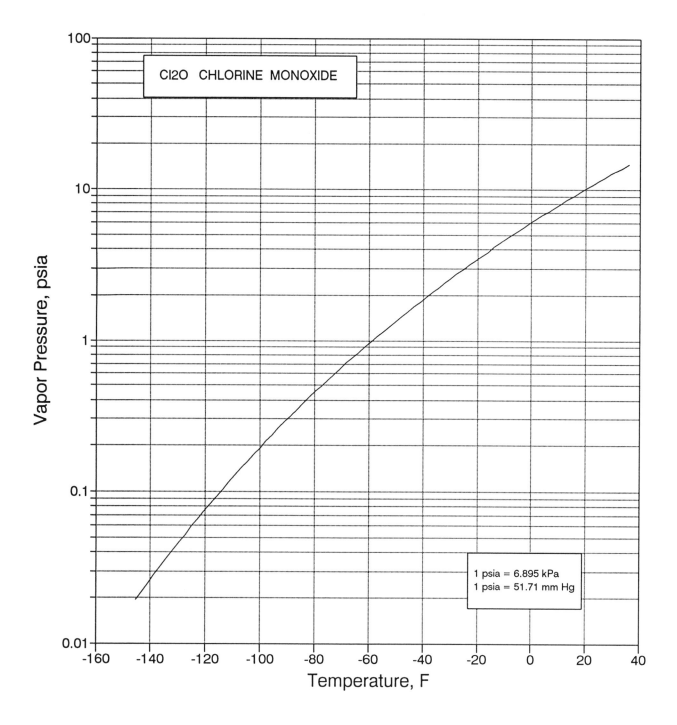

Cl2O  CHLORINE  MONOXIDE

1 psia = 6.895 kPa
1 psia = 51.71 mm Hg

Vapor Pressure, psia

Temperature, F

81

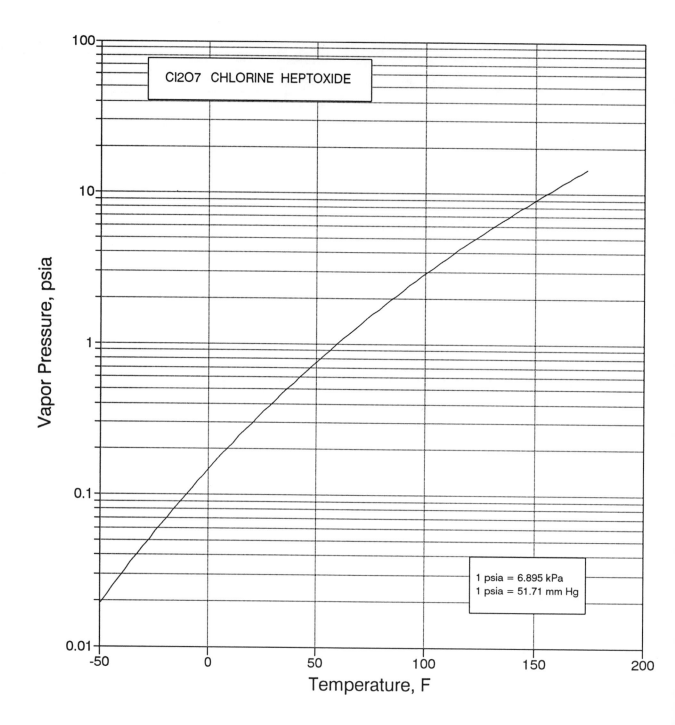

1 psia = 6.895 kPa
1 psia = 51.71 mm Hg

Cl2O7 CHLORINE HEPTOXIDE

Vapor Pressure, psia

Temperature, F

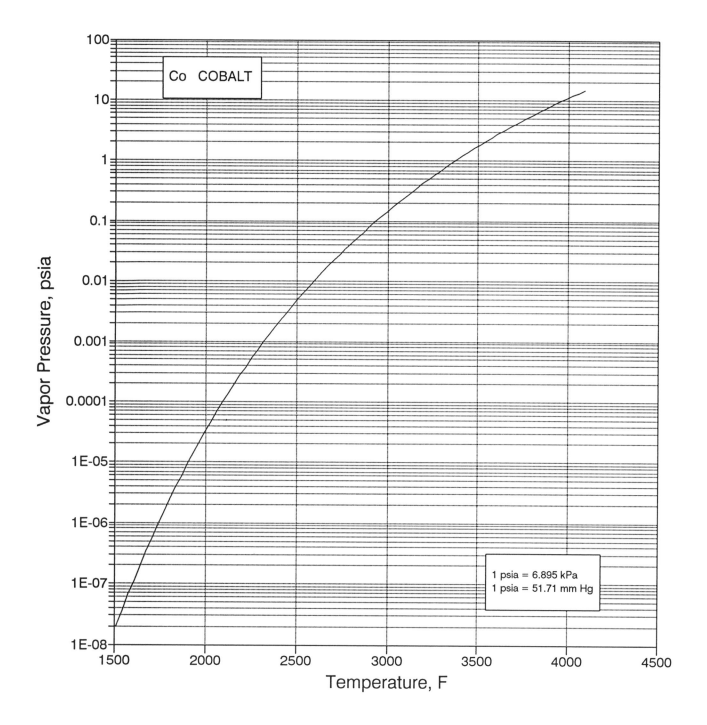

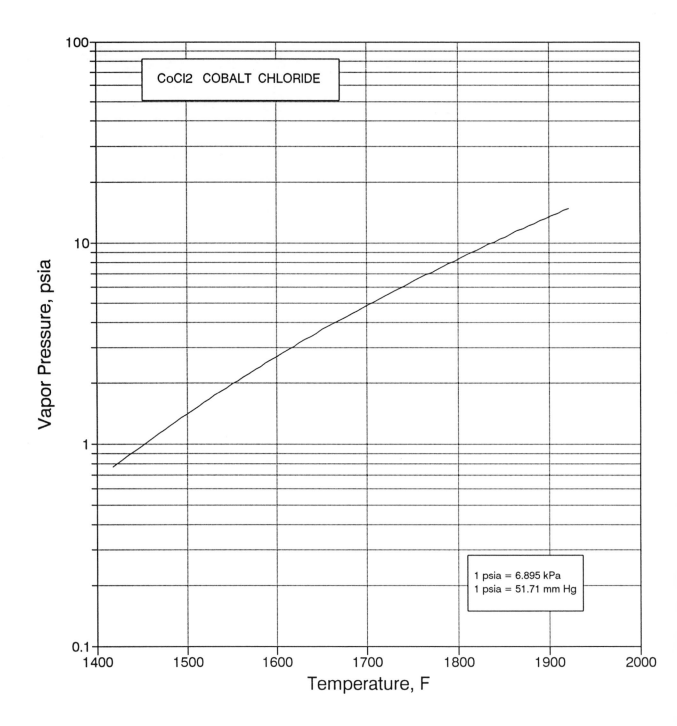

CoCl2  COBALT CHLORIDE

1 psia = 6.895 kPa
1 psia = 51.71 mm Hg

Vapor Pressure, psia

Temperature, F

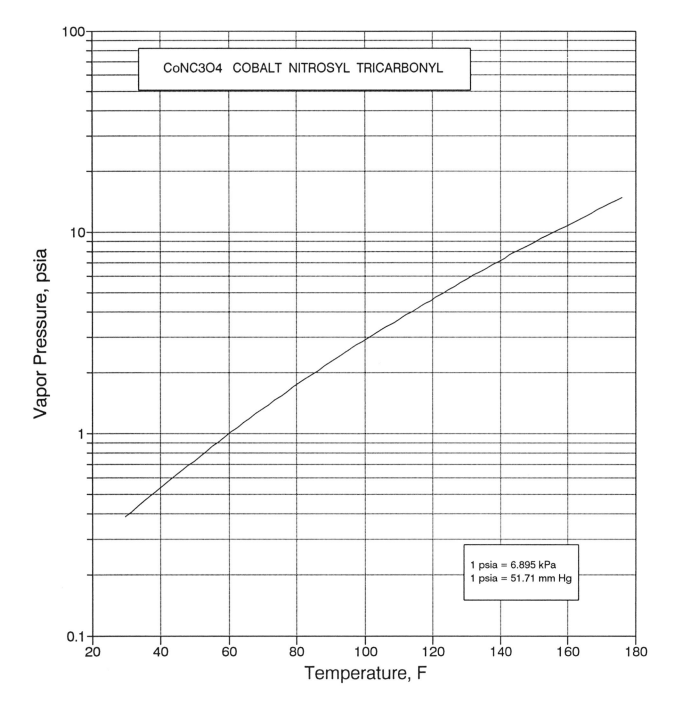

CoNC3O4   COBALT NITROSYL TRICARBONYL

Vapor Pressure, psia

Temperature, F

1 psia = 6.895 kPa
1 psia = 51.71 mm Hg

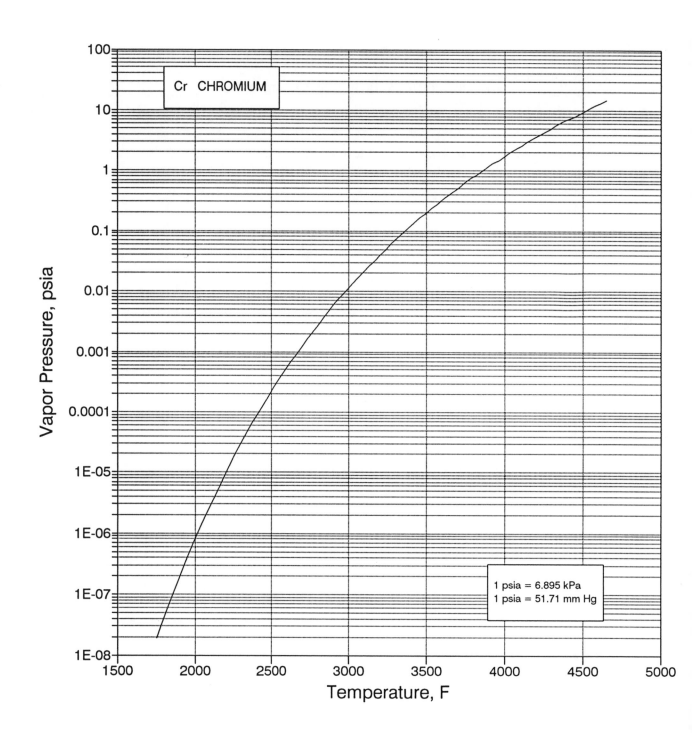

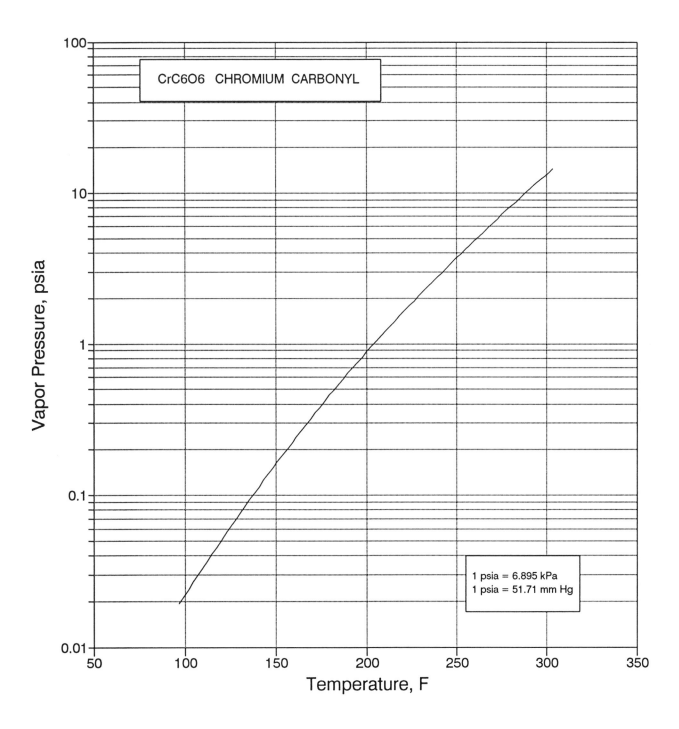

CrC6O6  CHROMIUM  CARBONYL

Vapor Pressure, psia

Temperature, F

1 psia = 6.895 kPa
1 psia = 51.71 mm Hg

87

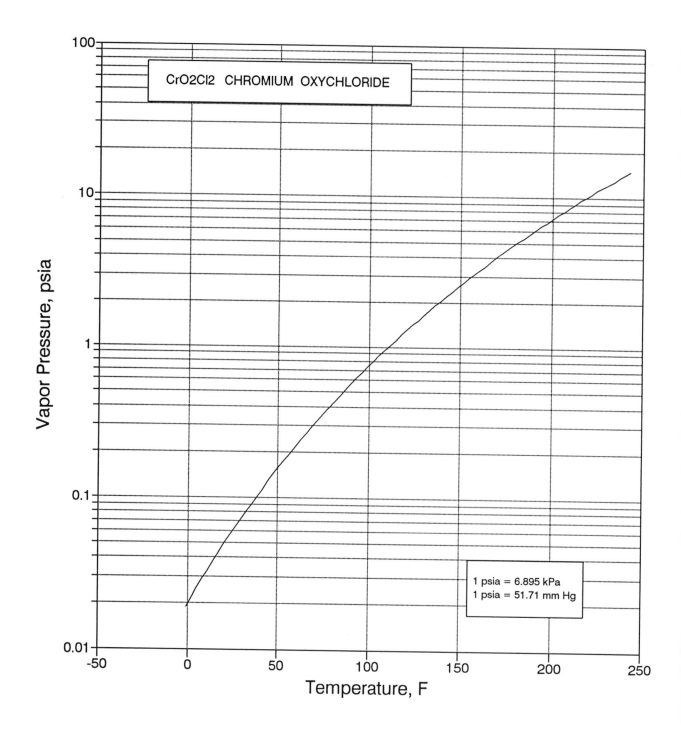

CrO2Cl2  CHROMIUM  OXYCHLORIDE

Vapor Pressure, psia

Temperature, F

1 psia = 6.895 kPa
1 psia = 51.71 mm Hg

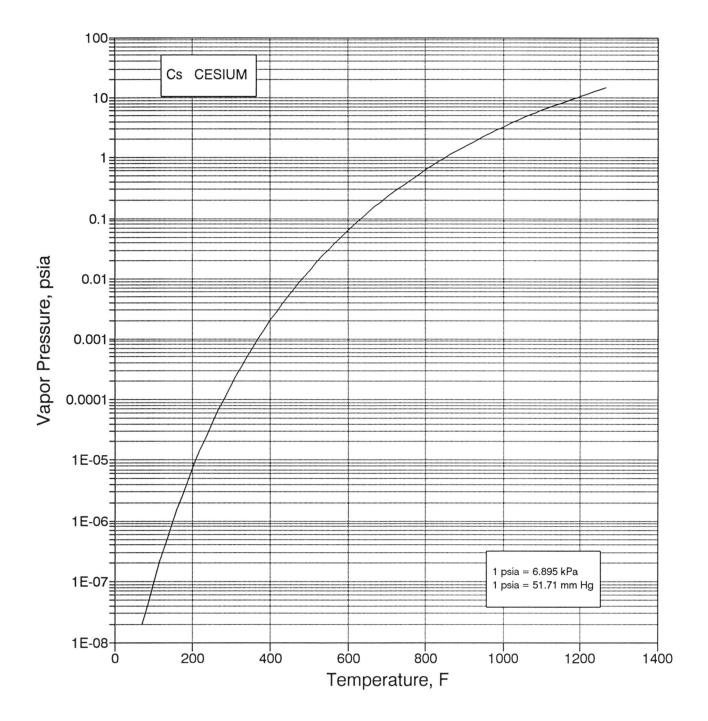

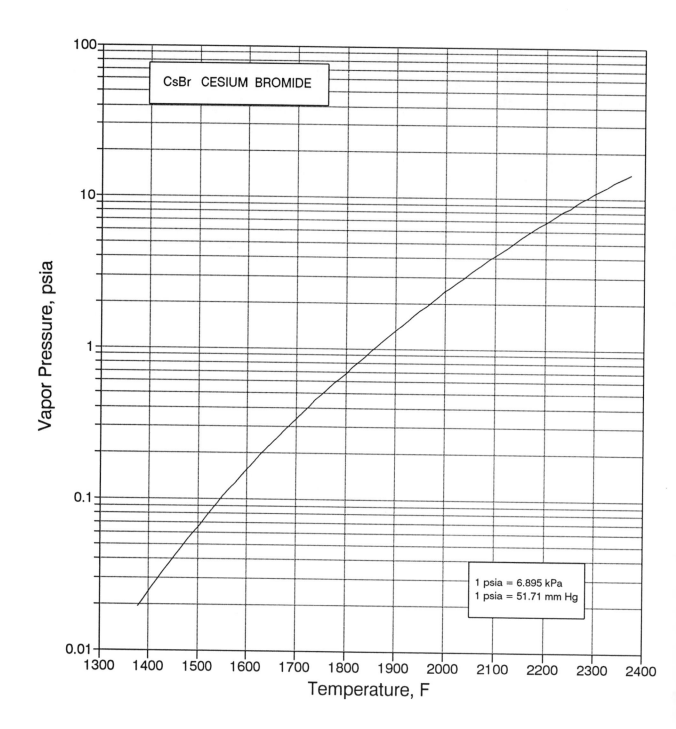

CsBr  CESIUM BROMIDE

1 psia = 6.895 kPa
1 psia = 51.71 mm Hg

Vapor Pressure, psia

Temperature, F

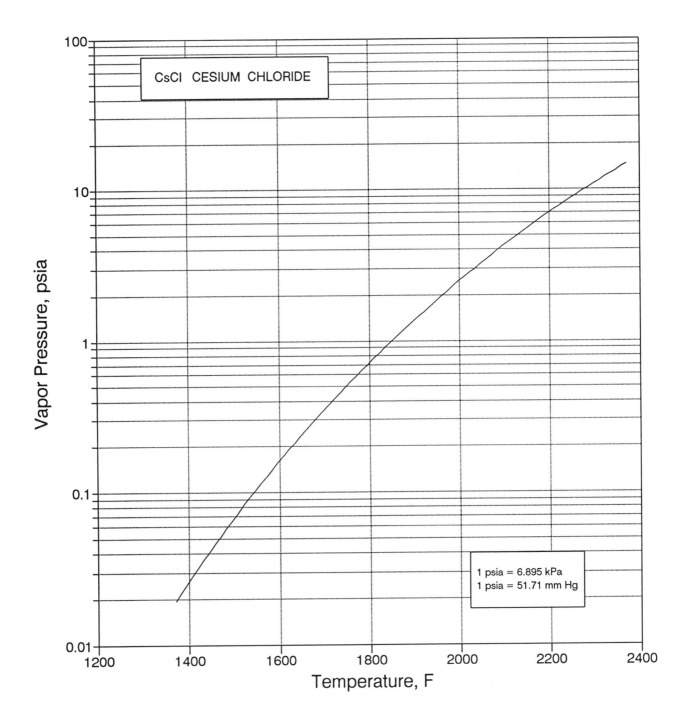

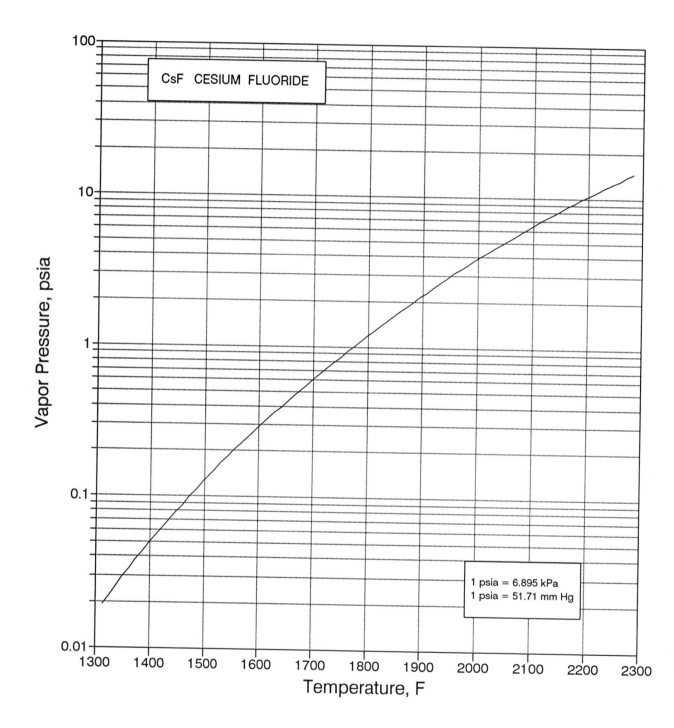

CsF  CESIUM FLUORIDE

Vapor Pressure, psia

Temperature, F

1 psia = 6.895 kPa
1 psia = 51.71 mm Hg

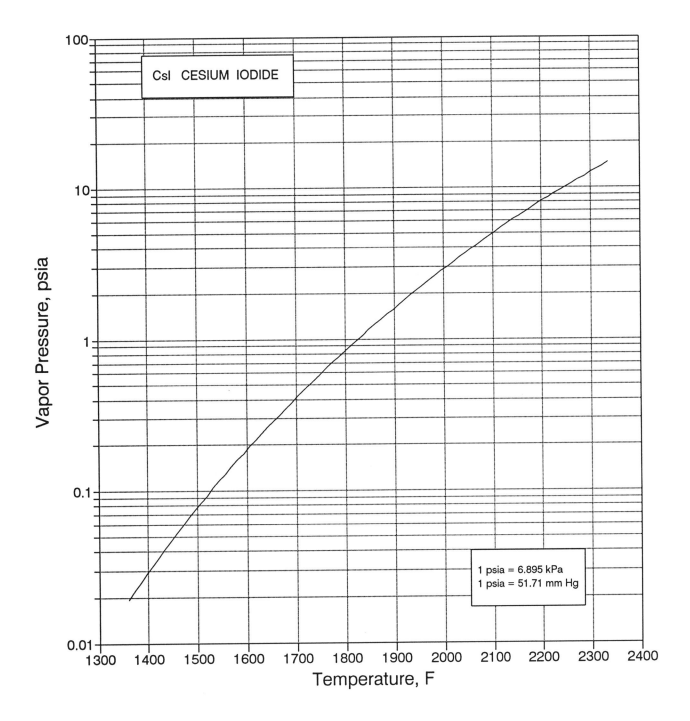

CsI  CESIUM IODIDE

1 psia = 6.895 kPa
1 psia = 51.71 mm Hg

Vapor Pressure, psia

Temperature, F

93

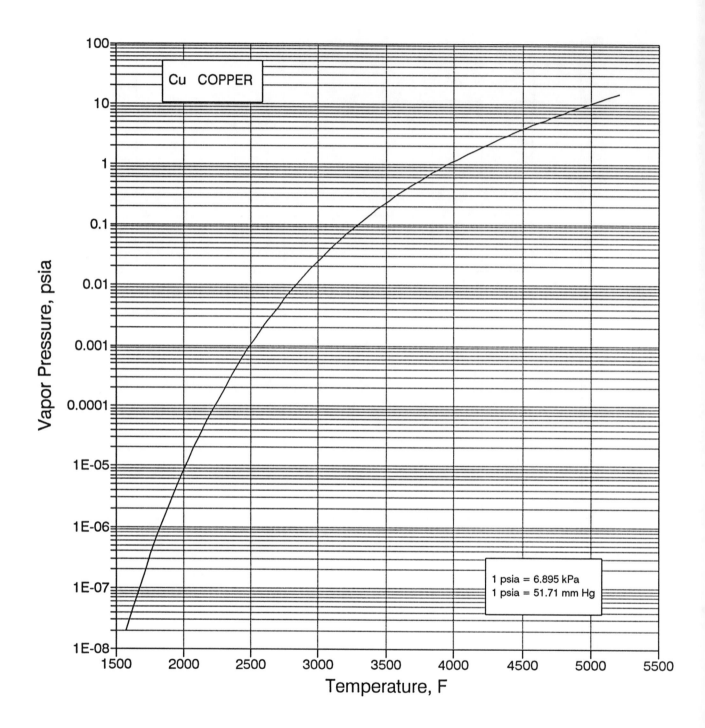

Cu  COPPER

Vapor Pressure, psia

Temperature, F

1 psia = 6.895 kPa
1 psia = 51.71 mm Hg

94

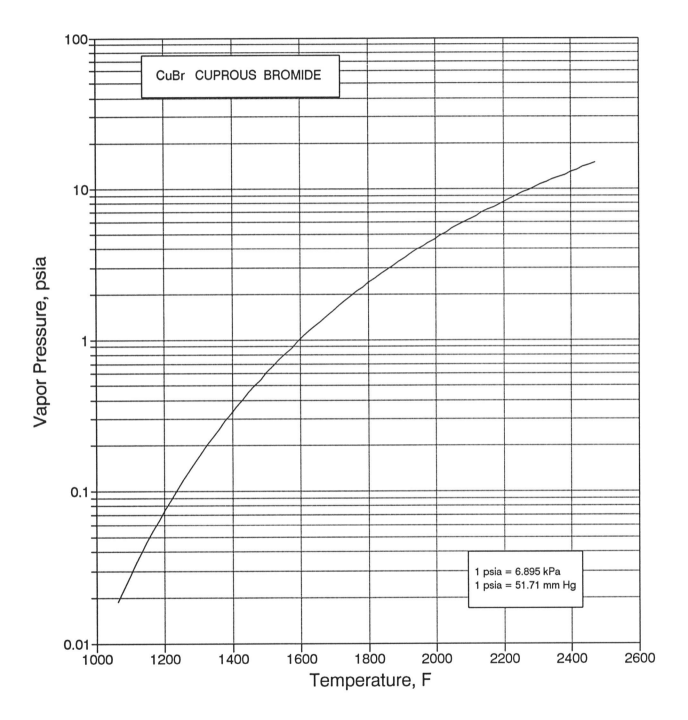

CuBr CUPROUS BROMIDE

Vapor Pressure, psia

Temperature, F

1 psia = 6.895 kPa
1 psia = 51.71 mm Hg

95

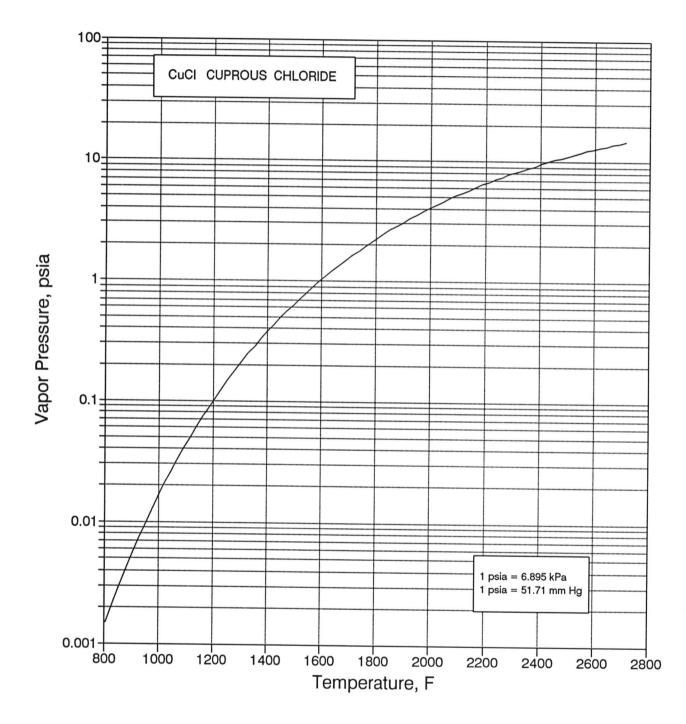

CuCl CUPROUS CHLORIDE

1 psia = 6.895 kPa
1 psia = 51.71 mm Hg

Vapor Pressure, psia

Temperature, F

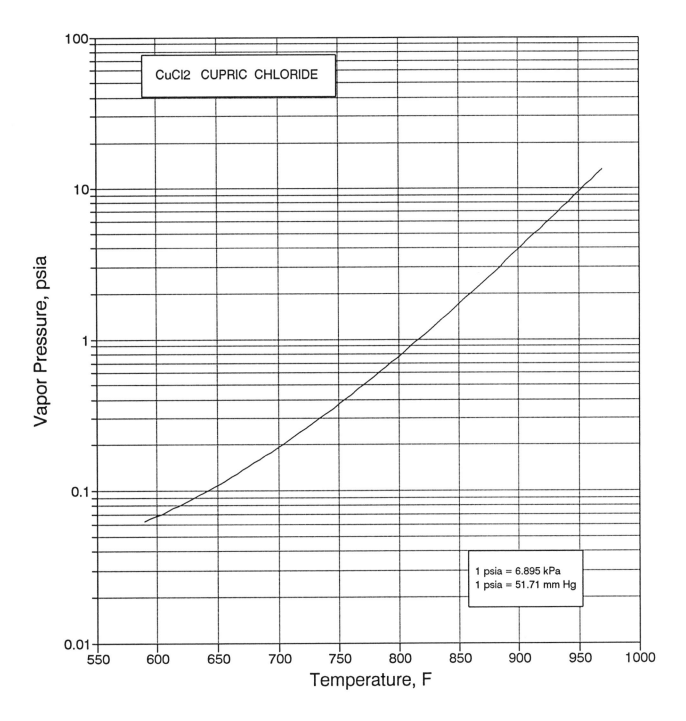

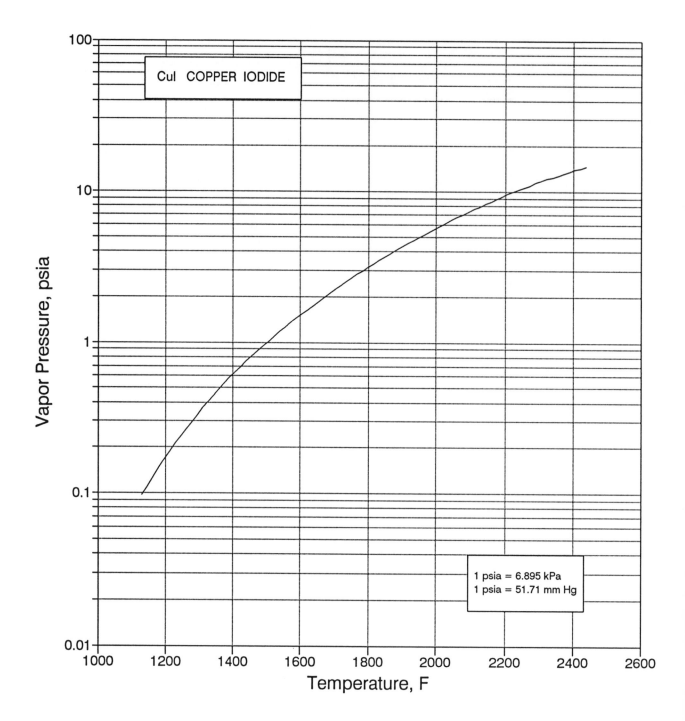

CuI  COPPER IODIDE

1 psia = 6.895 kPa
1 psia = 51.71 mm Hg

Vapor Pressure, psia

Temperature, F

98

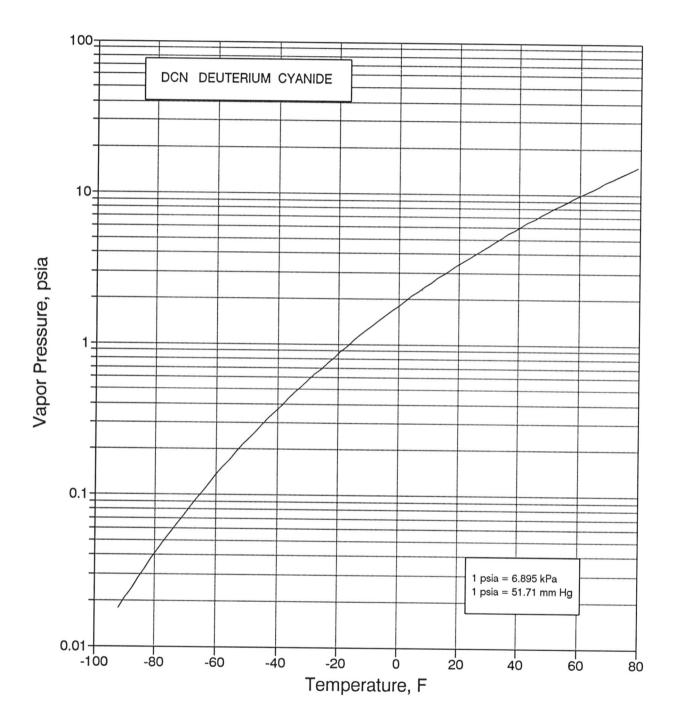

DCN DEUTERIUM CYANIDE

1 psia = 6.895 kPa
1 psia = 51.71 mm Hg

Vapor Pressure, psia

Temperature, F

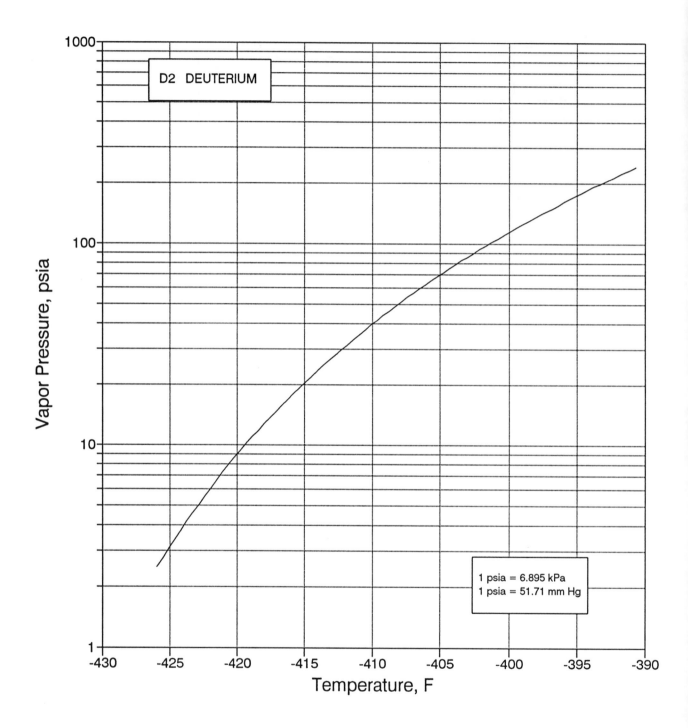

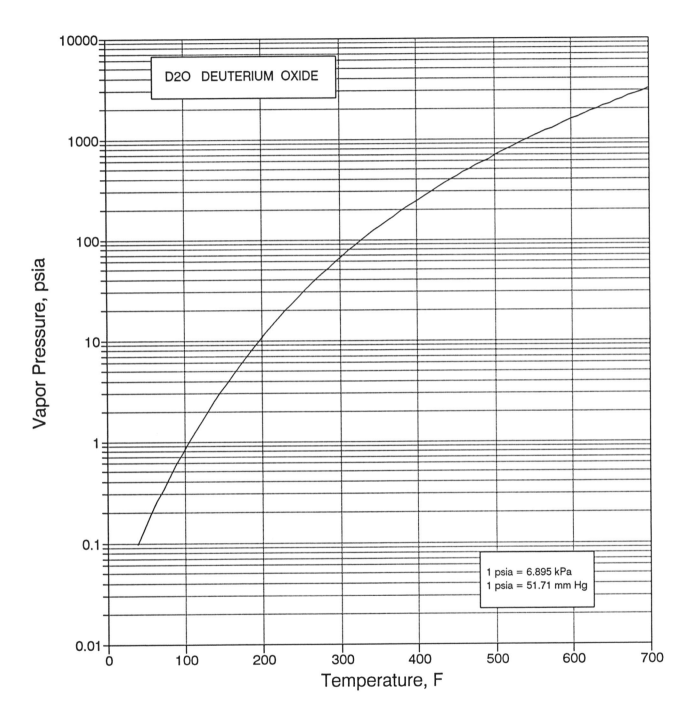

D2O  DEUTERIUM OXIDE

Vapor Pressure, psia

Temperature, F

1 psia = 6.895 kPa
1 psia = 51.71 mm Hg

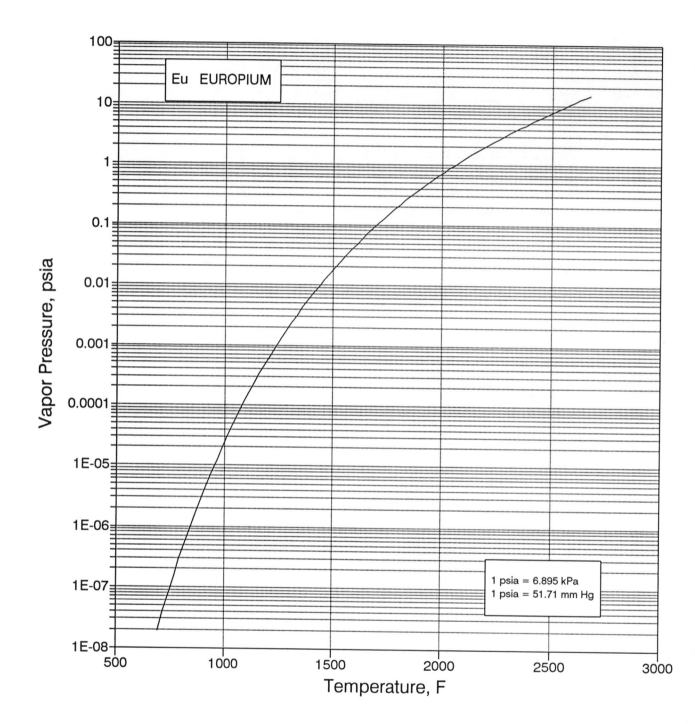

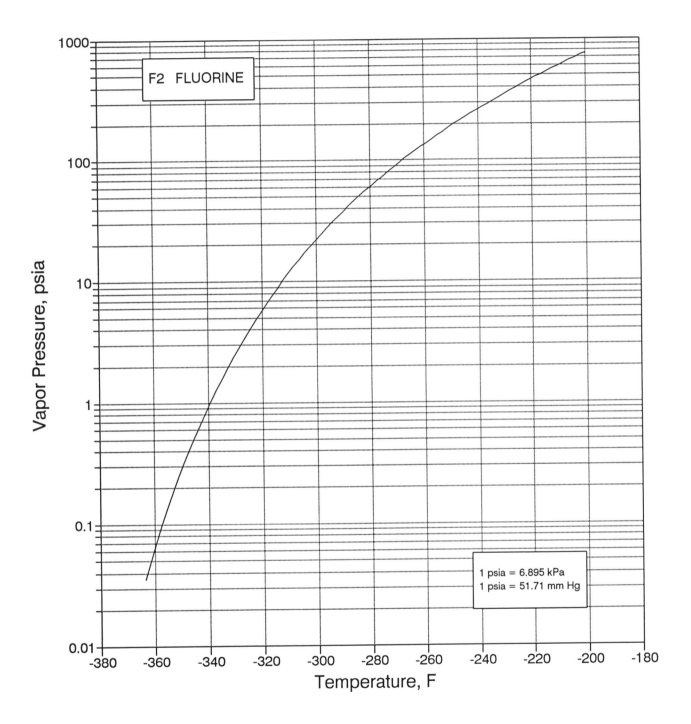

F2  FLUORINE

1 psia = 6.895 kPa
1 psia = 51.71 mm Hg

Vapor Pressure, psia

Temperature, F

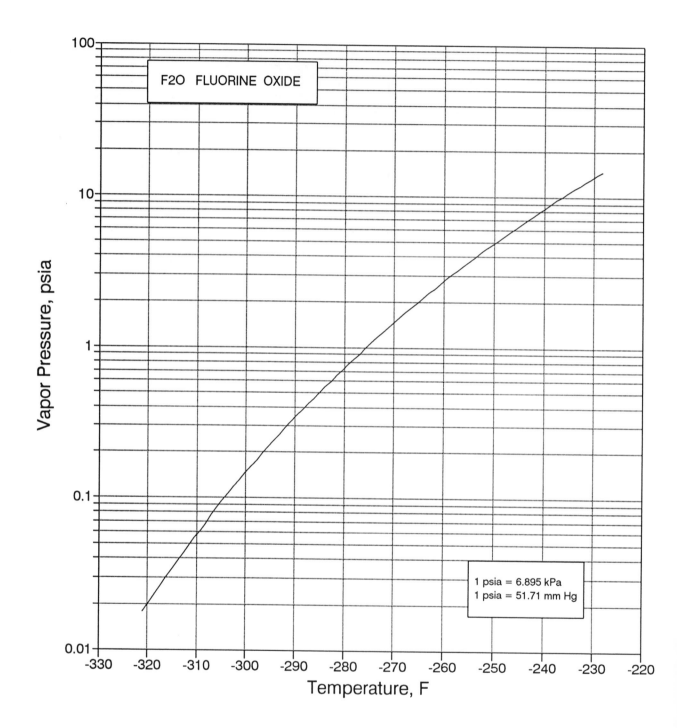

F2O  FLUORINE OXIDE

Vapor Pressure, psia

Temperature, F

1 psia = 6.895 kPa
1 psia = 51.71 mm Hg

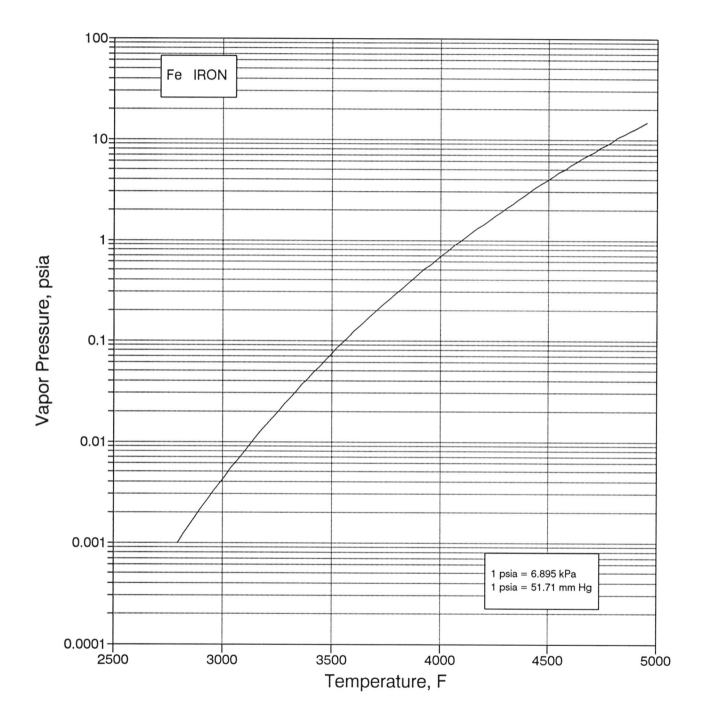

Fe IRON

Vapor Pressure, psia

Temperature, F

1 psia = 6.895 kPa
1 psia = 51.71 mm Hg

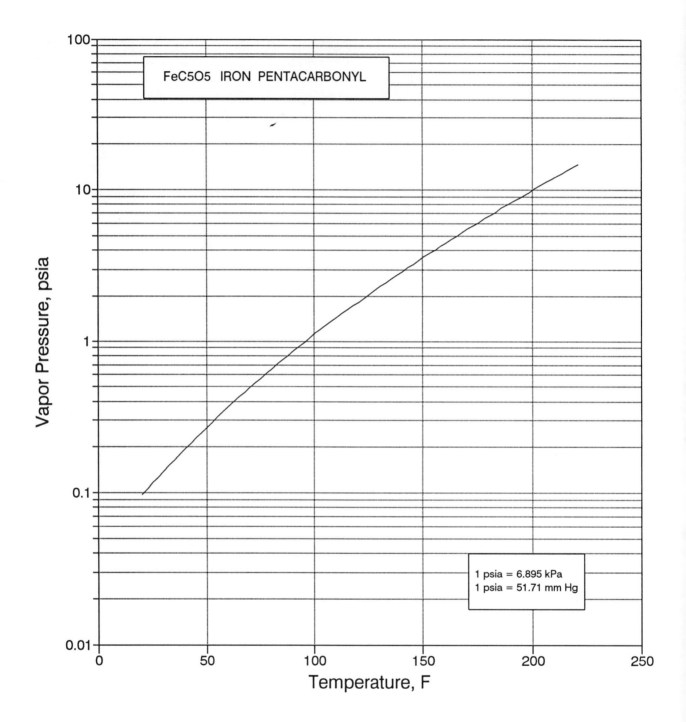

FeC5O5  IRON  PENTACARBONYL

Vapor Pressure, psia

Temperature, F

1 psia = 6.895 kPa
1 psia = 51.71 mm Hg

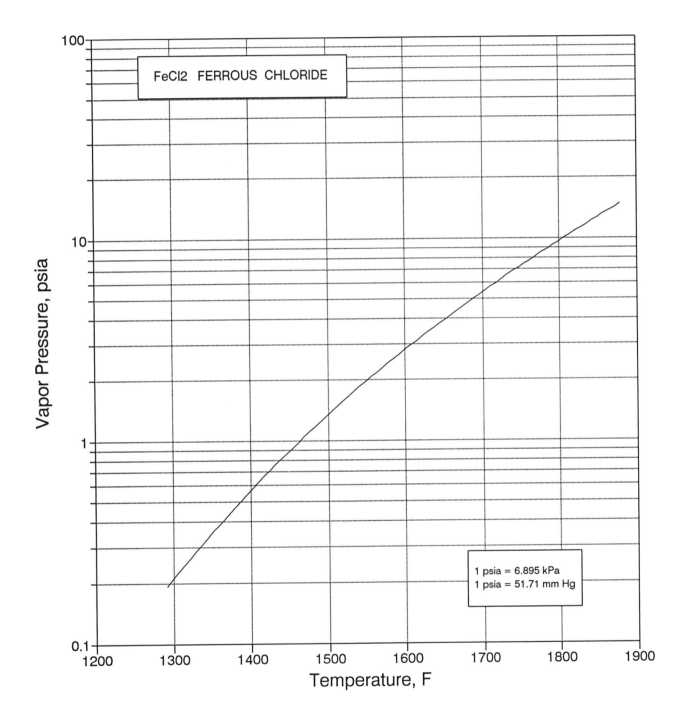

FeCl2 FERROUS CHLORIDE

Vapor Pressure, psia

Temperature, F

1 psia = 6.895 kPa
1 psia = 51.71 mm Hg

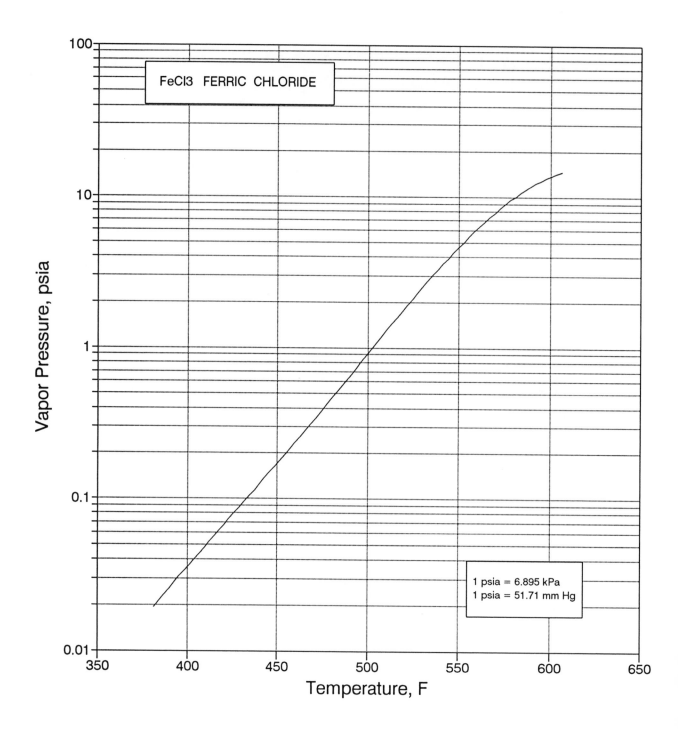

FeCl3  FERRIC CHLORIDE

Vapor Pressure, psia

Temperature, F

1 psia = 6.895 kPa
1 psia = 51.71 mm Hg

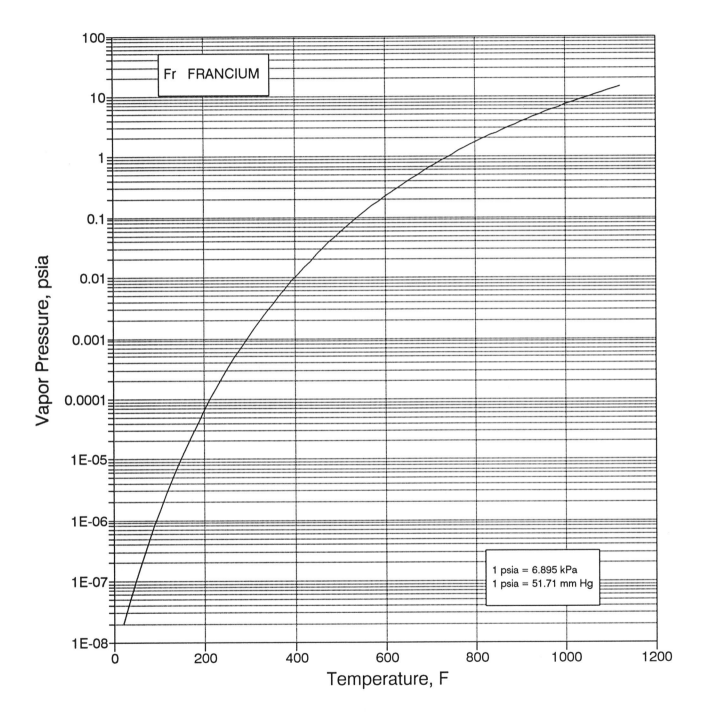

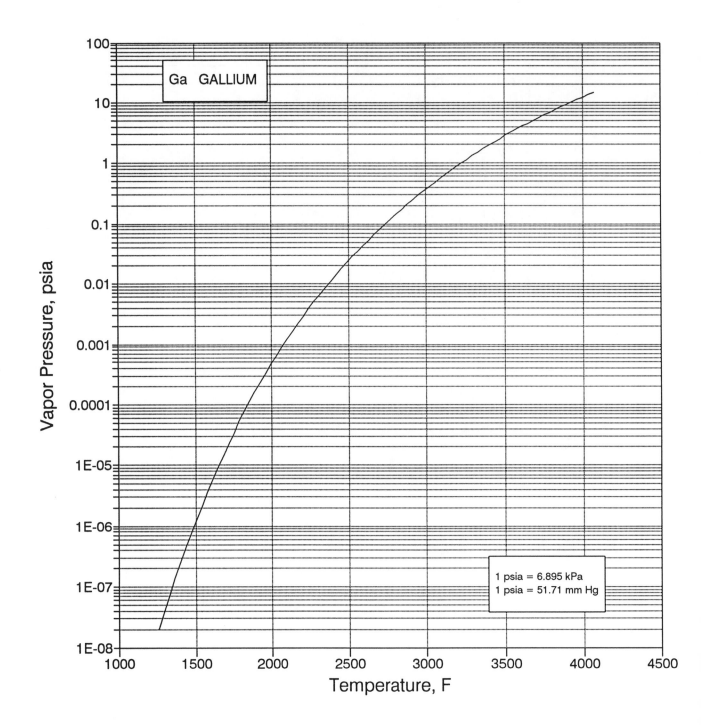

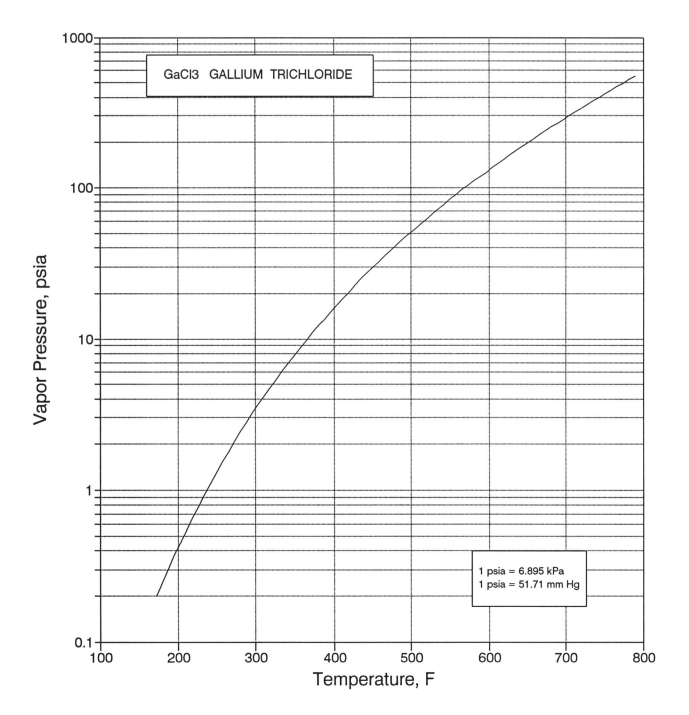

GaCl3 GALLIUM TRICHLORIDE

1 psia = 6.895 kPa
1 psia = 51.71 mm Hg

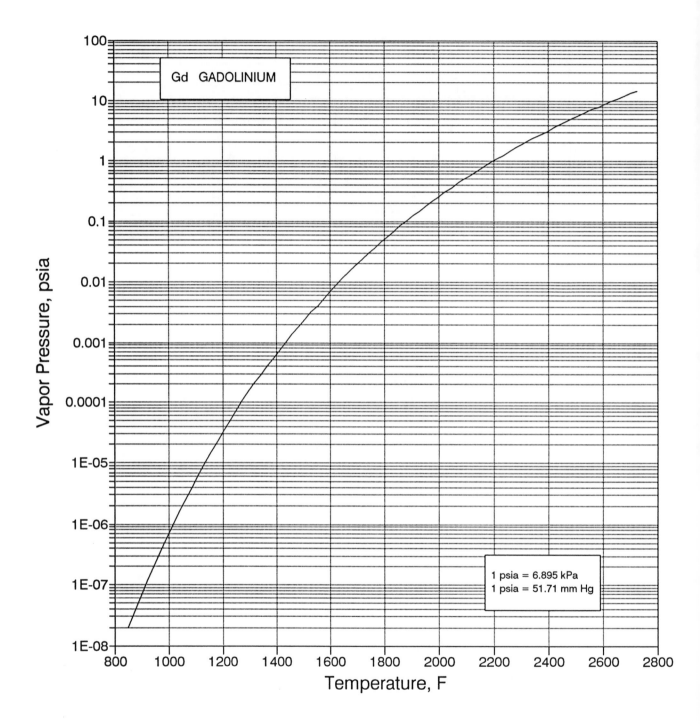

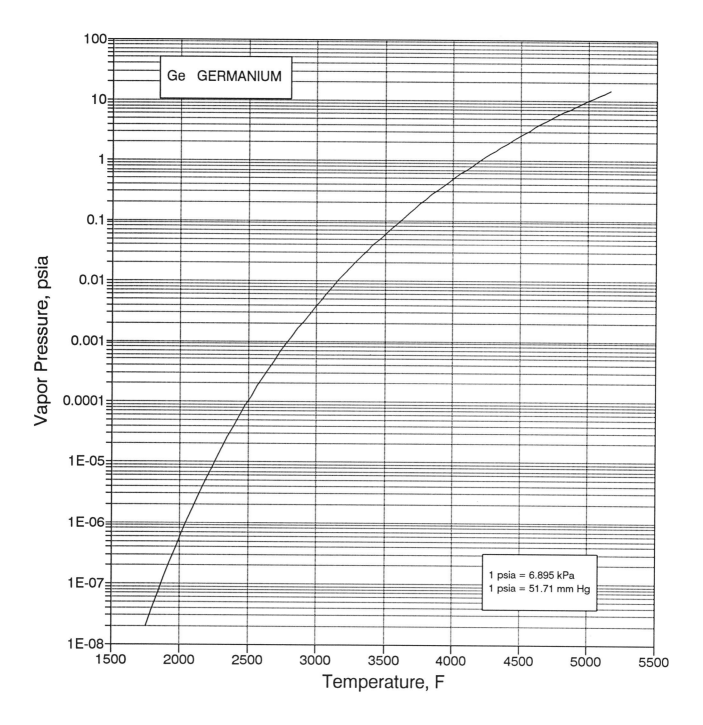

Ge GERMANIUM

Vapor Pressure, psia

Temperature, F

1 psia = 6.895 kPa
1 psia = 51.71 mm Hg

113

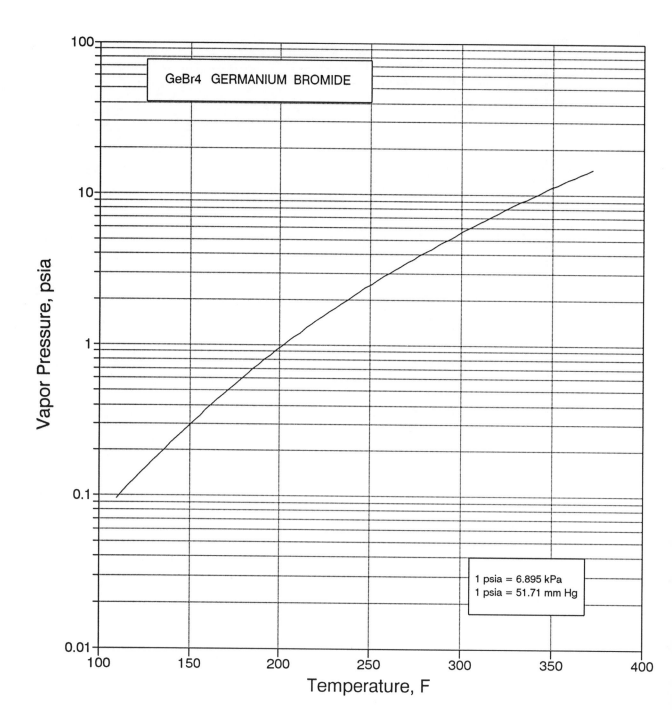

GeBr4  GERMANIUM  BROMIDE

1 psia = 6.895 kPa
1 psia = 51.71 mm Hg

Vapor Pressure, psia

Temperature, F

114

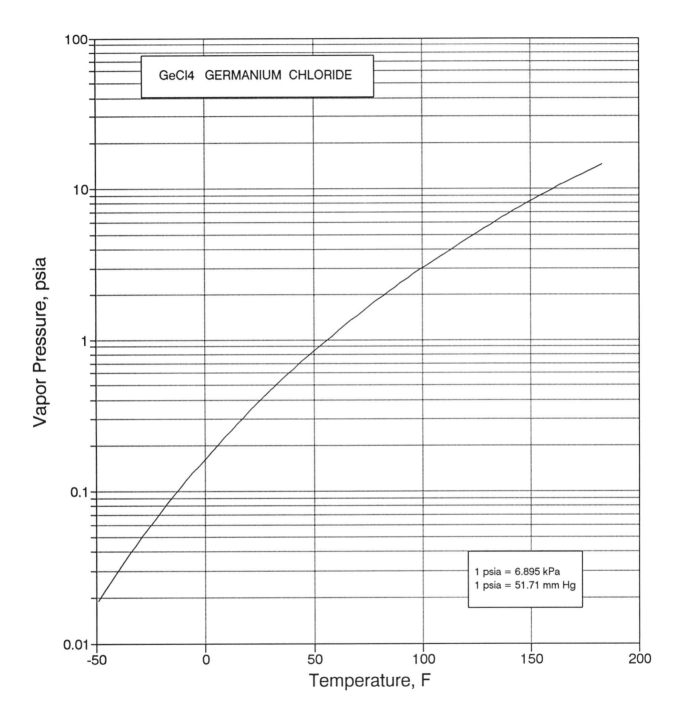

GeCl4  GERMANIUM CHLORIDE

Vapor Pressure, psia

Temperature, F

1 psia = 6.895 kPa
1 psia = 51.71 mm Hg

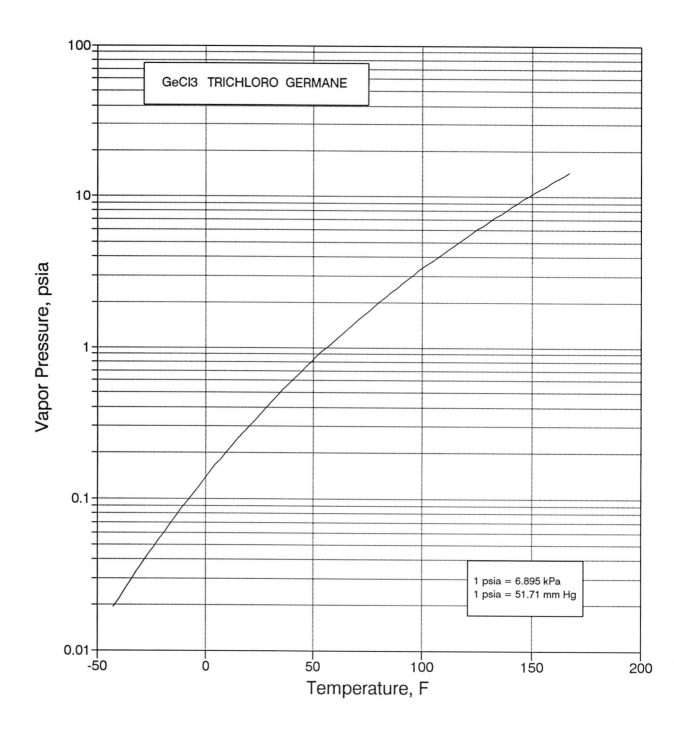

GeCl3 TRICHLORO GERMANE

1 psia = 6.895 kPa
1 psia = 51.71 mm Hg

Vapor Pressure, psia

Temperature, F

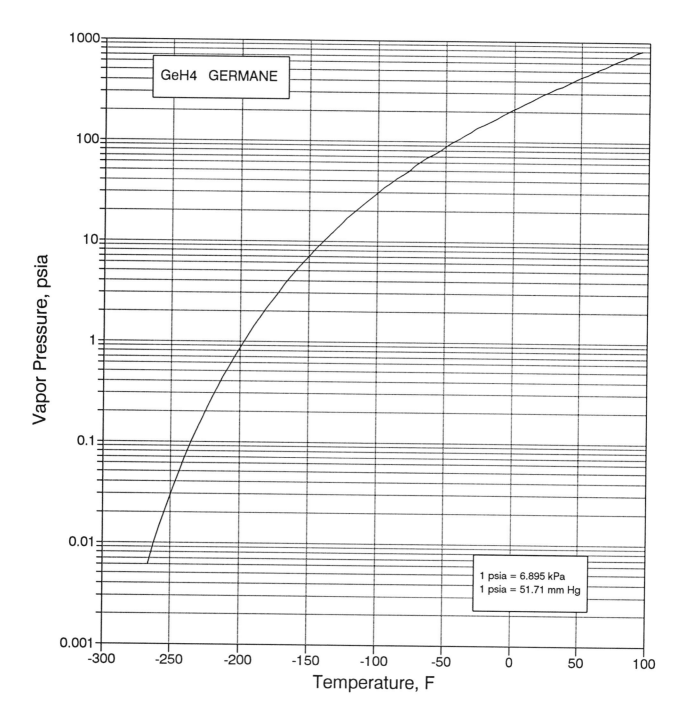

GeH4 GERMANE

Vapor Pressure, psia

Temperature, F

1 psia = 6.895 kPa
1 psia = 51.71 mm Hg

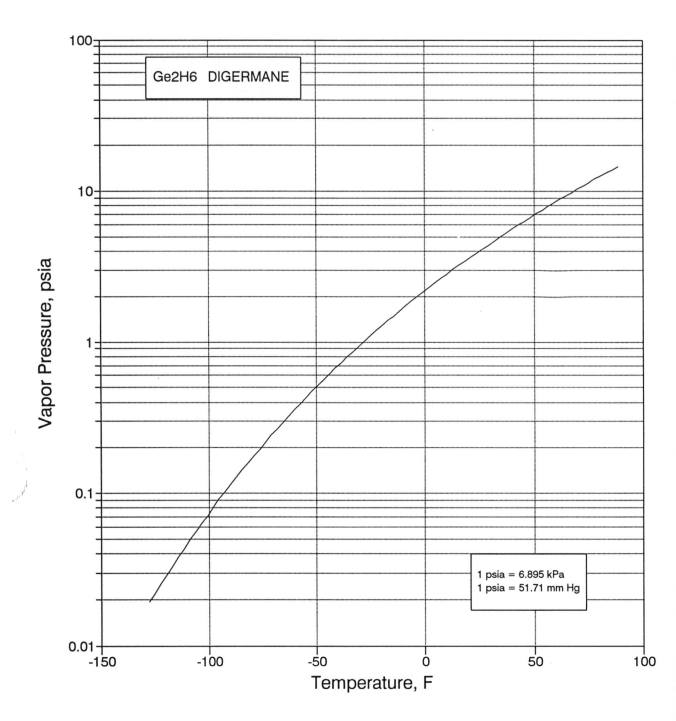

Ge2H6   DIGERMANE

Vapor Pressure, psia

Temperature, F

1 psia = 6.895 kPa
1 psia = 51.71 mm Hg

118

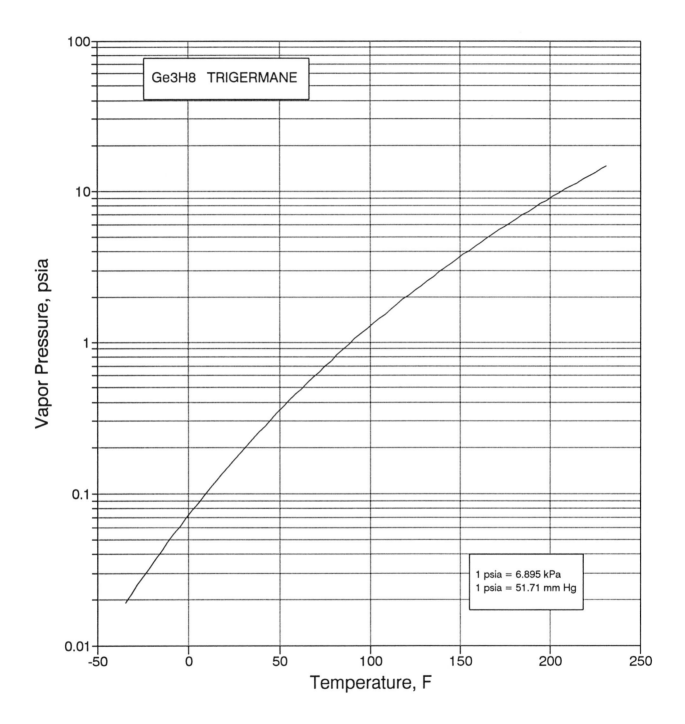

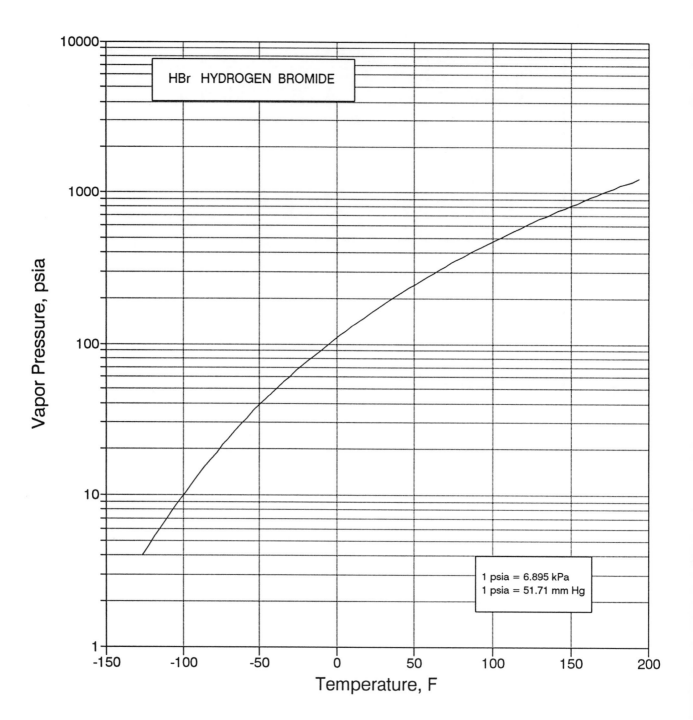

HBr  HYDROGEN BROMIDE

Vapor Pressure, psia

Temperature, F

1 psia = 6.895 kPa
1 psia = 51.71 mm Hg

120

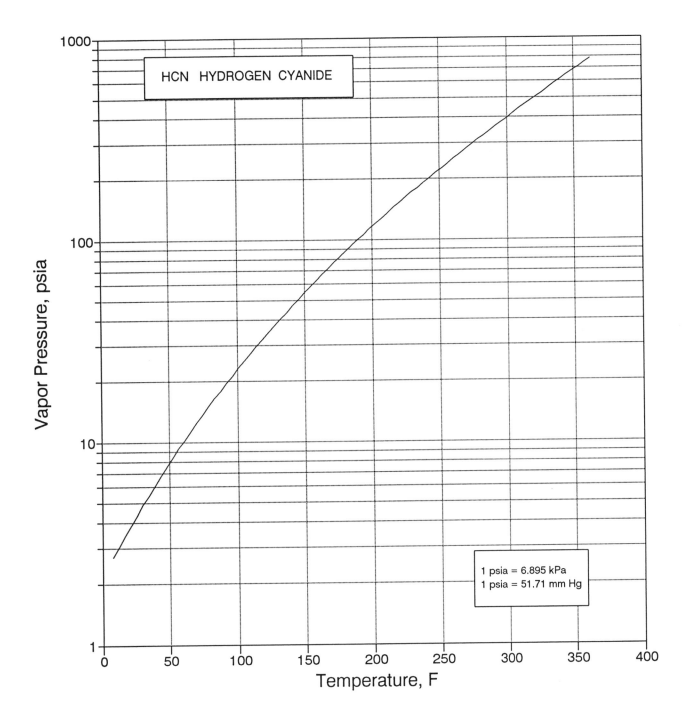

HCN HYDROGEN CYANIDE

1 psia = 6.895 kPa
1 psia = 51.71 mm Hg

Vapor Pressure, psia

Temperature, F

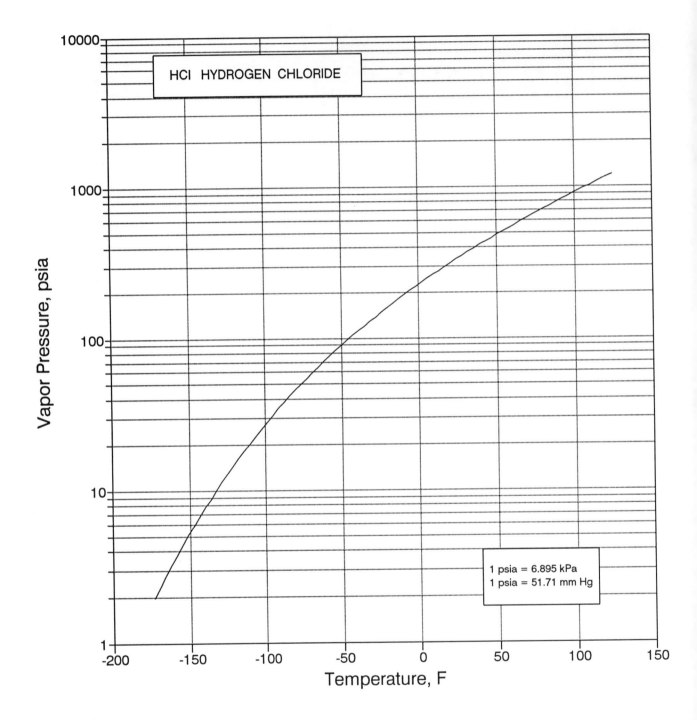

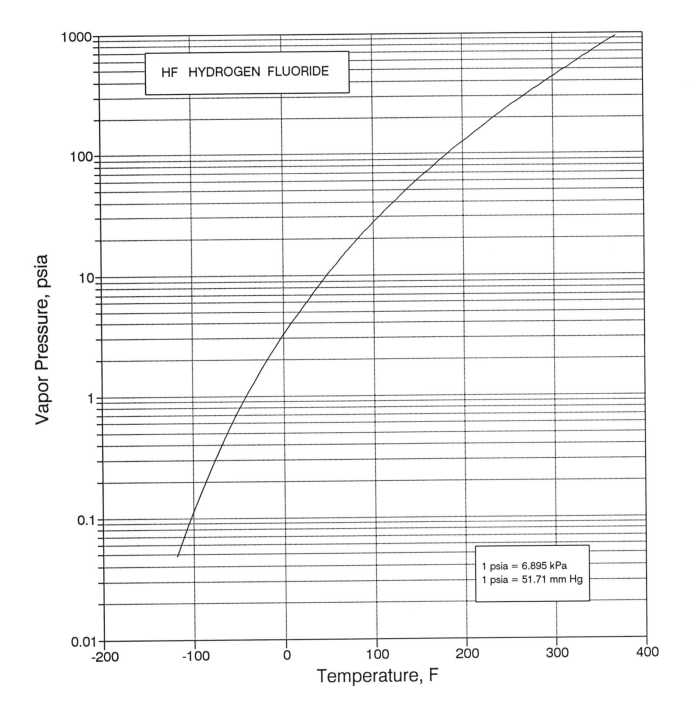

HF HYDROGEN FLUORIDE

1 psia = 6.895 kPa
1 psia = 51.71 mm Hg

Vapor Pressure, psia

Temperature, F

123

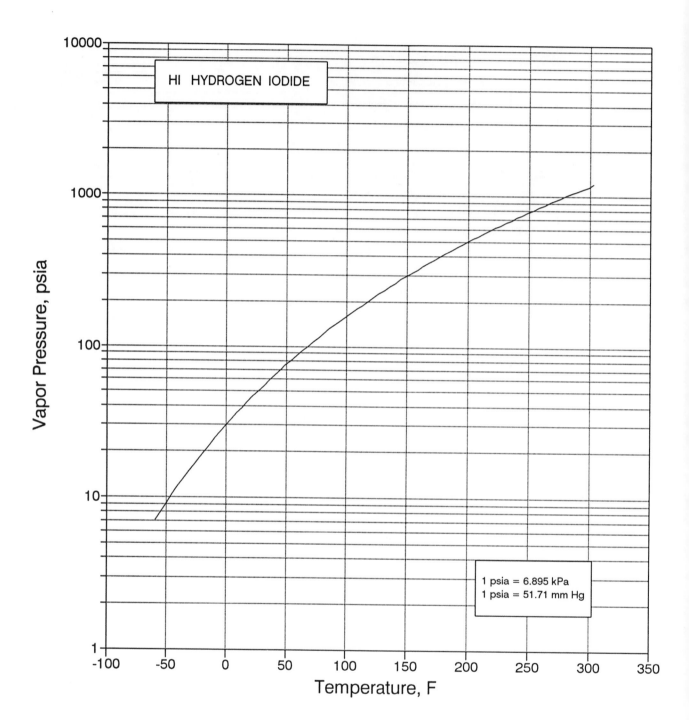

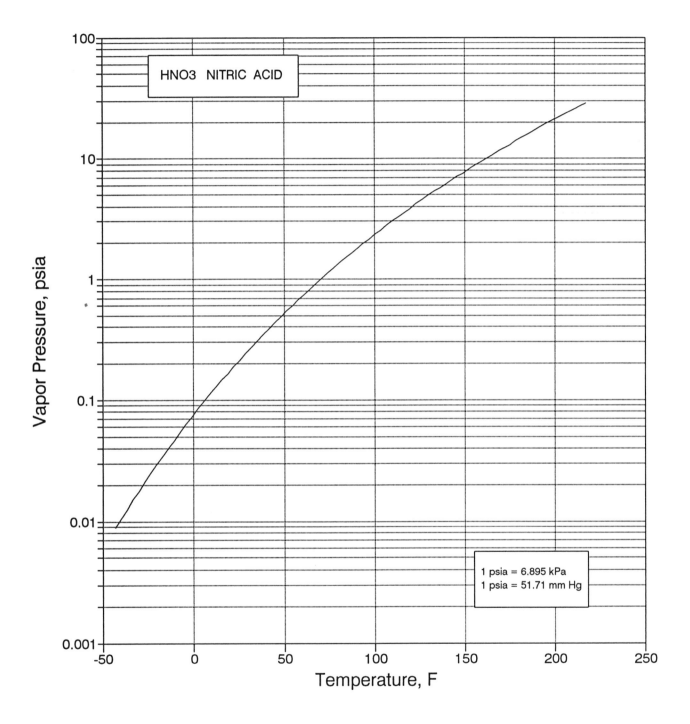

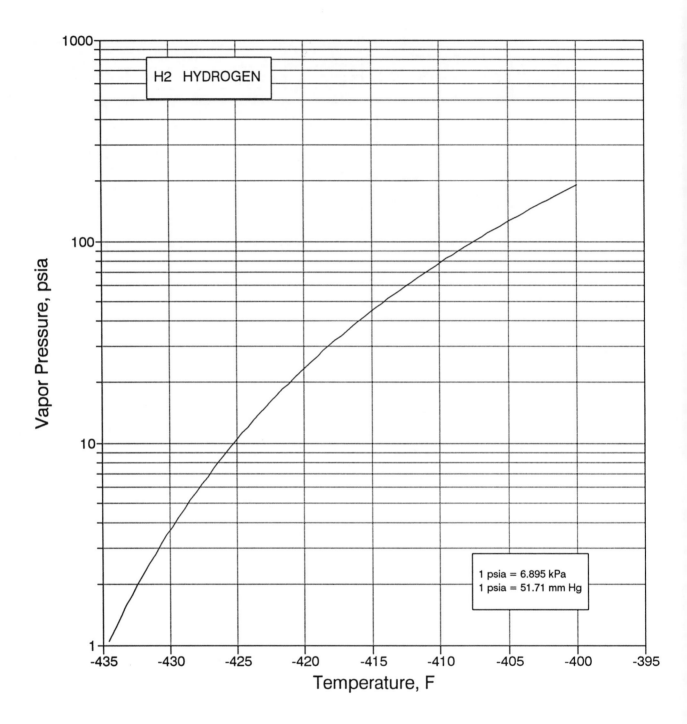

H2 HYDROGEN

Vapor Pressure, psia

Temperature, F

1 psia = 6.895 kPa
1 psia = 51.71 mm Hg

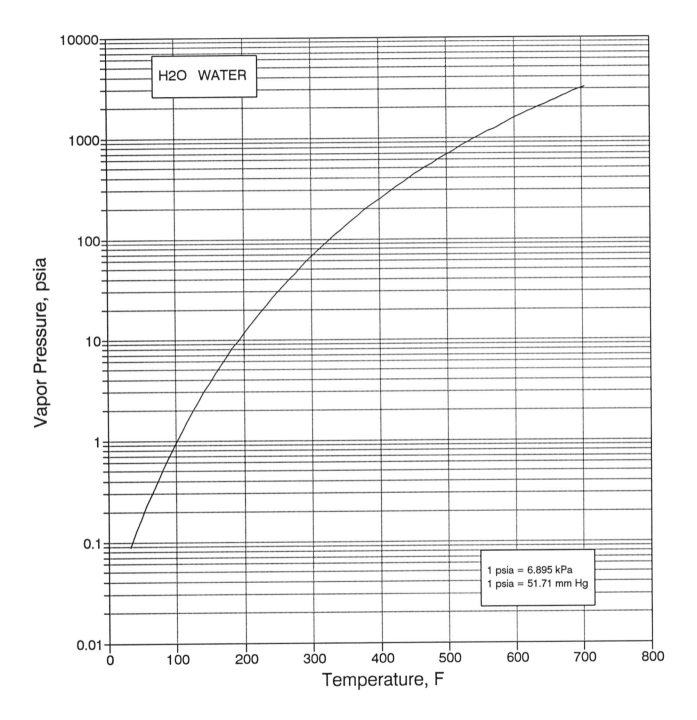

H2O   WATER

1 psia = 6.895 kPa
1 psia = 51.71 mm Hg

Vapor Pressure, psia

Temperature, F

127

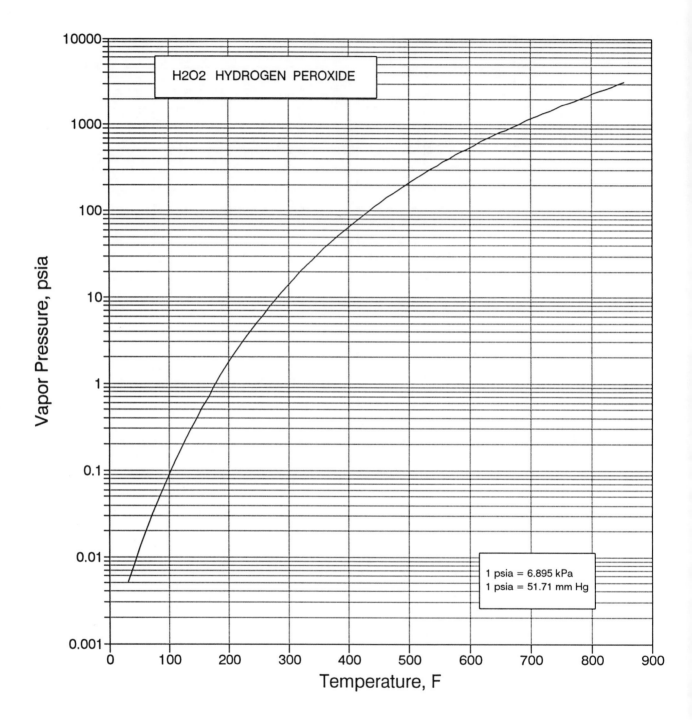

H2O2  HYDROGEN PEROXIDE

1 psia = 6.895 kPa
1 psia = 51.71 mm Hg

128

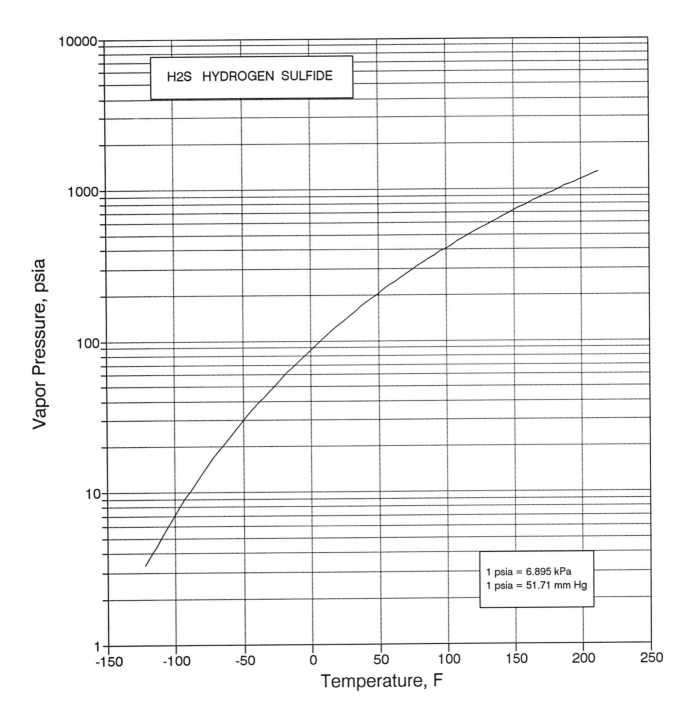

H2S  HYDROGEN SULFIDE

Vapor Pressure, psia

Temperature, F

1 psia = 6.895 kPa
1 psia = 51.71 mm Hg

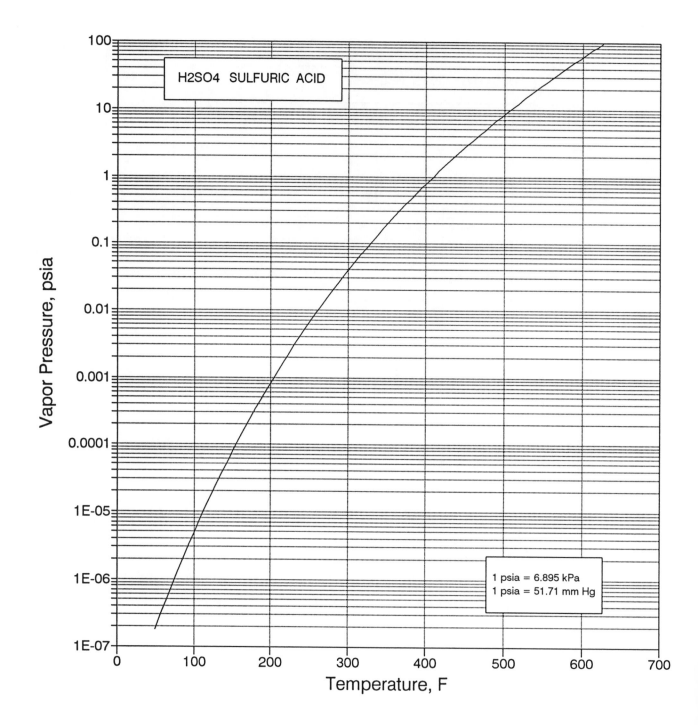

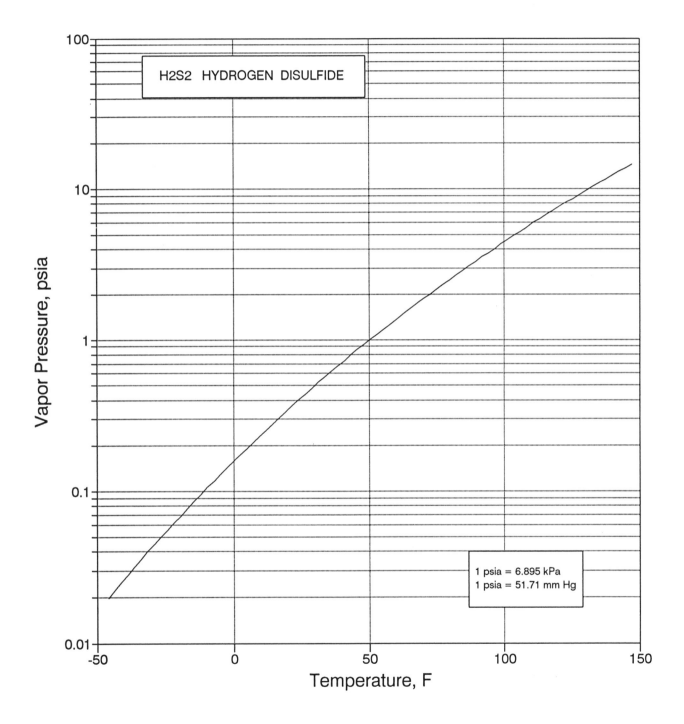

1 psia = 6.895 kPa
1 psia = 51.71 mm Hg

H2S2 HYDROGEN DISULFIDE

Vapor Pressure, psia

Temperature, F

131

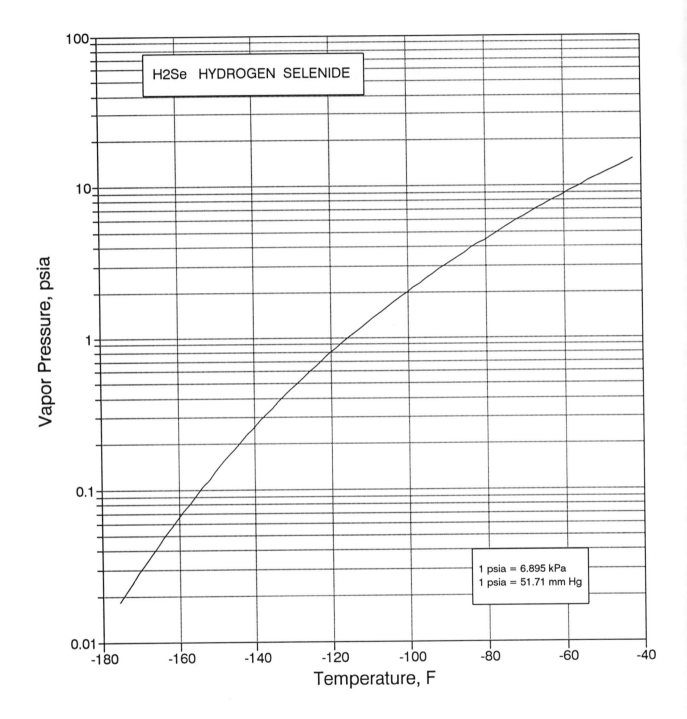

H2Se  HYDROGEN SELENIDE

Vapor Pressure, psia

Temperature, F

1 psia = 6.895 kPa
1 psia = 51.71 mm Hg

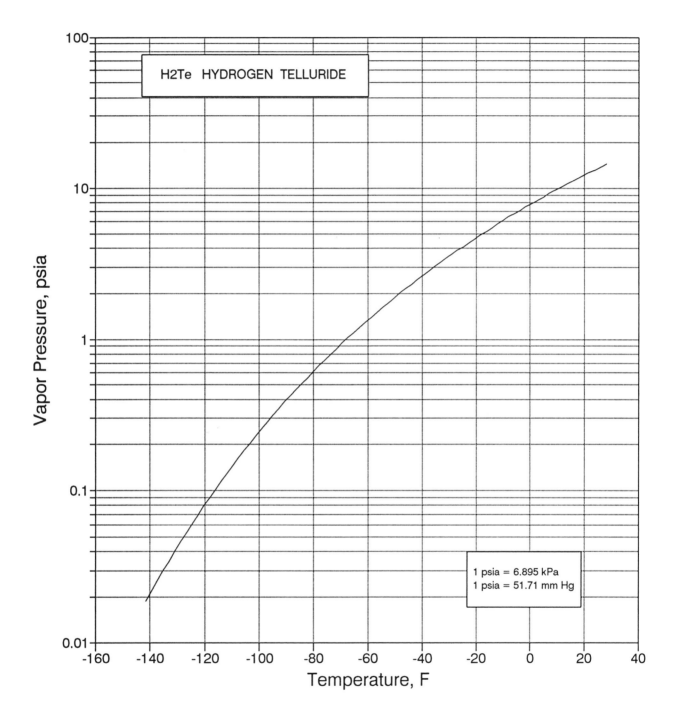

Vapor Pressure, psia

Temperature, F

H2Te  HYDROGEN TELLURIDE

1 psia = 6.895 kPa
1 psia = 51.71 mm Hg

133

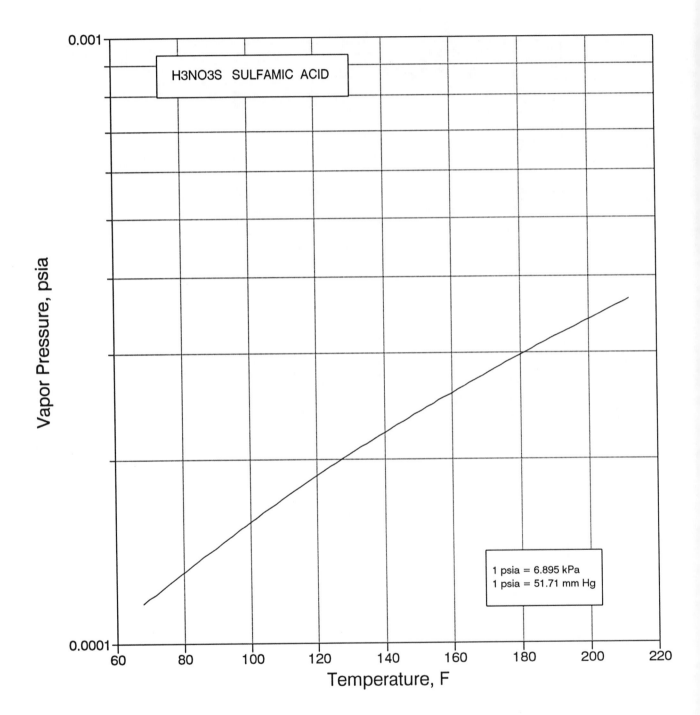

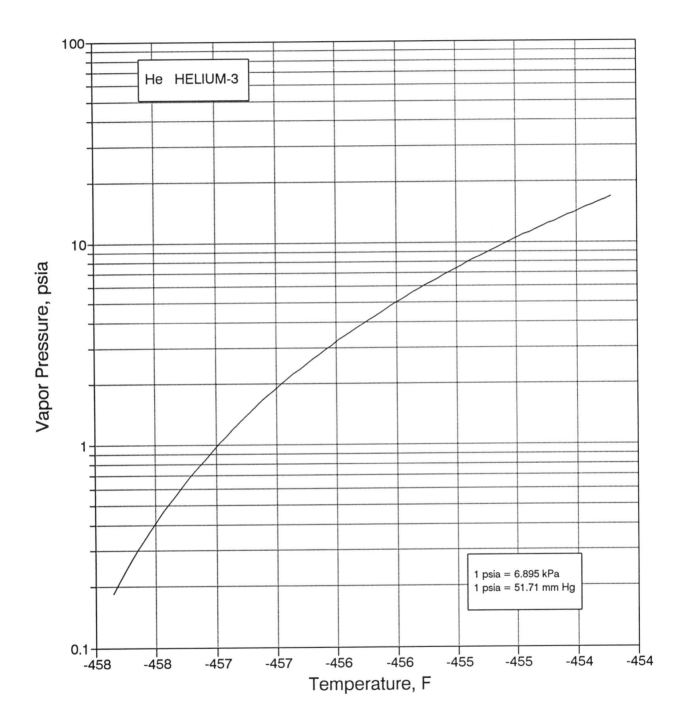

He HELIUM-3

Vapor Pressure, psia

Temperature, F

1 psia = 6.895 kPa
1 psia = 51.71 mm Hg

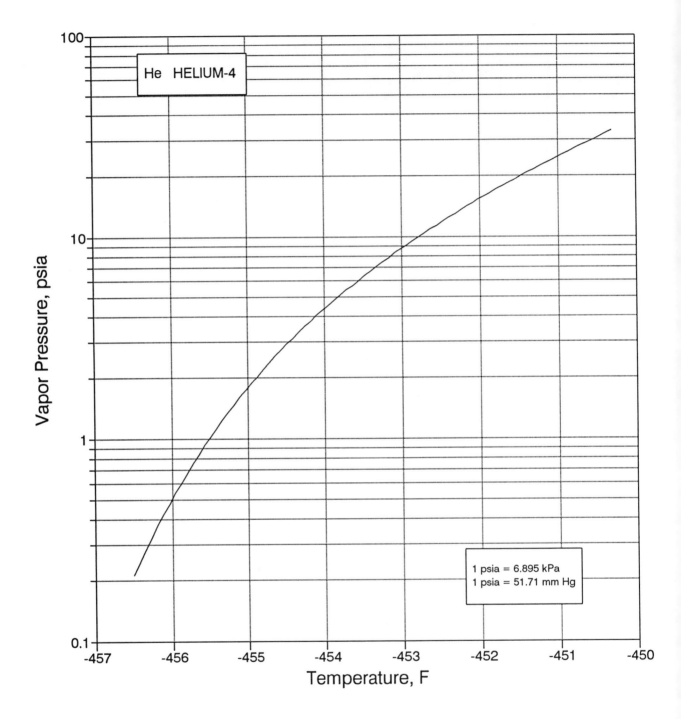

Vapor Pressure, psia

He   HELIUM-4

1 psia = 6.895 kPa
1 psia = 51.71 mm Hg

Temperature, F

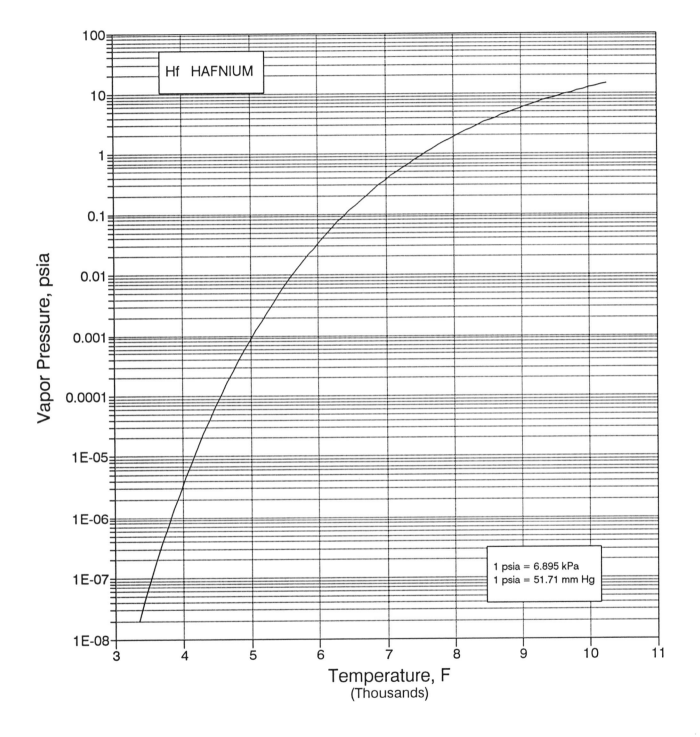

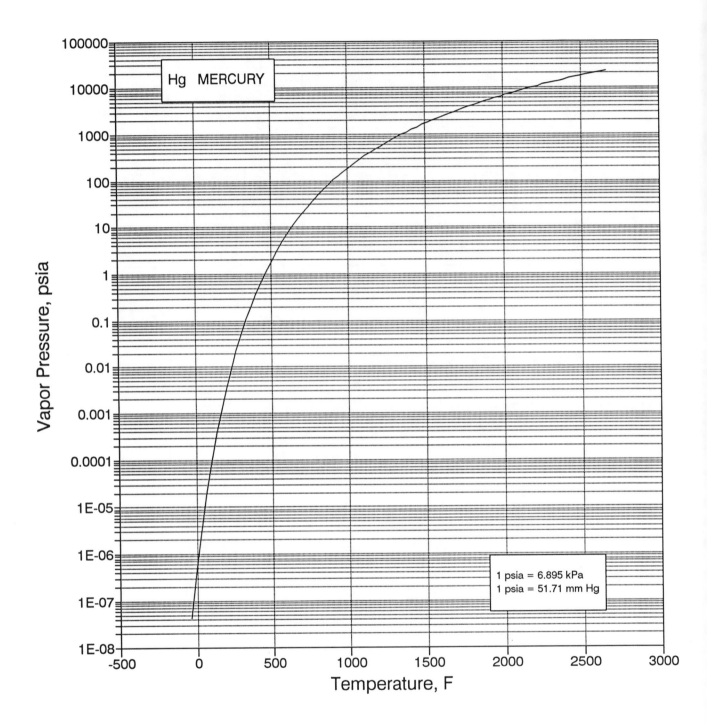

Hg   MERCURY

1 psia = 6.895 kPa
1 psia = 51.71 mm Hg

Vapor Pressure, psia

Temperature, F

138

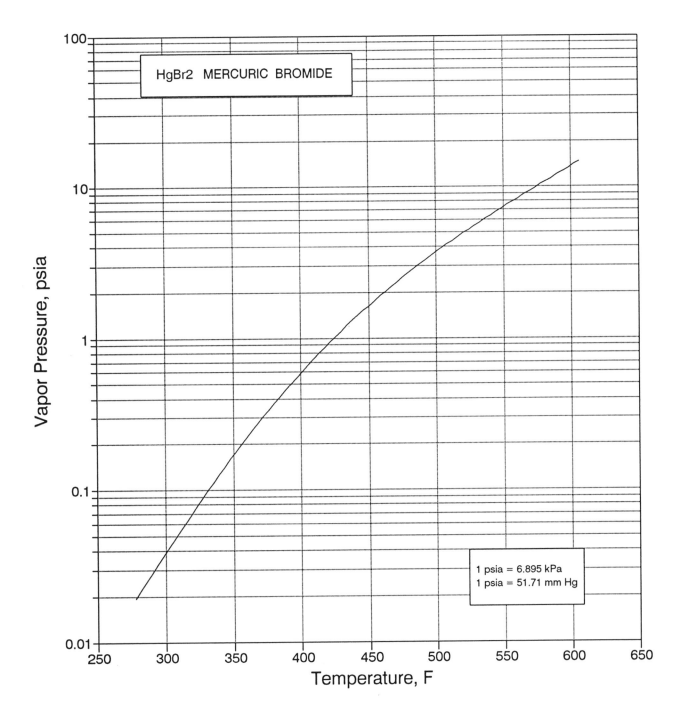

HgBr2  MERCURIC  BROMIDE

Vapor Pressure, psia

Temperature, F

1 psia = 6.895 kPa
1 psia = 51.71 mm Hg

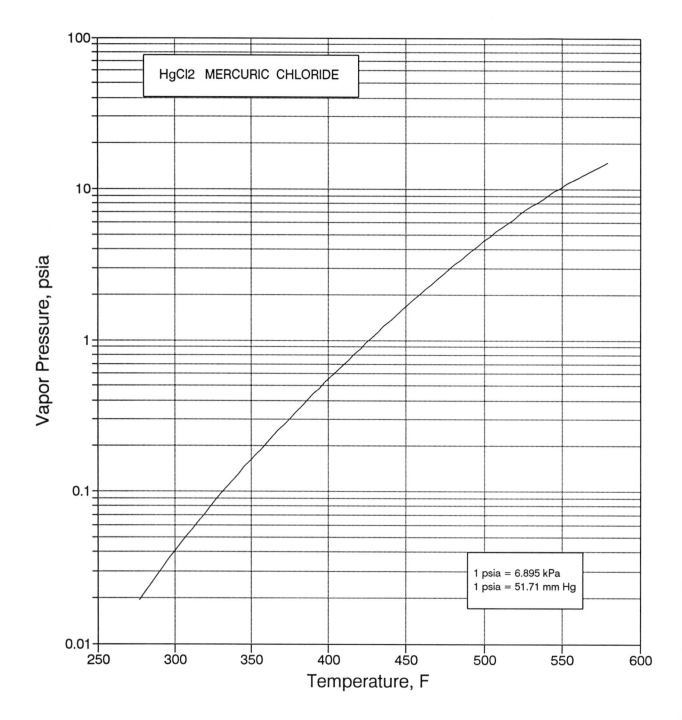

HgCl2  MERCURIC CHLORIDE

Vapor Pressure, psia

Temperature, F

1 psia = 6.895 kPa
1 psia = 51.71 mm Hg

140

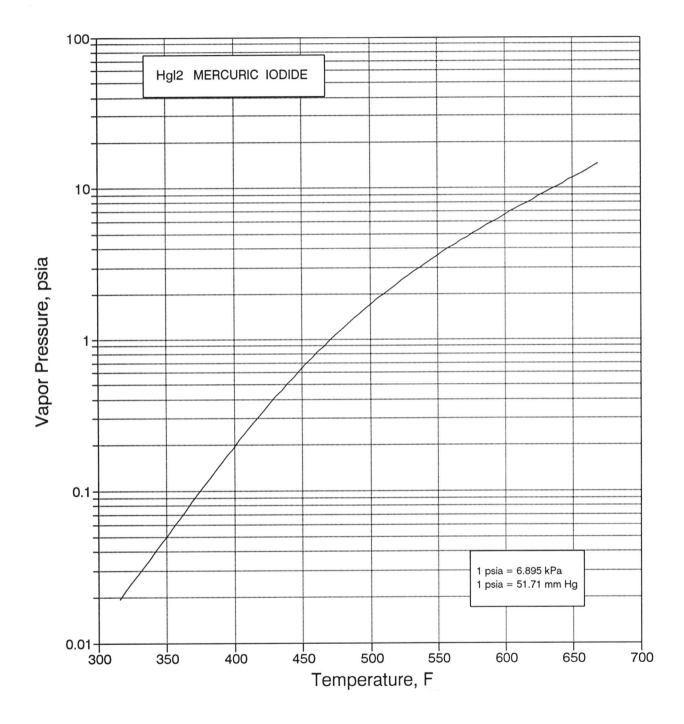

HgI2 MERCURIC IODIDE

Vapor Pressure, psia

Temperature, F

1 psia = 6.895 kPa
1 psia = 51.71 mm Hg

141

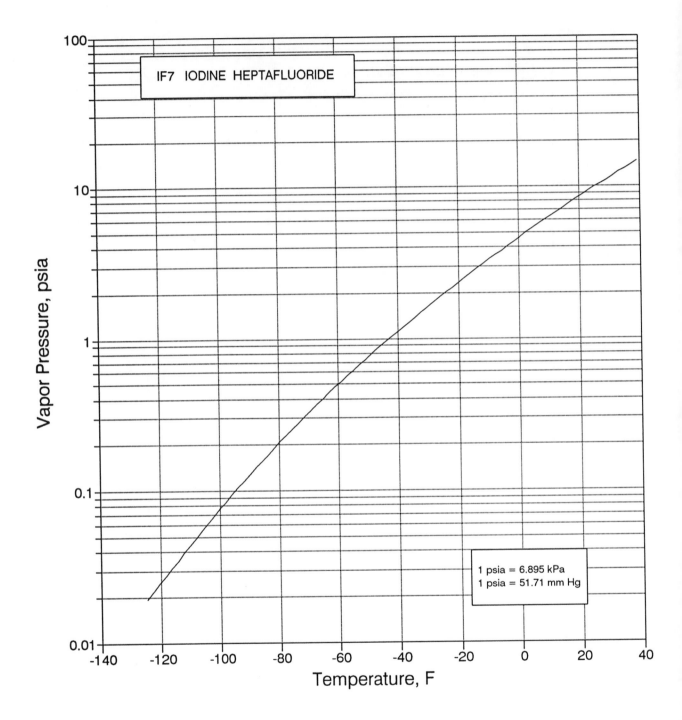

IF7  IODINE HEPTAFLUORIDE

Vapor Pressure, psia

Temperature, F

1 psia = 6.895 kPa
1 psia = 51.71 mm Hg

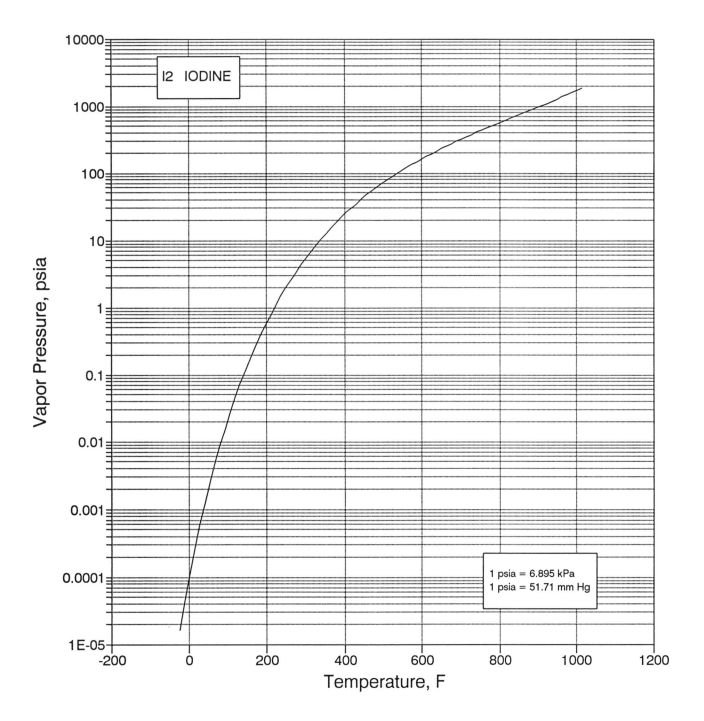

I2 IODINE

1 psia = 6.895 kPa
1 psia = 51.71 mm Hg

Vapor Pressure, psia

Temperature, F

143

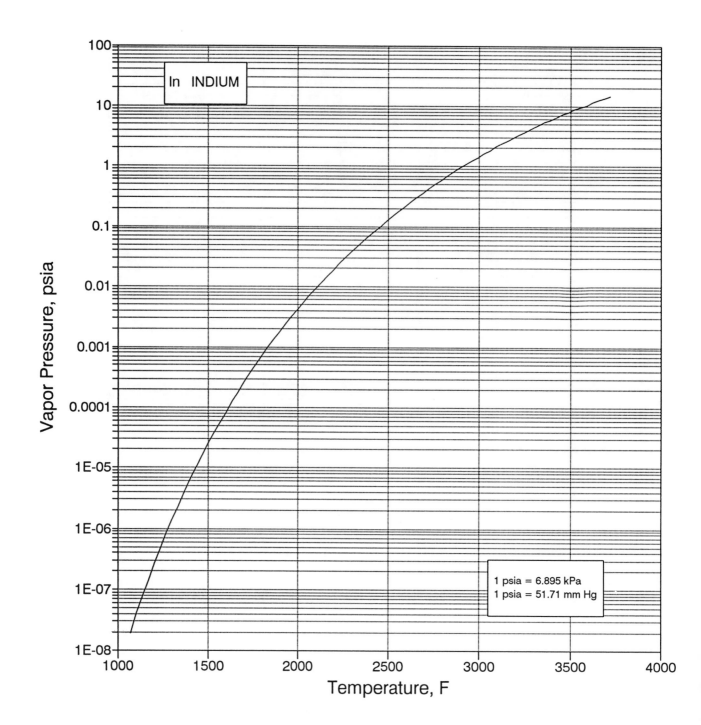

144

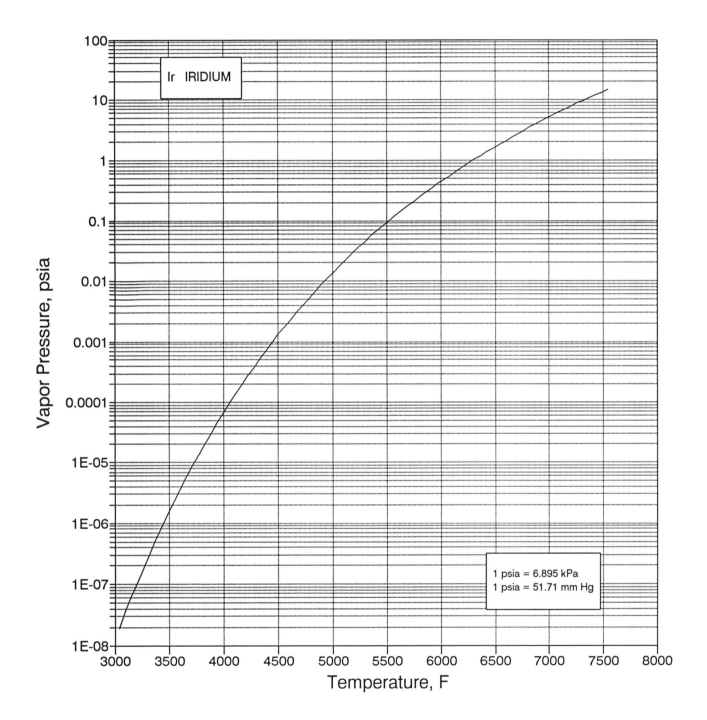

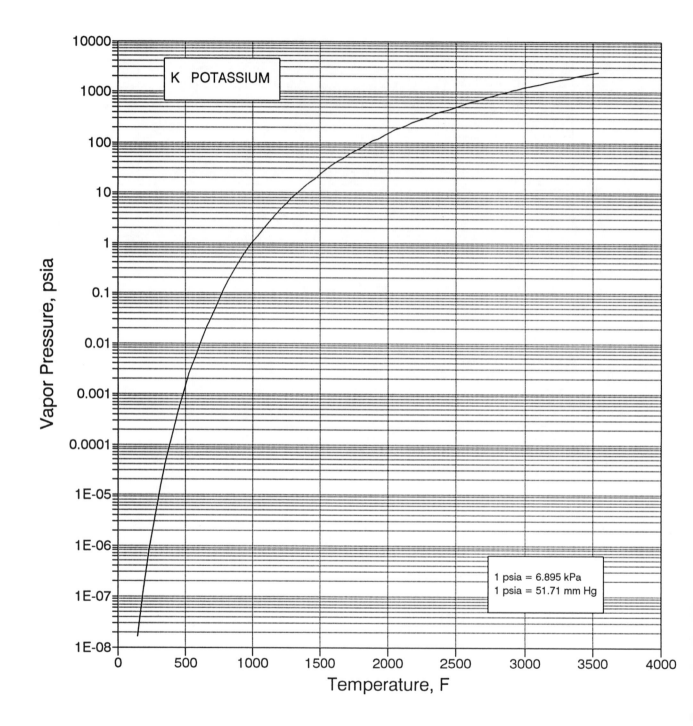

K  POTASSIUM

1 psia = 6.895 kPa
1 psia = 51.71 mm Hg

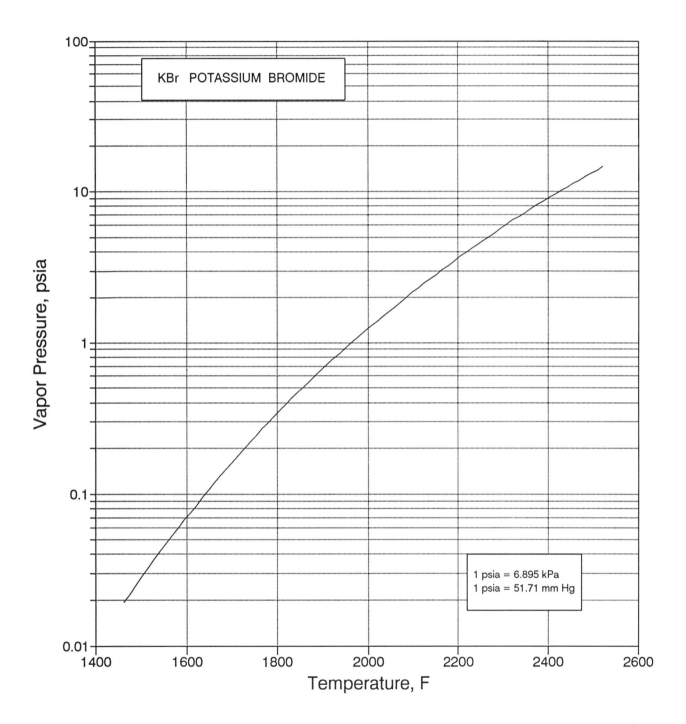

KBr  POTASSIUM BROMIDE

Vapor Pressure, psia

Temperature, F

1 psia = 6.895 kPa
1 psia = 51.71 mm Hg

147

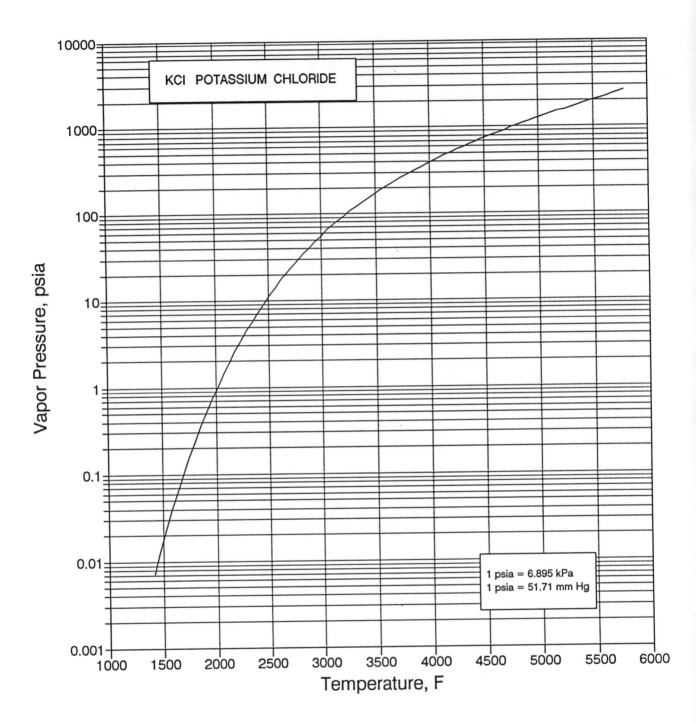

KCl POTASSIUM CHLORIDE

Vapor Pressure, psia

Temperature, F

1 psia = 6.895 kPa
1 psia = 51.71 mm Hg

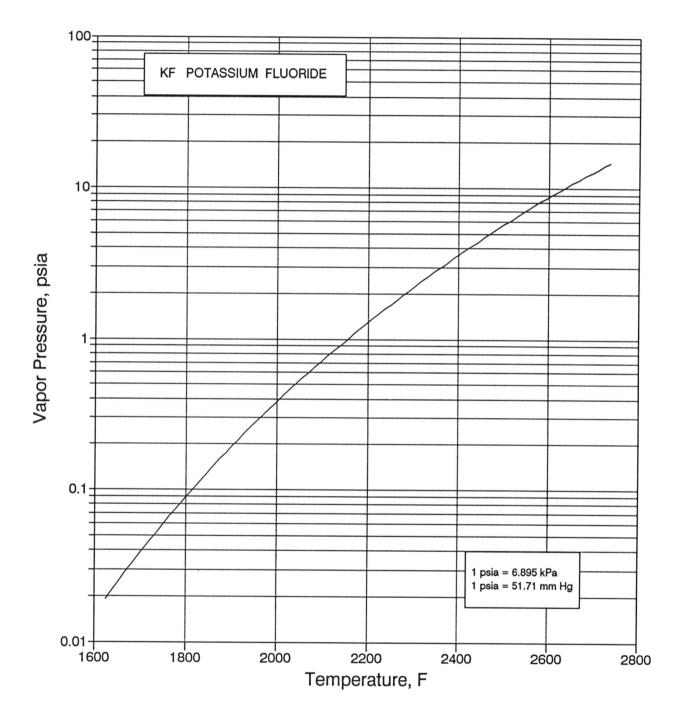

KF POTASSIUM FLUORIDE

Vapor Pressure, psia

Temperature, F

1 psia = 6.895 kPa
1 psia = 51.71 mm Hg

149

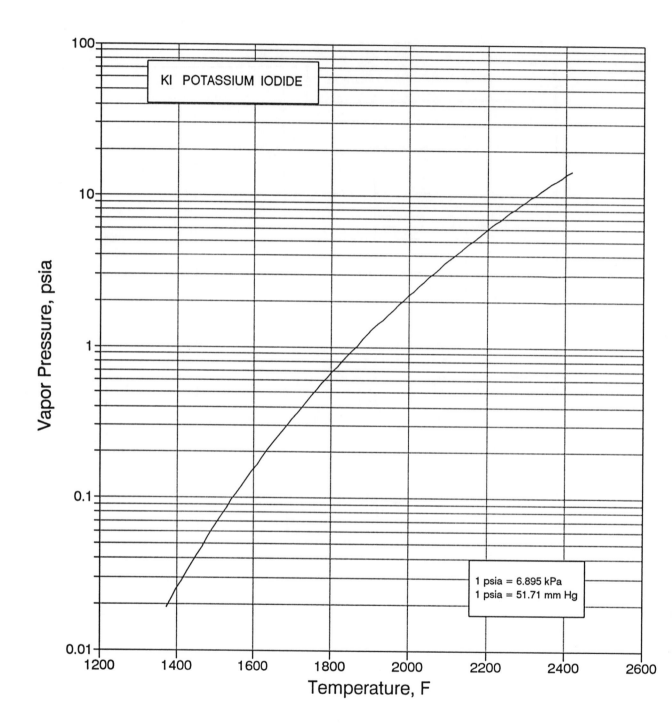

KI  POTASSIUM IODIDE

Vapor Pressure, psia

Temperature, F

1 psia = 6.895 kPa
1 psia = 51.71 mm Hg

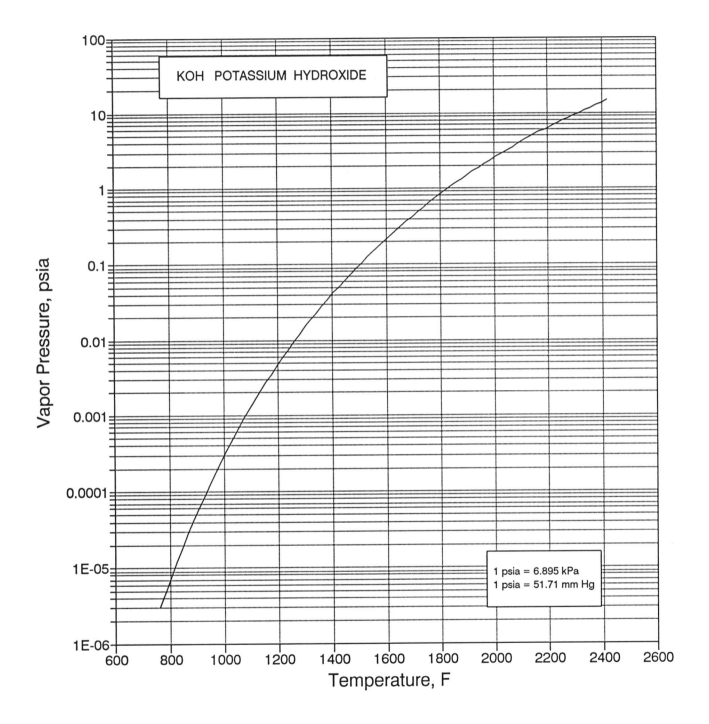

KOH POTASSIUM HYDROXIDE

1 psia = 6.895 kPa
1 psia = 51.71 mm Hg

Vapor Pressure, psia

Temperature, F

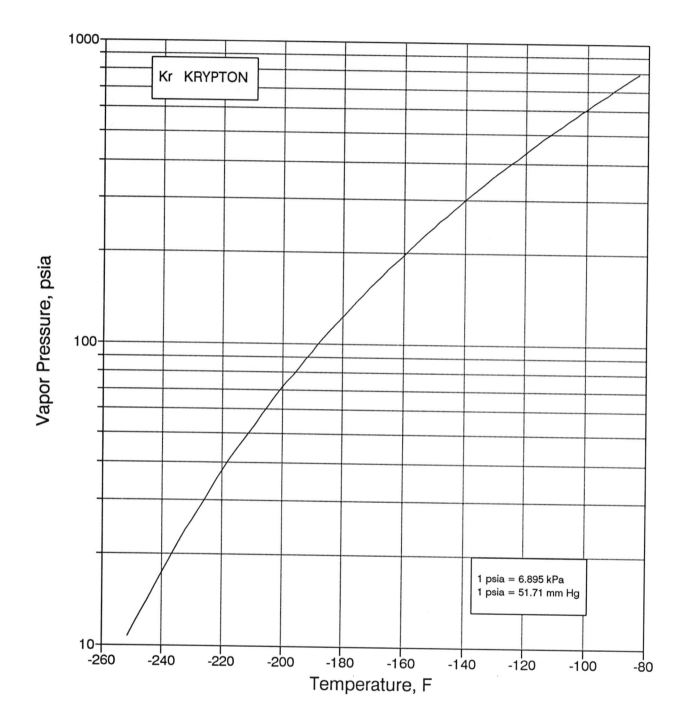

Kr  KRYPTON

Vapor Pressure, psia

Temperature, F

1 psia = 6.895 kPa
1 psia = 51.71 mm Hg

152

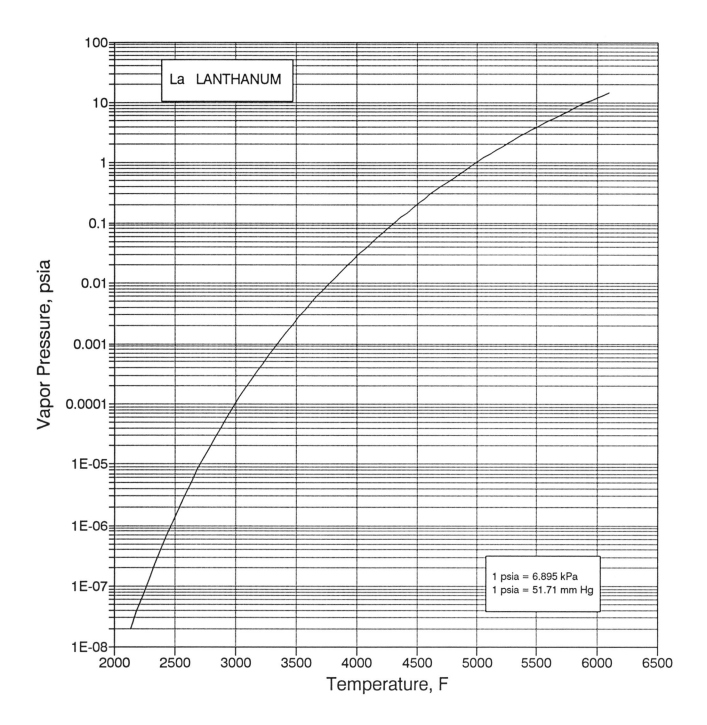

La LANTHANUM

1 psia = 6.895 kPa
1 psia = 51.71 mm Hg

Vapor Pressure, psia

Temperature, F

153

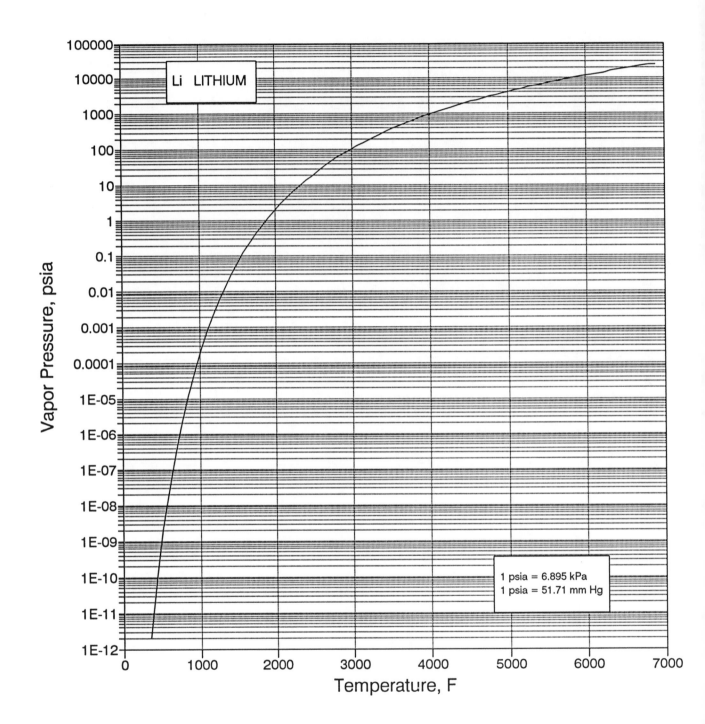

Li   LITHIUM

Vapor Pressure, psia

Temperature, F

1 psia = 6.895 kPa
1 psia = 51.71 mm Hg

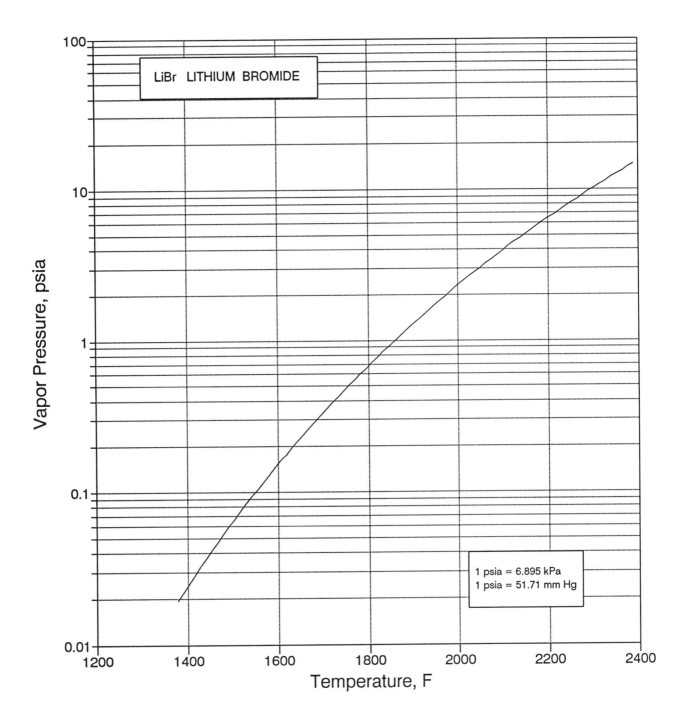

LiBr LITHIUM BROMIDE

1 psia = 6.895 kPa
1 psia = 51.71 mm Hg

Vapor Pressure, psia

Temperature, F

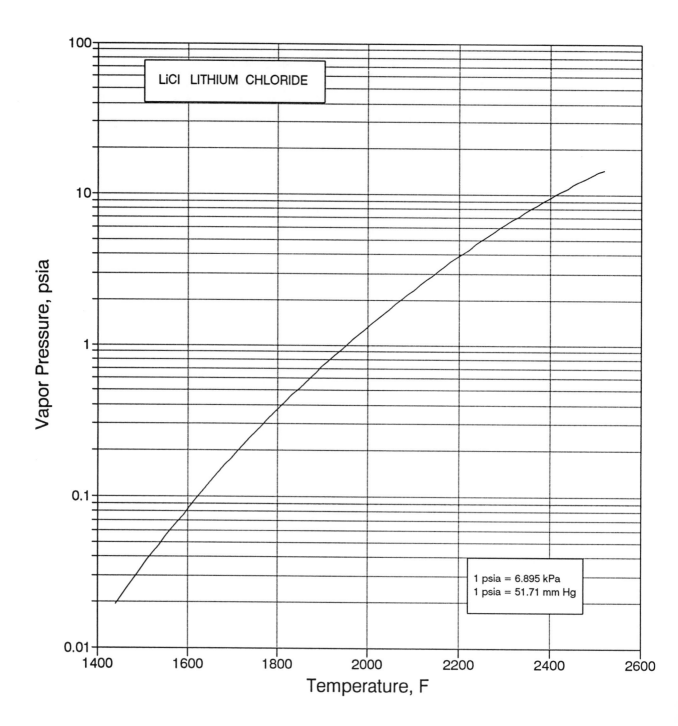

LiCl LITHIUM CHLORIDE

Vapor Pressure, psia

Temperature, F

1 psia = 6.895 kPa
1 psia = 51.71 mm Hg

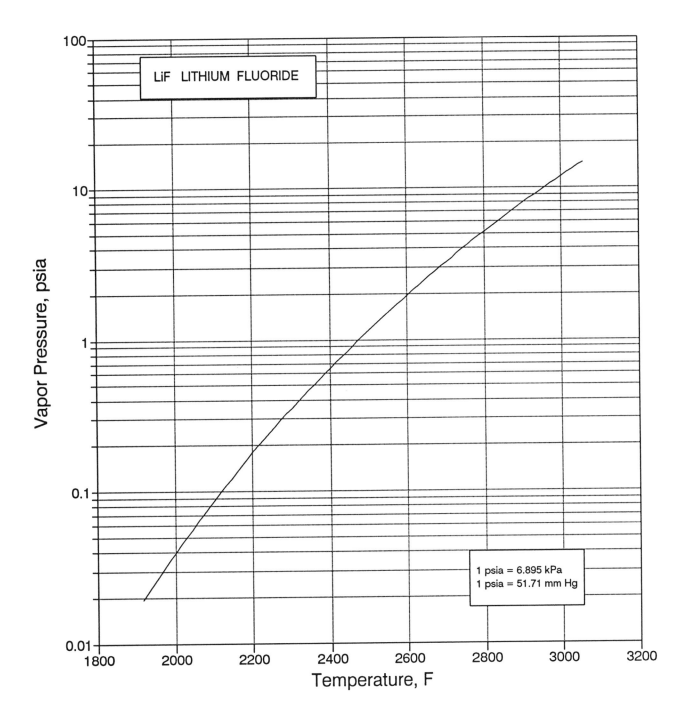

LiF  LITHIUM FLUORIDE

Vapor Pressure, psia

Temperature, F

1 psia = 6.895 kPa
1 psia = 51.71 mm Hg

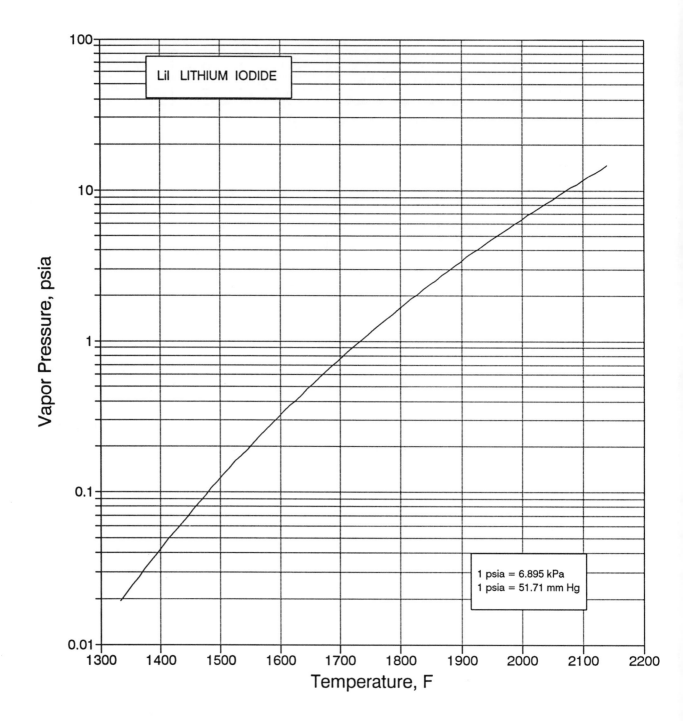

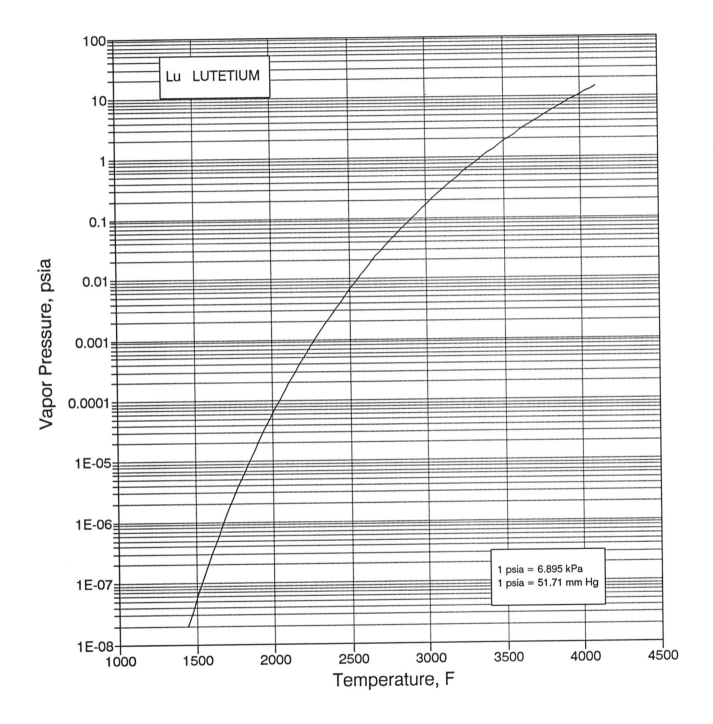

Lu  LUTETIUM

1 psia = 6.895 kPa
1 psia = 51.71 mm Hg

Vapor Pressure, psia

Temperature, F

159

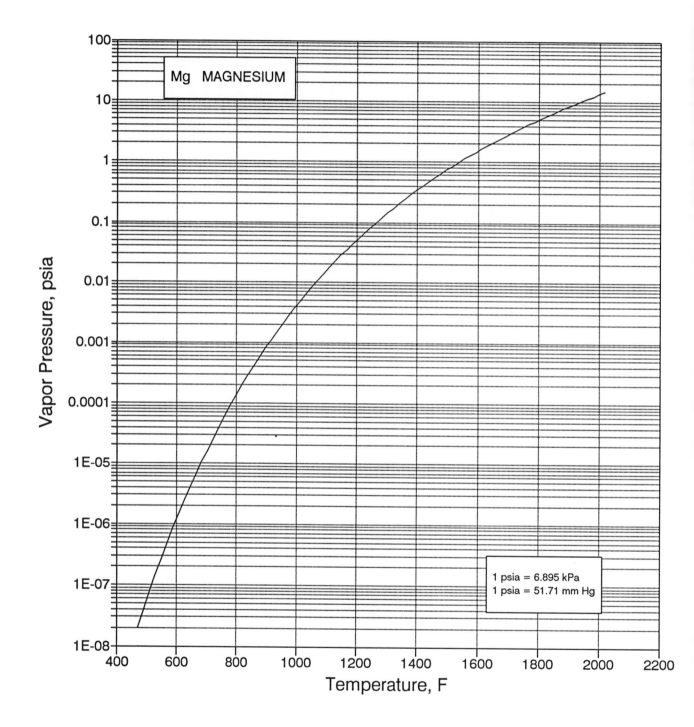

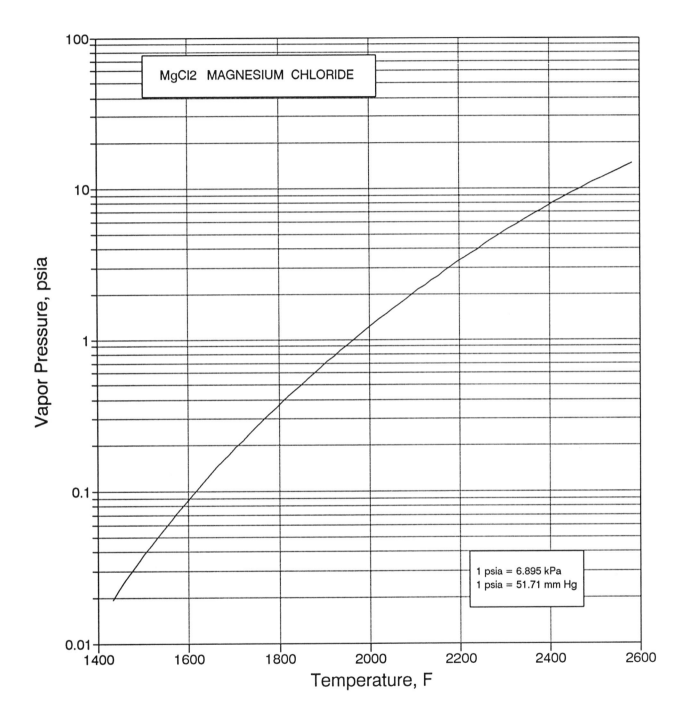

MgCl2  MAGNESIUM CHLORIDE

Vapor Pressure, psia

Temperature, F

1 psia = 6.895 kPa
1 psia = 51.71 mm Hg

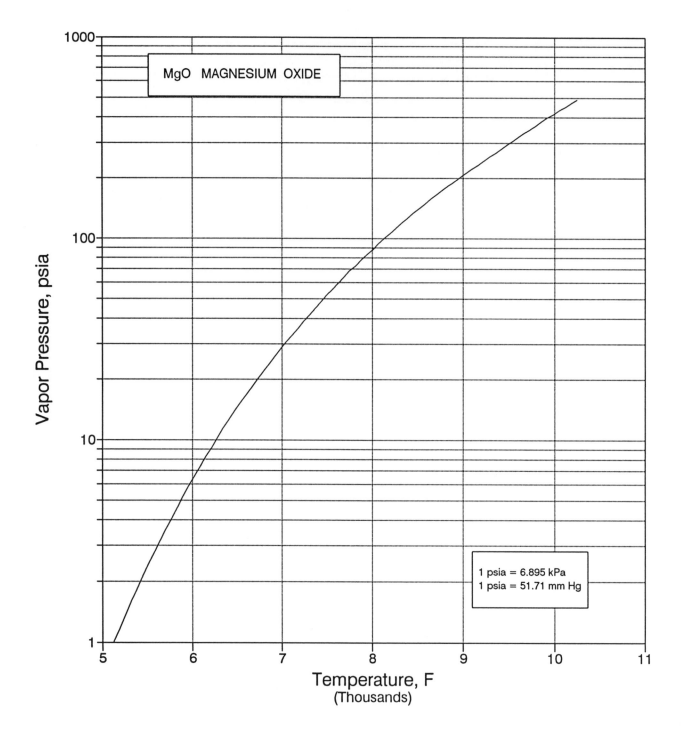

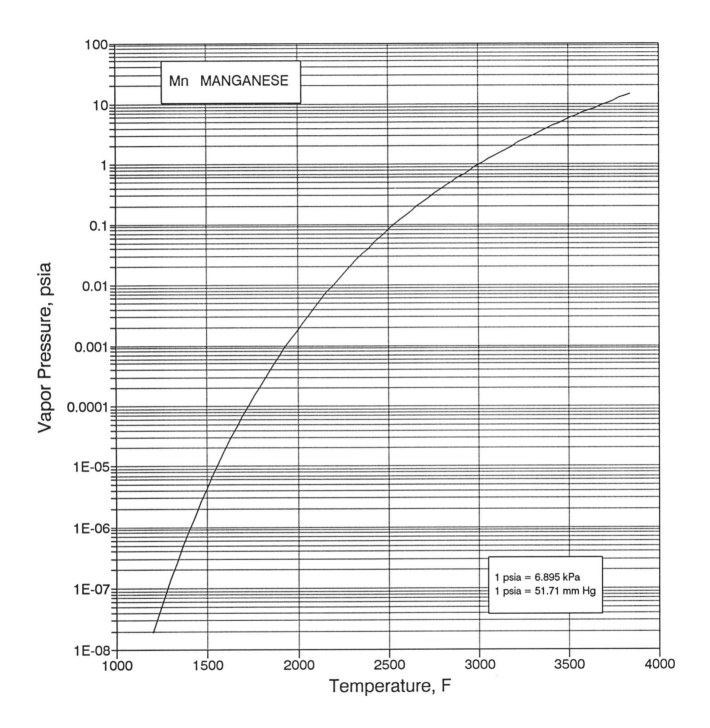

Mn MANGANESE

1 psia = 6.895 kPa
1 psia = 51.71 mm Hg

Vapor Pressure, psia

Temperature, F

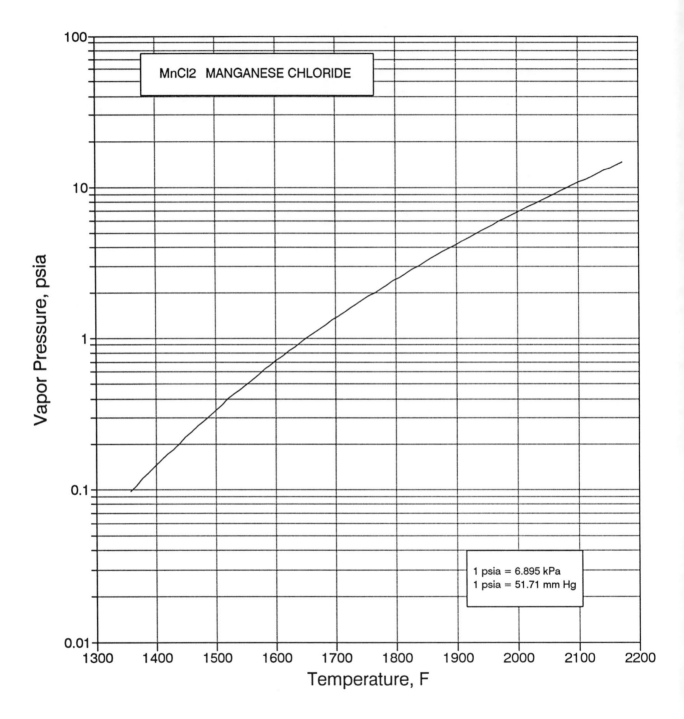

MnCl2 MANGANESE CHLORIDE

1 psia = 6.895 kPa
1 psia = 51.71 mm Hg

Vapor Pressure, psia

Temperature, F

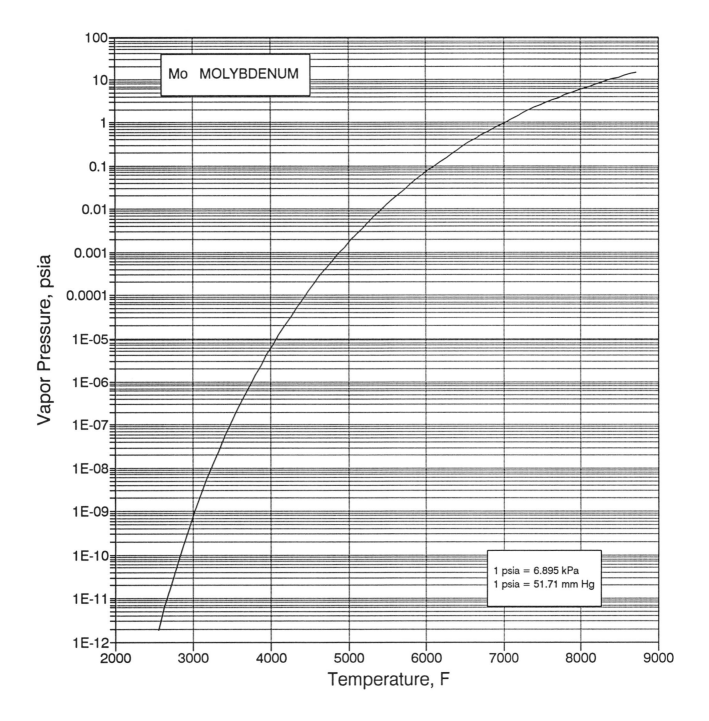

The graph shows Vapor Pressure (psia) vs Temperature (F) for Mo MOLYBDENUM.

Mo MOLYBDENUM

1 psia = 6.895 kPa
1 psia = 51.71 mm Hg

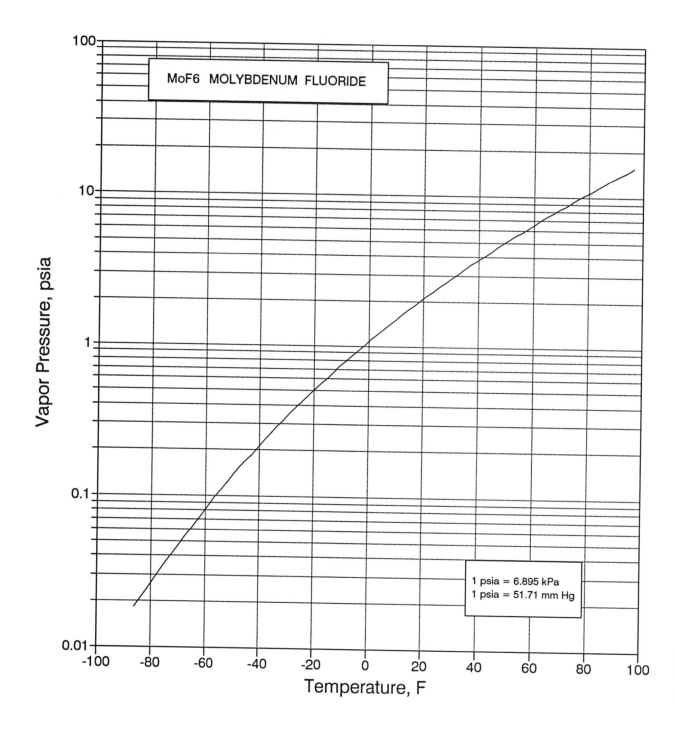

MoF6 MOLYBDENUM FLUORIDE

1 psia = 6.895 kPa
1 psia = 51.71 mm Hg

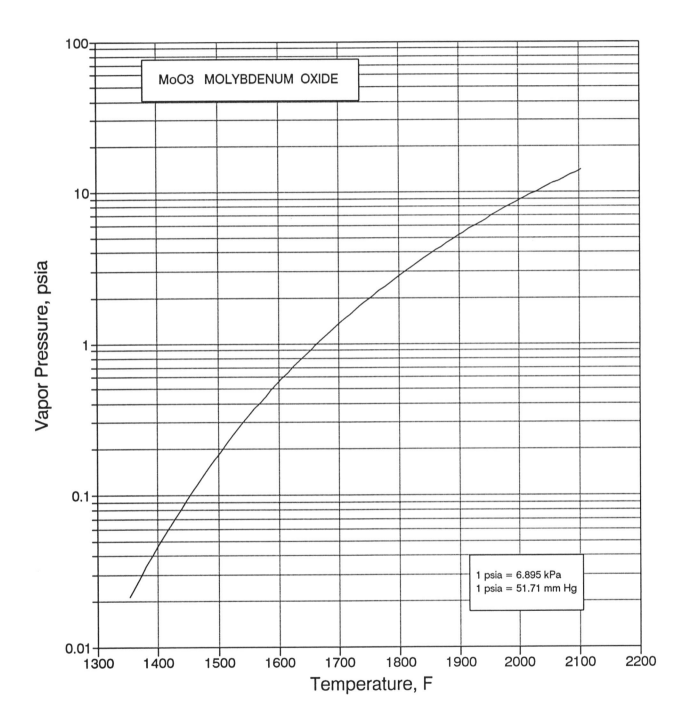

MoO3  MOLYBDENUM OXIDE

Vapor Pressure, psia

Temperature, F

1 psia = 6.895 kPa
1 psia = 51.71 mm Hg

167

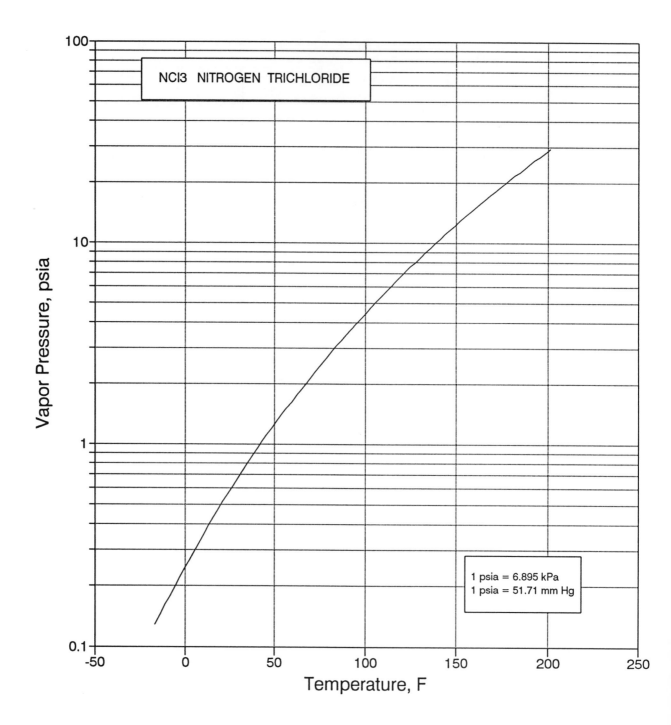

NCl3 NITROGEN TRICHLORIDE

Vapor Pressure, psia

Temperature, F

1 psia = 6.895 kPa
1 psia = 51.71 mm Hg

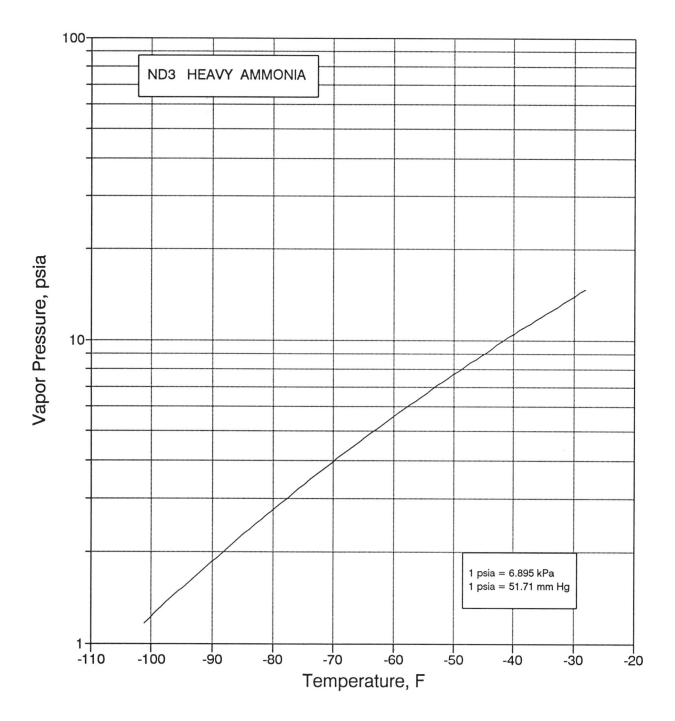

ND3  HEAVY AMMONIA

1 psia = 6.895 kPa
1 psia = 51.71 mm Hg

Vapor Pressure, psia

Temperature, F

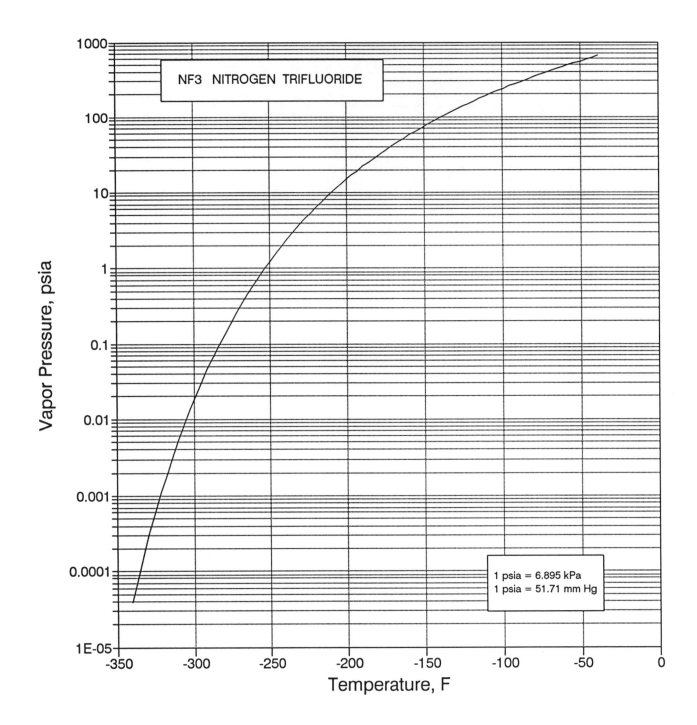

NF3   NITROGEN  TRIFLUORIDE

Vapor Pressure, psia

Temperature, F

1 psia = 6.895 kPa
1 psia = 51.71 mm Hg

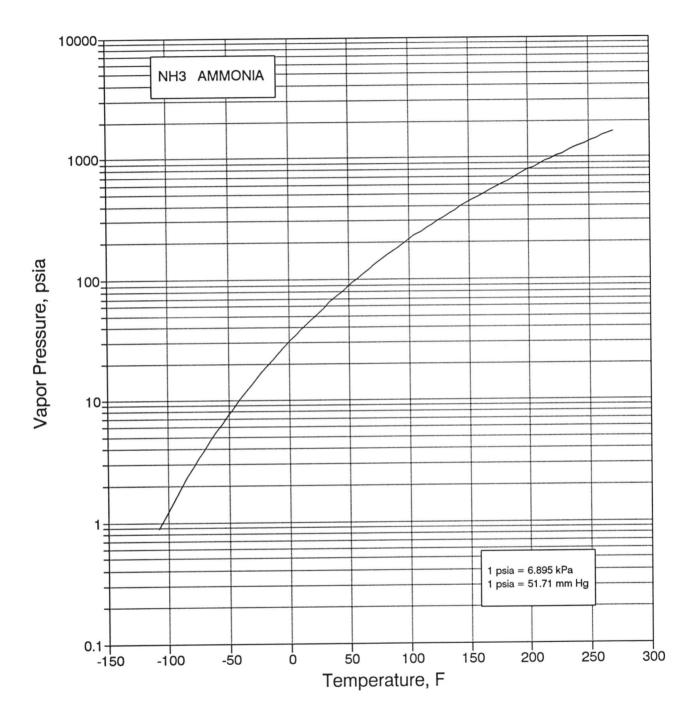

NH3   AMMONIA

1 psia = 6.895 kPa
1 psia = 51.71 mm Hg

Vapor Pressure, psia

Temperature, F

171

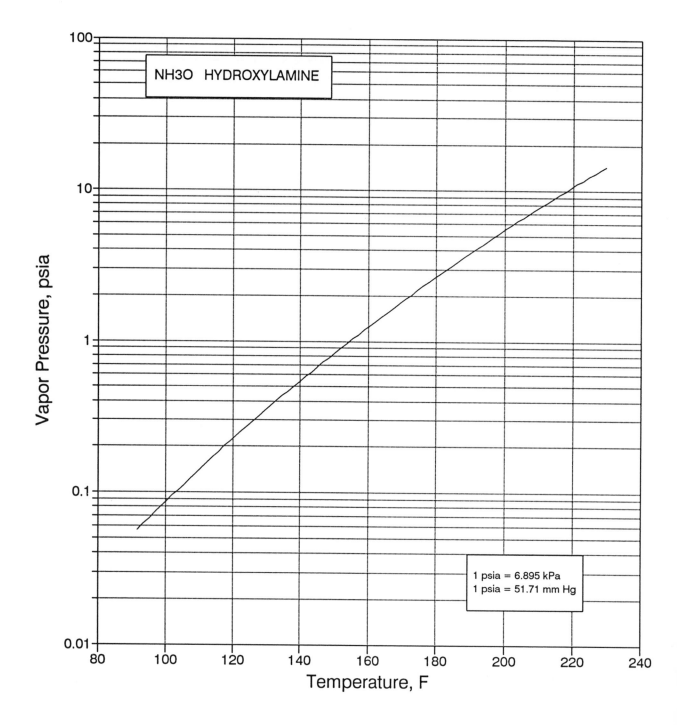

NH3O  HYDROXYLAMINE

Vapor Pressure, psia

Temperature, F

1 psia = 6.895 kPa
1 psia = 51.71 mm Hg

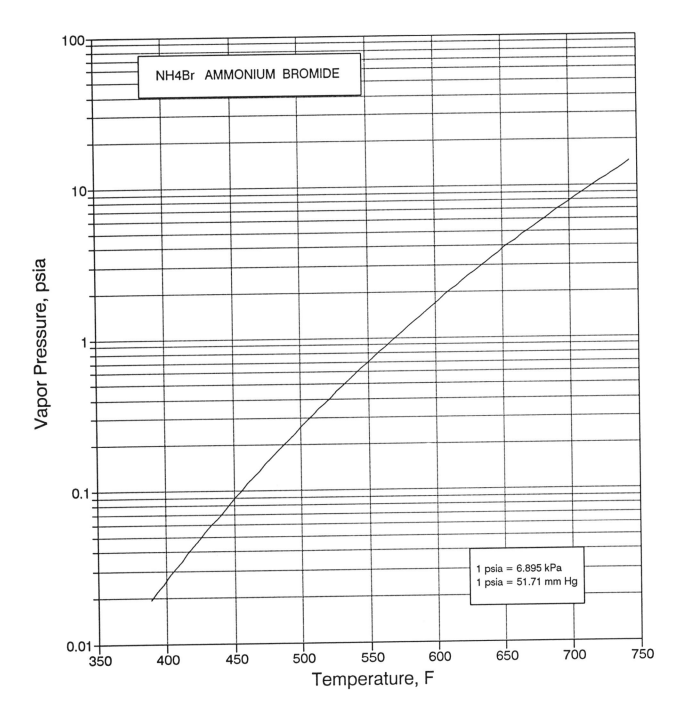

NH4Br  AMMONIUM  BROMIDE

Vapor Pressure, psia

Temperature, F

1 psia = 6.895 kPa
1 psia = 51.71 mm Hg

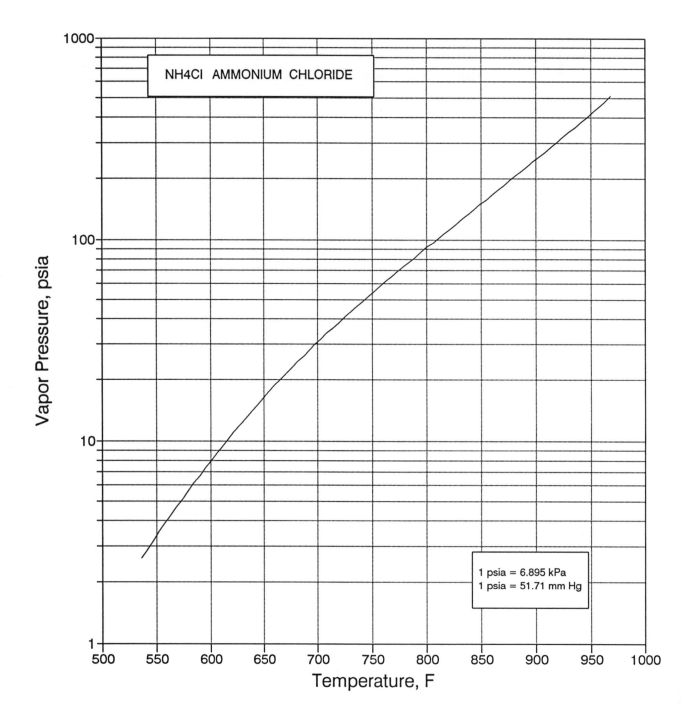

NH4Cl  AMMONIUM  CHLORIDE

1 psia = 6.895 kPa
1 psia = 51.71 mm Hg

Vapor Pressure, psia

Temperature, F

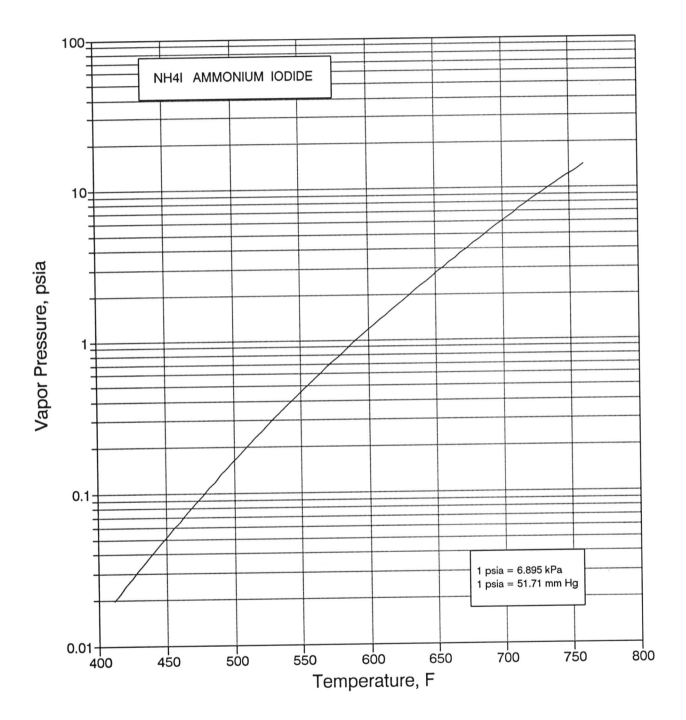

NH4I  AMMONIUM  IODIDE

Vapor Pressure, psia

Temperature, F

1 psia = 6.895 kPa
1 psia = 51.71 mm Hg

175

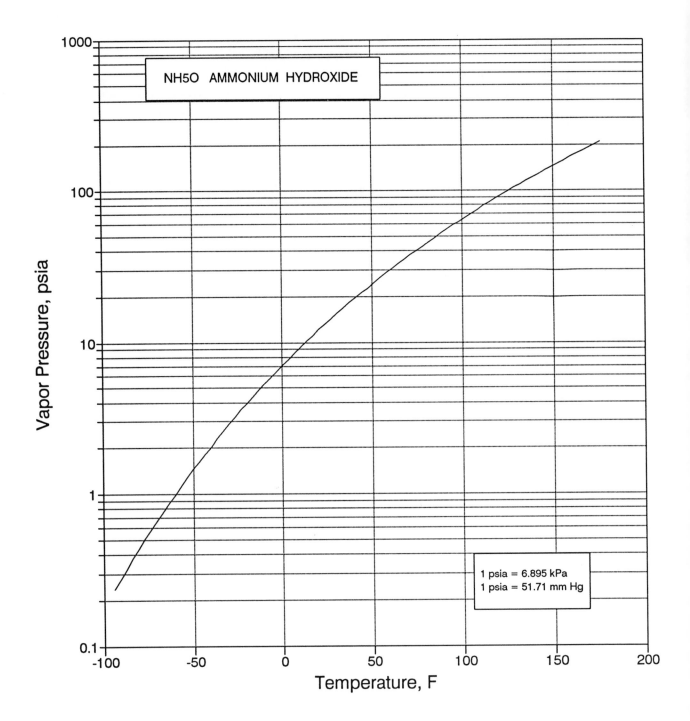

NH5O  AMMONIUM HYDROXIDE

Vapor Pressure, psia

Temperature, F

1 psia = 6.895 kPa
1 psia = 51.71 mm Hg

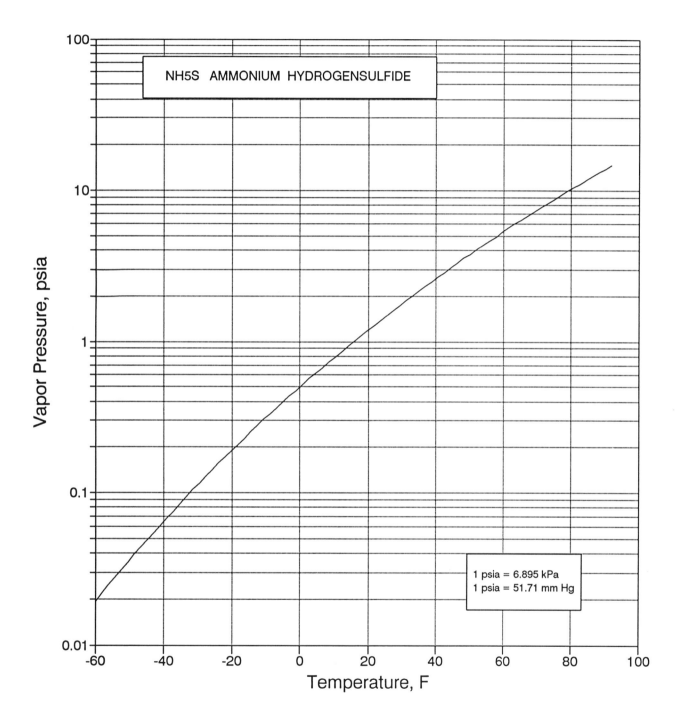

NH5S  AMMONIUM HYDROGENSULFIDE

Vapor Pressure, psia

Temperature, F

1 psia = 6.895 kPa
1 psia = 51.71 mm Hg

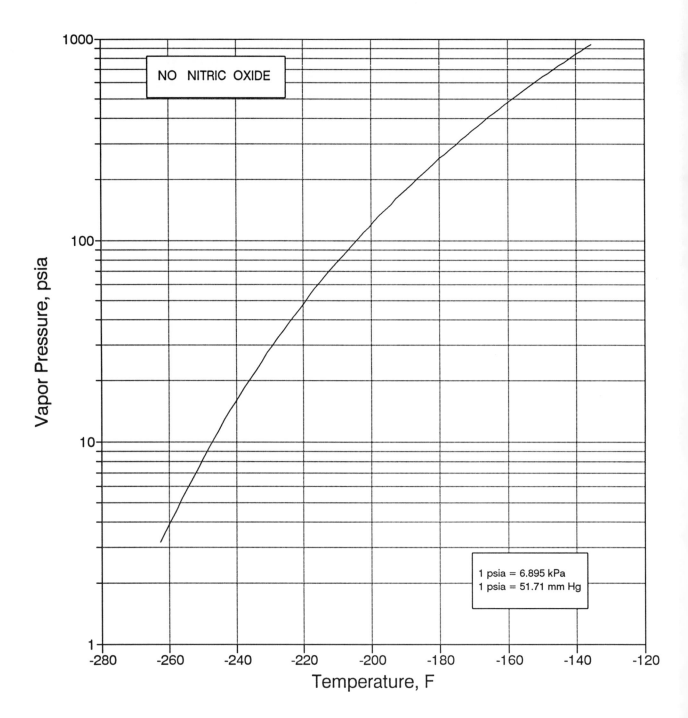

NO NITRIC OXIDE

1 psia = 6.895 kPa
1 psia = 51.71 mm Hg

Vapor Pressure, psia

Temperature, F

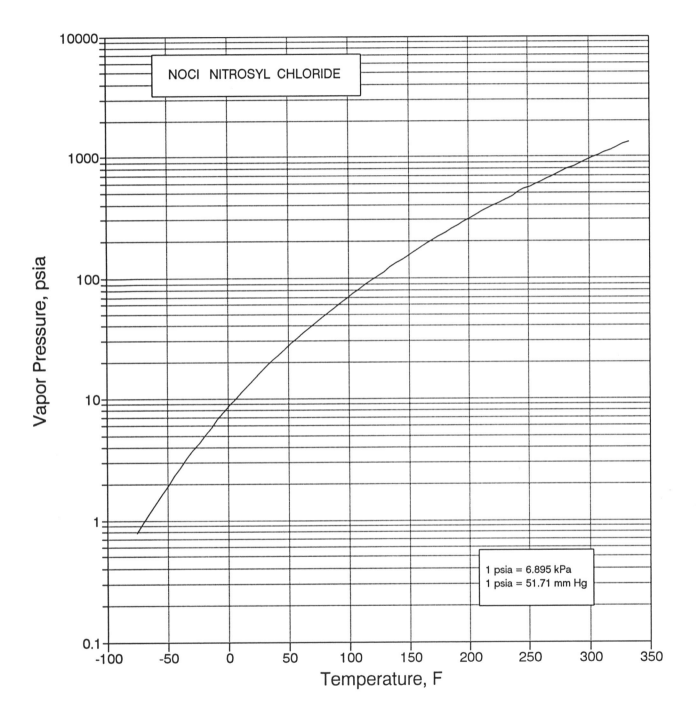

NOCl  NITROSYL CHLORIDE

Vapor Pressure, psia

Temperature, F

1 psia = 6.895 kPa
1 psia = 51.71 mm Hg

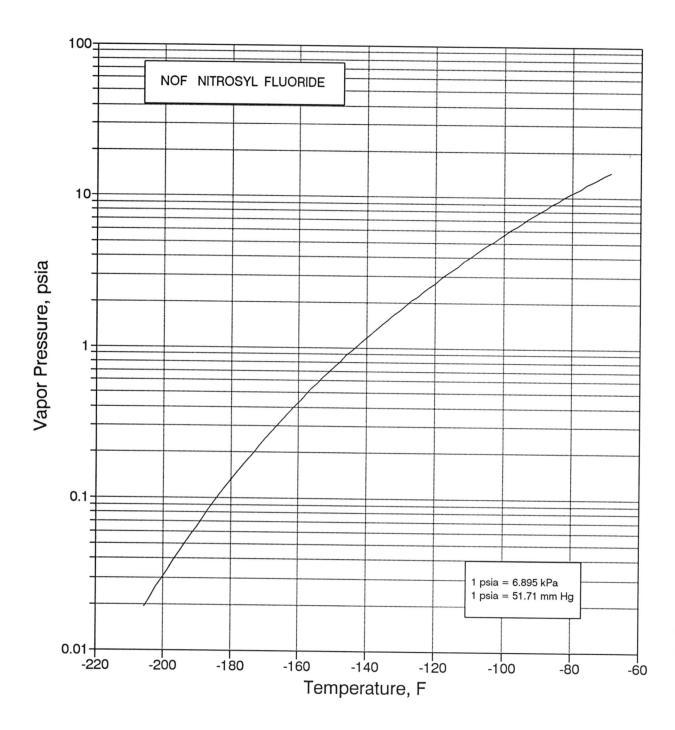

NOF  NITROSYL FLUORIDE

Vapor Pressure, psia

Temperature, F

1 psia = 6.895 kPa
1 psia = 51.71 mm Hg

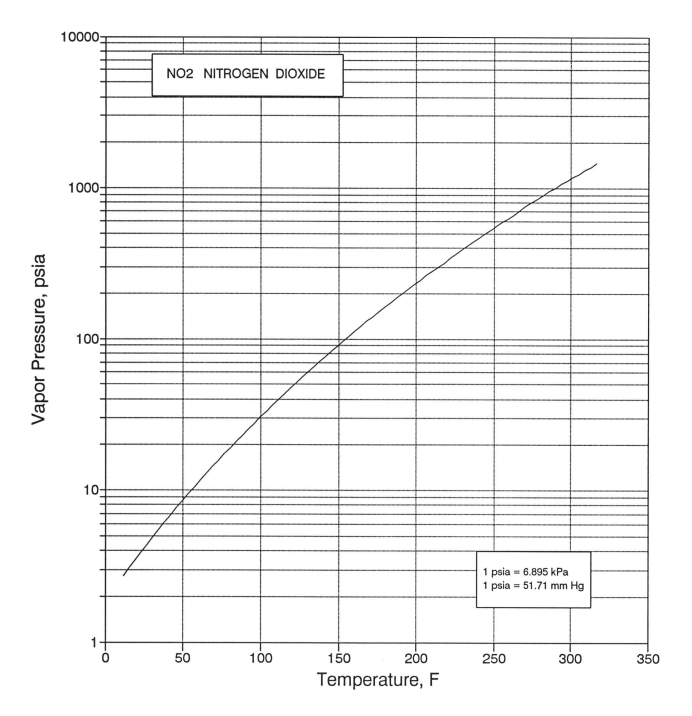

NO2 NITROGEN DIOXIDE

1 psia = 6.895 kPa
1 psia = 51.71 mm Hg

Vapor Pressure, psia

Temperature, F

181

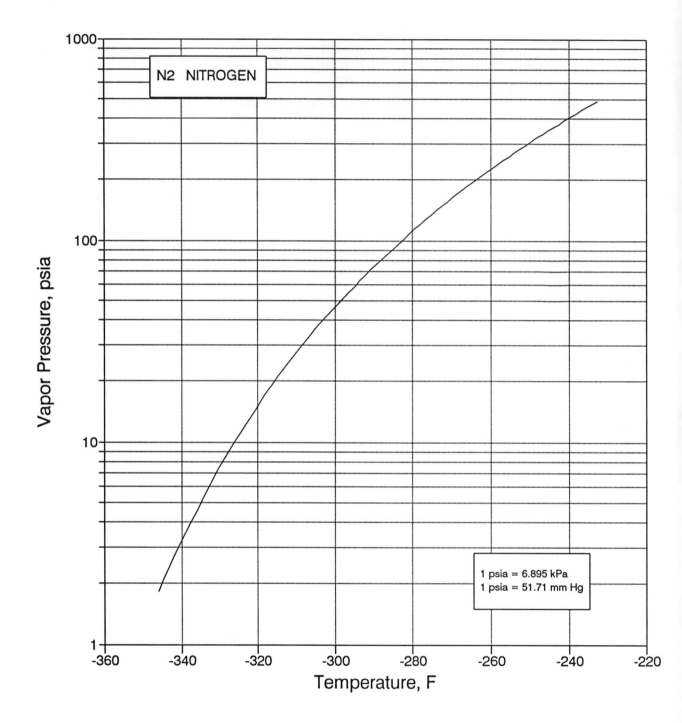

1 psia = 6.895 kPa
1 psia = 51.71 mm Hg

N2 NITROGEN

Vapor Pressure, psia

Temperature, F

182

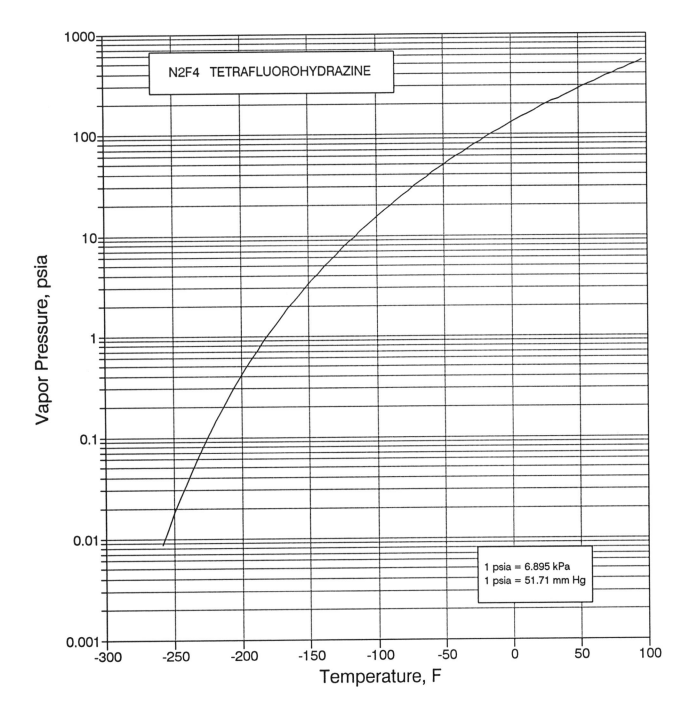

N2F4   TETRAFLUOROHYDRAZINE

1 psia = 6.895 kPa
1 psia = 51.71 mm Hg

Vapor Pressure, psia

Temperature, F

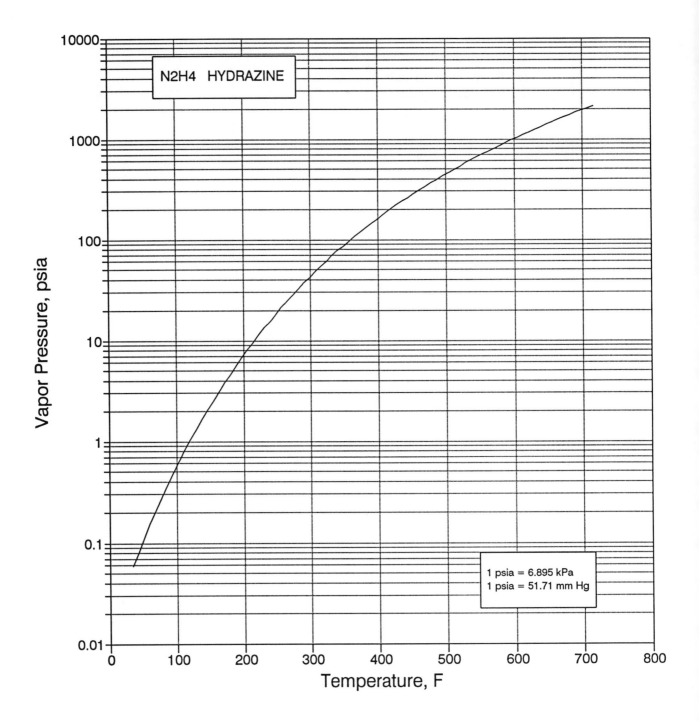

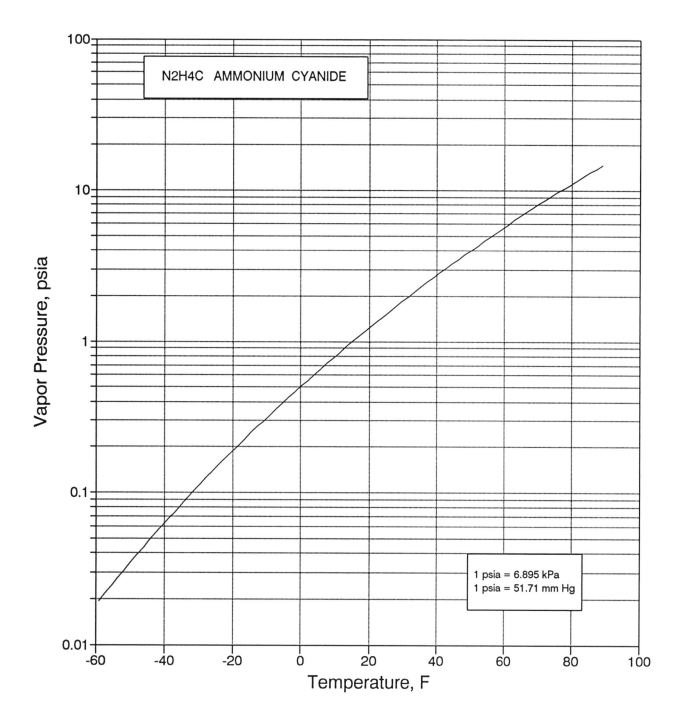

N2H4C  AMMONIUM CYANIDE

Vapor Pressure, psia

Temperature, F

1 psia = 6.895 kPa
1 psia = 51.71 mm Hg

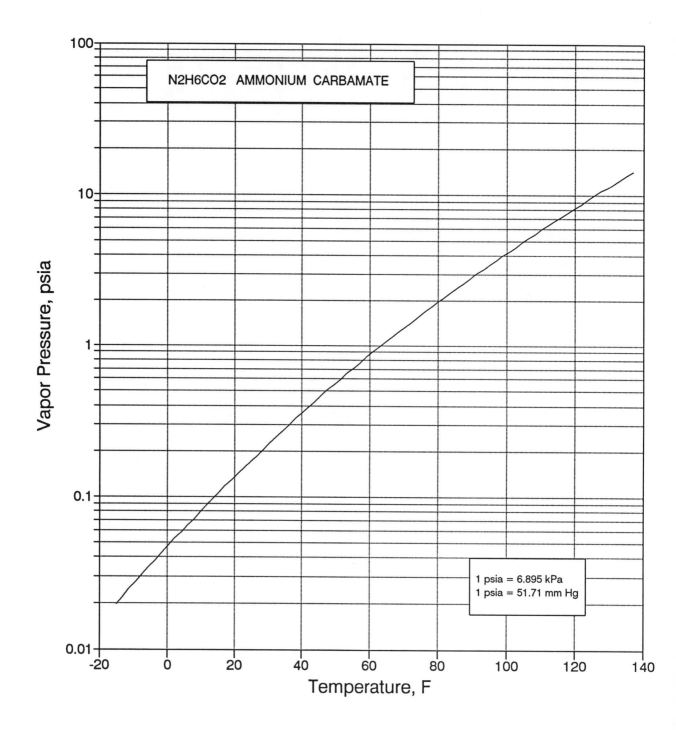

Vapor Pressure, psia

Temperature, F

N2H6CO2  AMMONIUM CARBAMATE

1 psia = 6.895 kPa
1 psia = 51.71 mm Hg

186

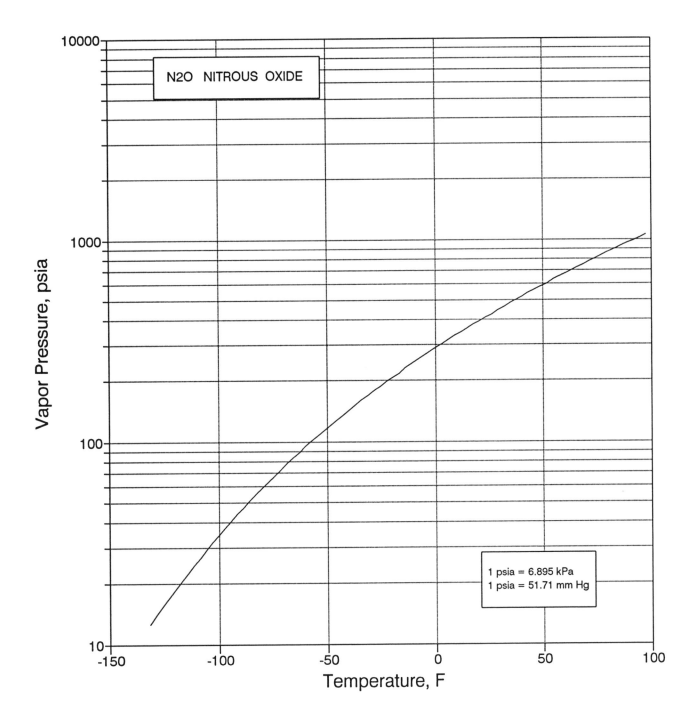

N2O  NITROUS OXIDE

Vapor Pressure, psia

10000

1000

100

10

1 psia = 6.895 kPa
1 psia = 51.71 mm Hg

-150    -100    -50    0    50    100

Temperature, F

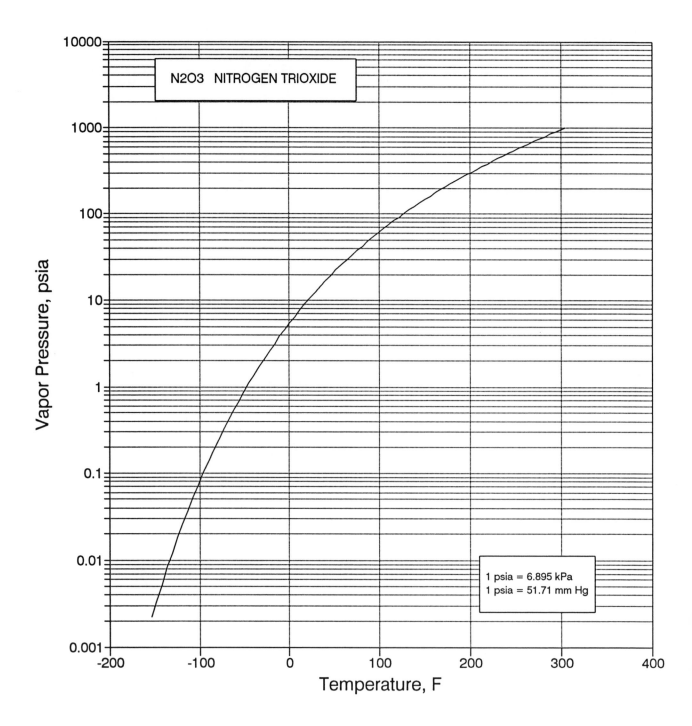

N2O3   NITROGEN TRIOXIDE

1 psia = 6.895 kPa
1 psia = 51.71 mm Hg

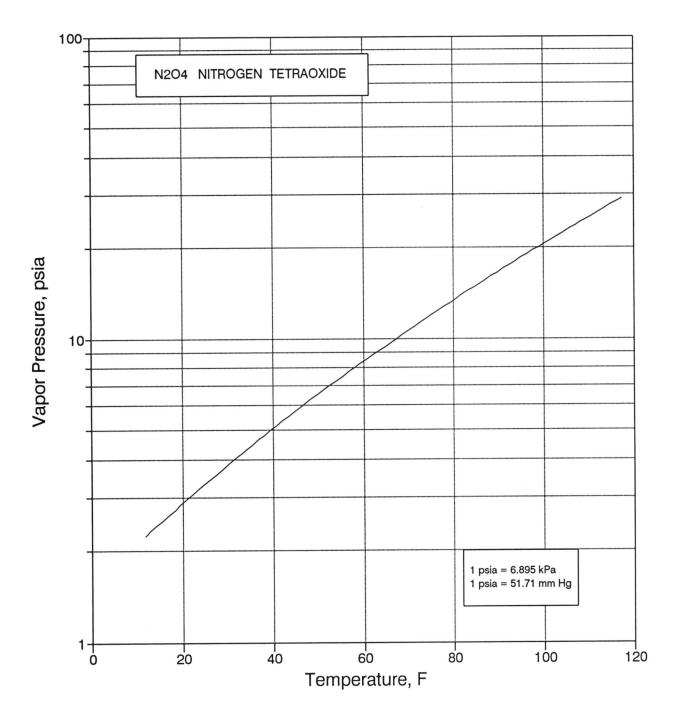

N2O4  NITROGEN TETRAOXIDE

Vapor Pressure, psia

Temperature, F

1 psia = 6.895 kPa
1 psia = 51.71 mm Hg

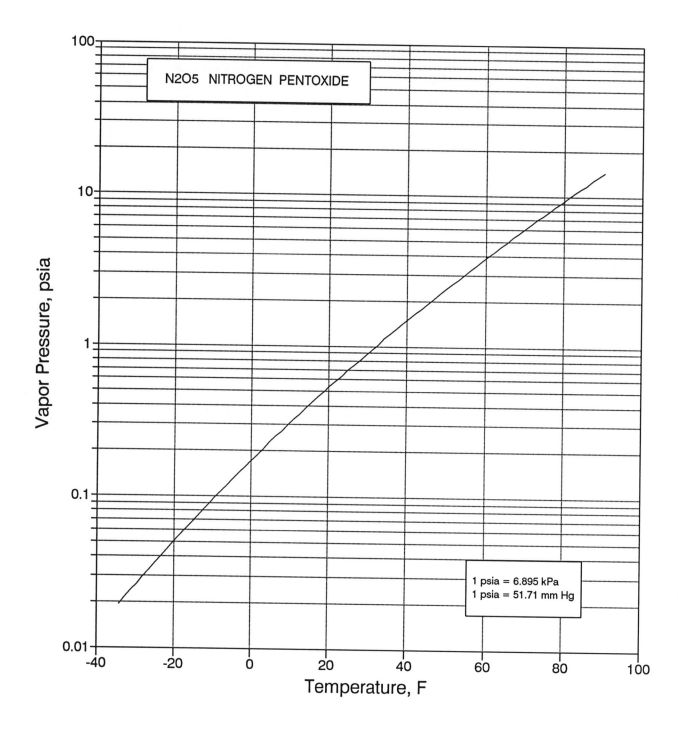

N2O5  NITROGEN PENTOXIDE

Vapor Pressure, psia

1 psia = 6.895 kPa
1 psia = 51.71 mm Hg

Temperature, F

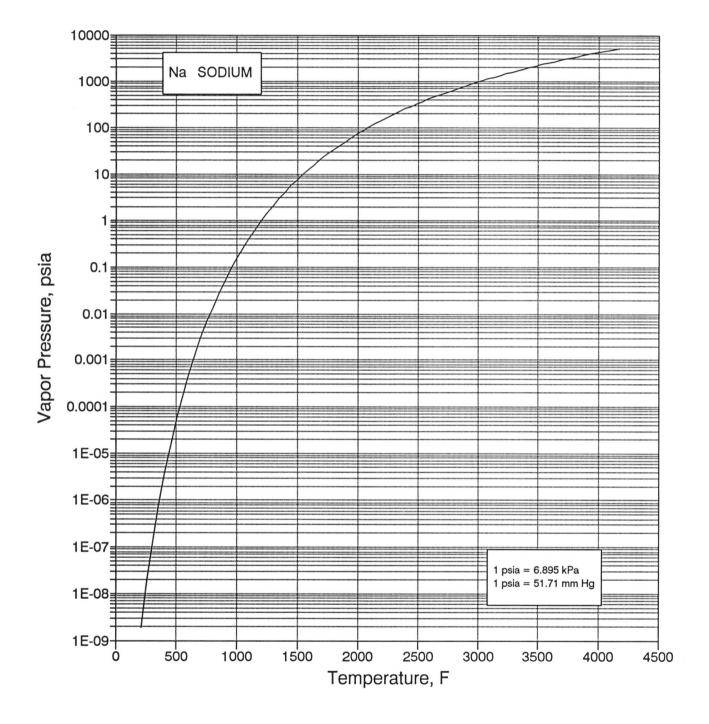

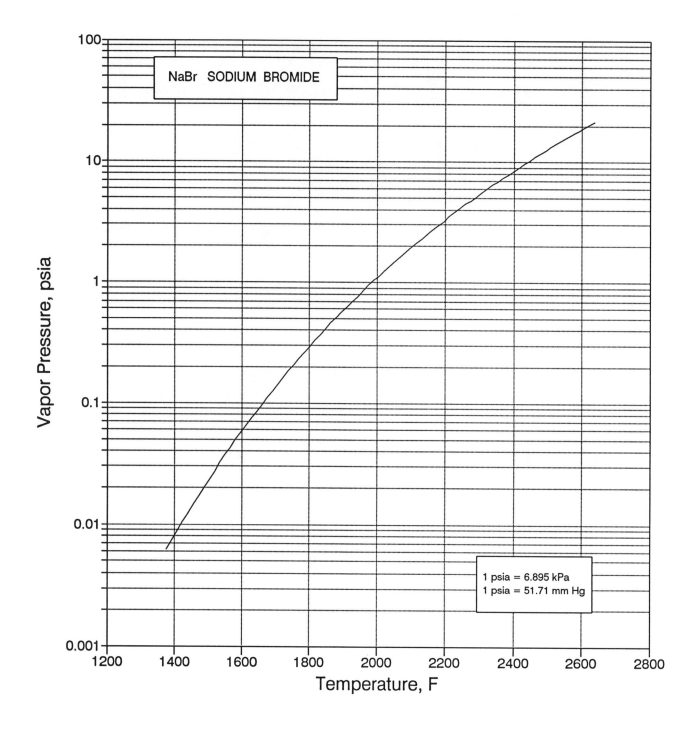

NaBr SODIUM BROMIDE

Vapor Pressure, psia

Temperature, F

1 psia = 6.895 kPa
1 psia = 51.71 mm Hg

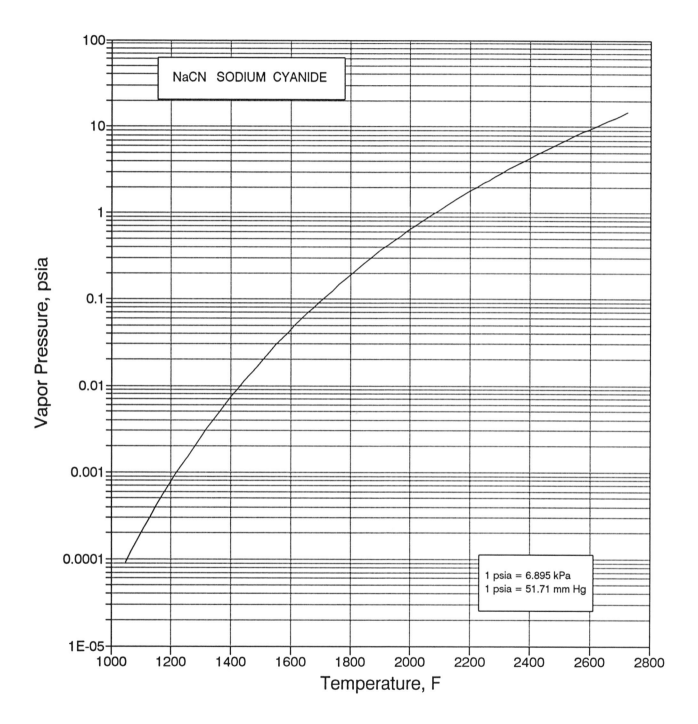

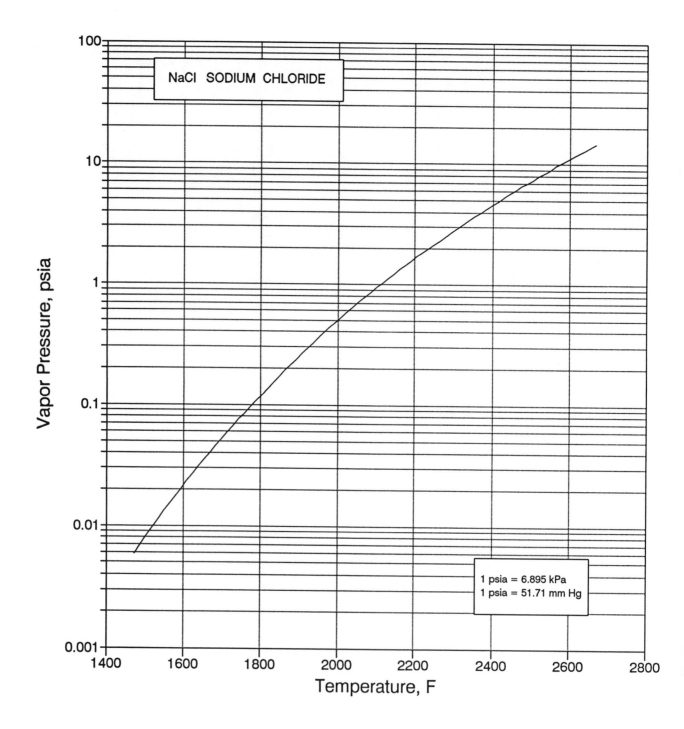

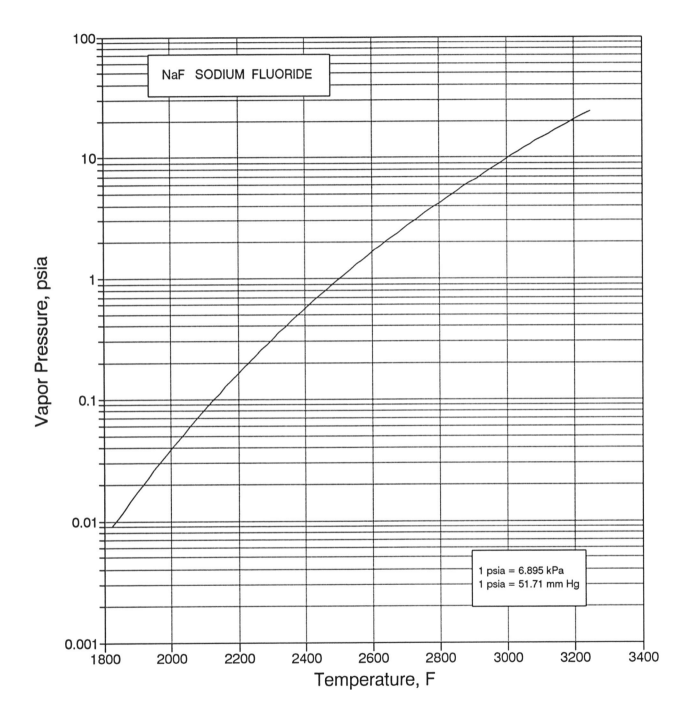

NaF  SODIUM FLUORIDE

1 psia = 6.895 kPa
1 psia = 51.71 mm Hg

Vapor Pressure, psia

Temperature, F

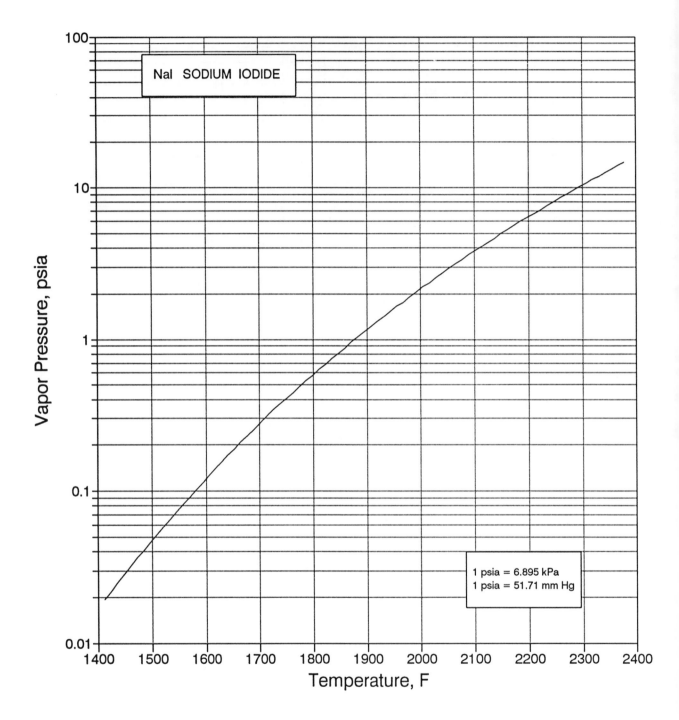

NaI  SODIUM  IODIDE

1 psia = 6.895 kPa
1 psia = 51.71 mm Hg

Vapor Pressure, psia

Temperature, F

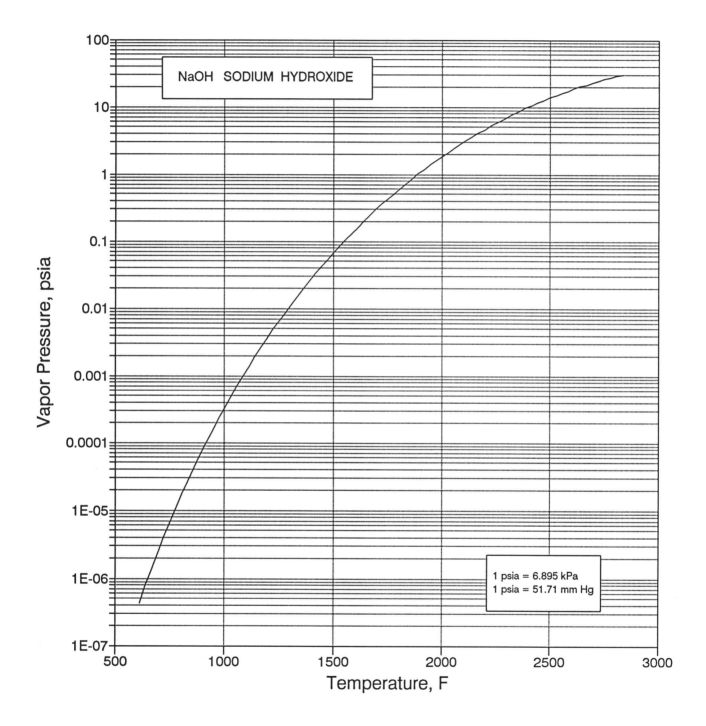

NaOH SODIUM HYDROXIDE

1 psia = 6.895 kPa
1 psia = 51.71 mm Hg

Vapor Pressure, psia

Temperature, F

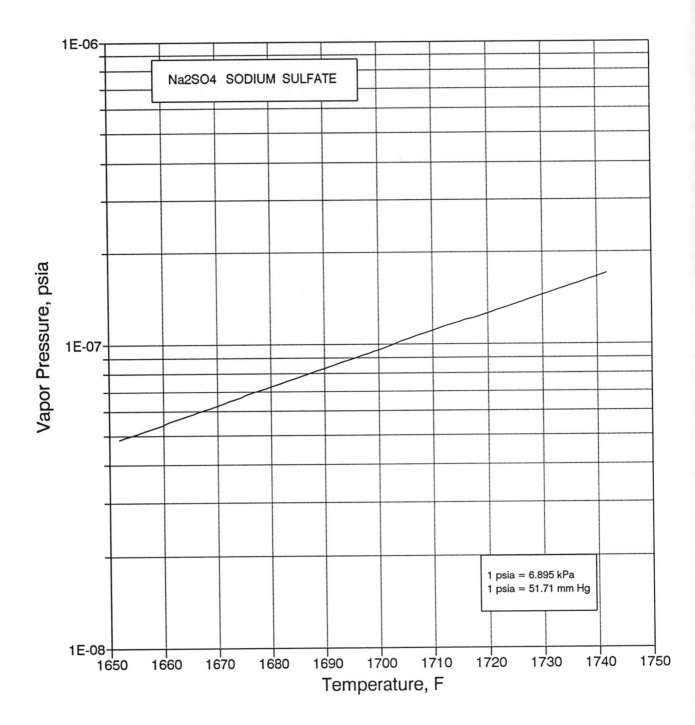

Na2SO4  SODIUM SULFATE

Vapor Pressure, psia

1E-06

1E-07

1E-08

1 psia = 6.895 kPa
1 psia = 51.71 mm Hg

Temperature, F

1650   1660   1670   1680   1690   1700   1710   1720   1730   1740   1750

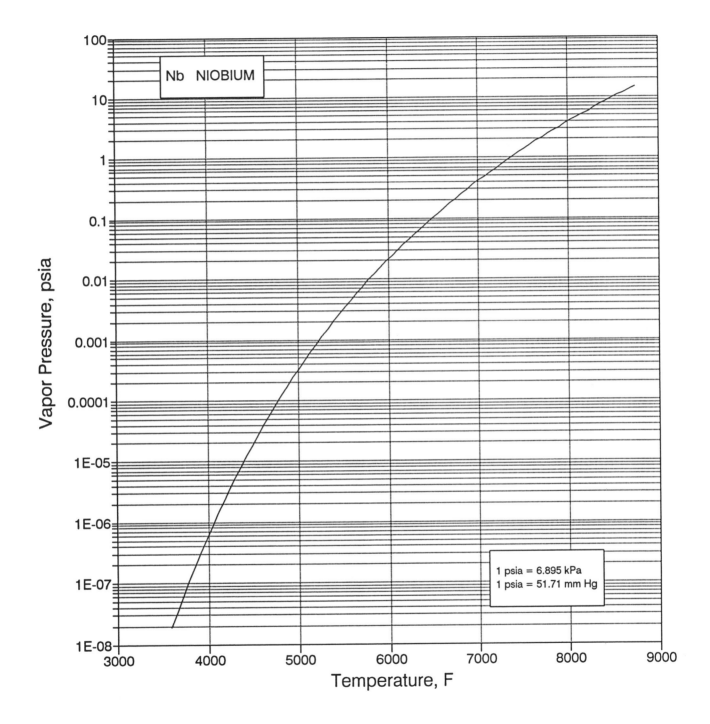

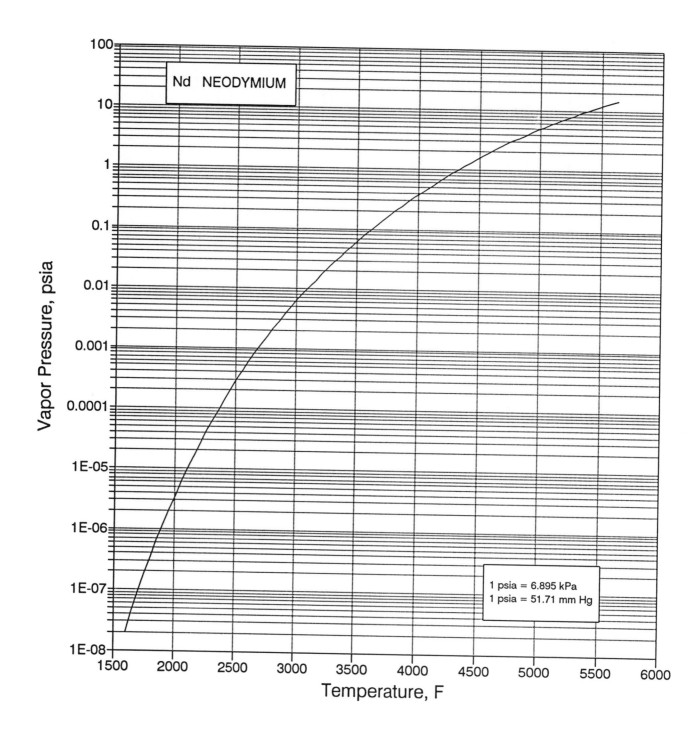

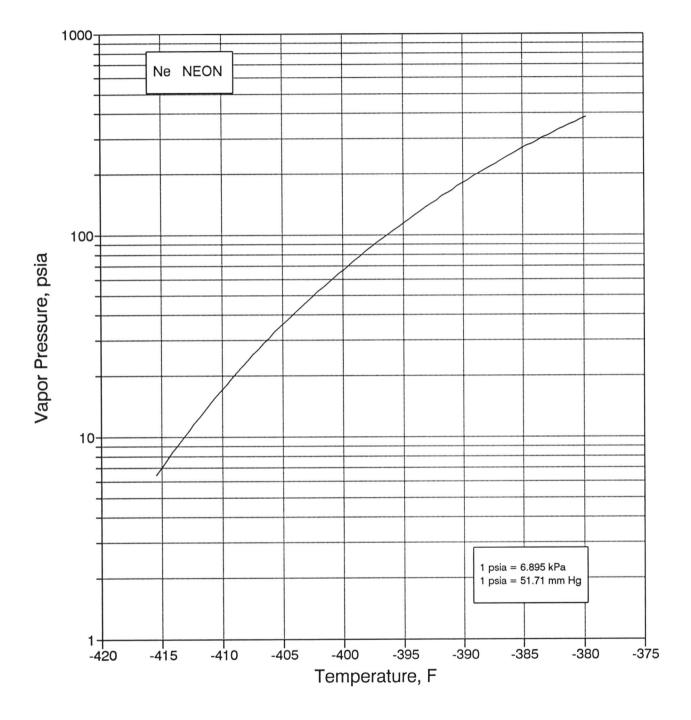

Ne NEON

Vapor Pressure, psia

Temperature, F

1 psia = 6.895 kPa
1 psia = 51.71 mm Hg

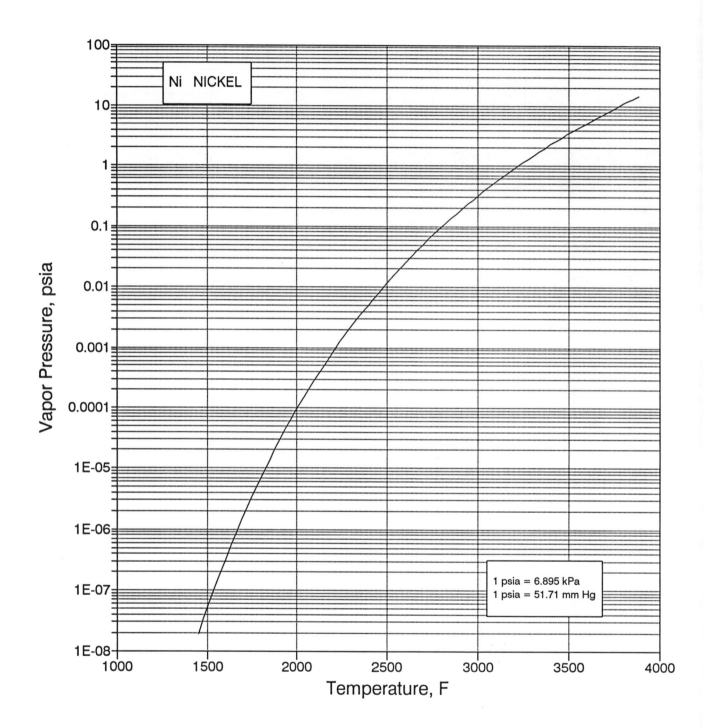

Ni  NICKEL

Vapor Pressure, psia

Temperature, F

1 psia = 6.895 kPa
1 psia = 51.71 mm Hg

202

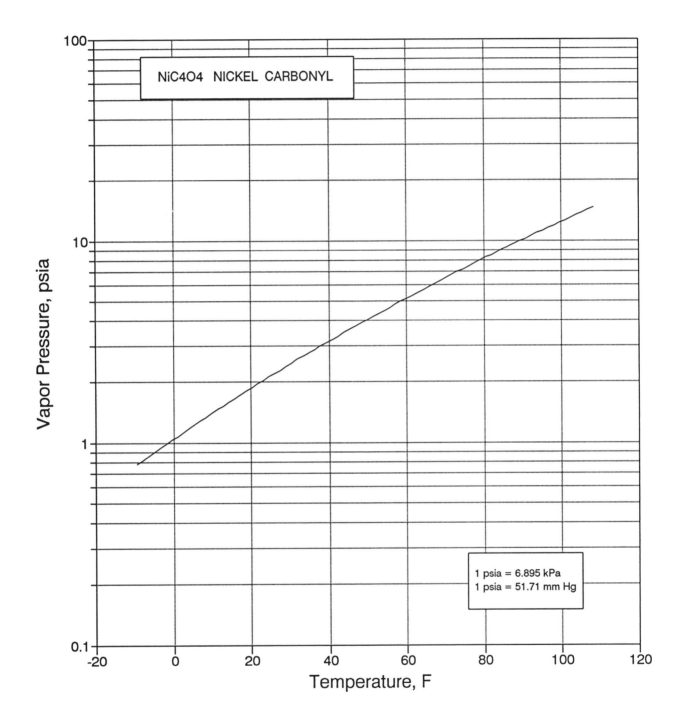

NiC4O4  NICKEL CARBONYL

Vapor Pressure, psia

Temperature, F

1 psia = 6.895 kPa
1 psia = 51.71 mm Hg

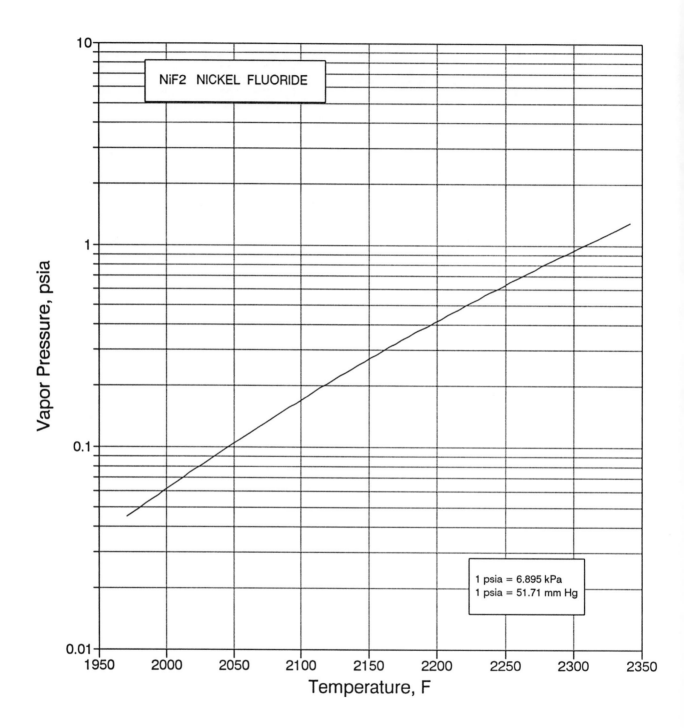

NiF2  NICKEL FLUORIDE

Vapor Pressure, psia

Temperature, F

1 psia = 6.895 kPa
1 psia = 51.71 mm Hg

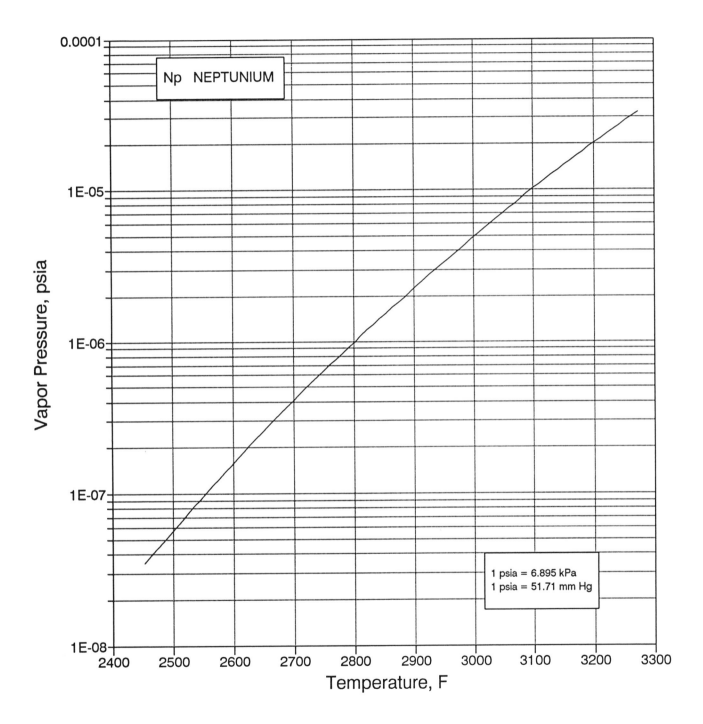

Np NEPTUNIUM

Vapor Pressure, psia

Temperature, F

1 psia = 6.895 kPa
1 psia = 51.71 mm Hg

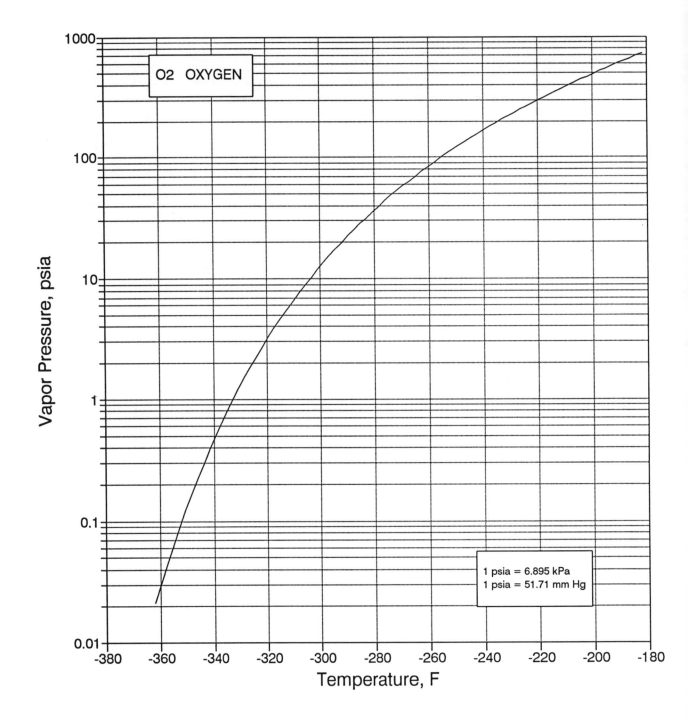

O2   OXYGEN

Vapor Pressure, psia

Temperature, F

1 psia = 6.895 kPa
1 psia = 51.71 mm Hg

206

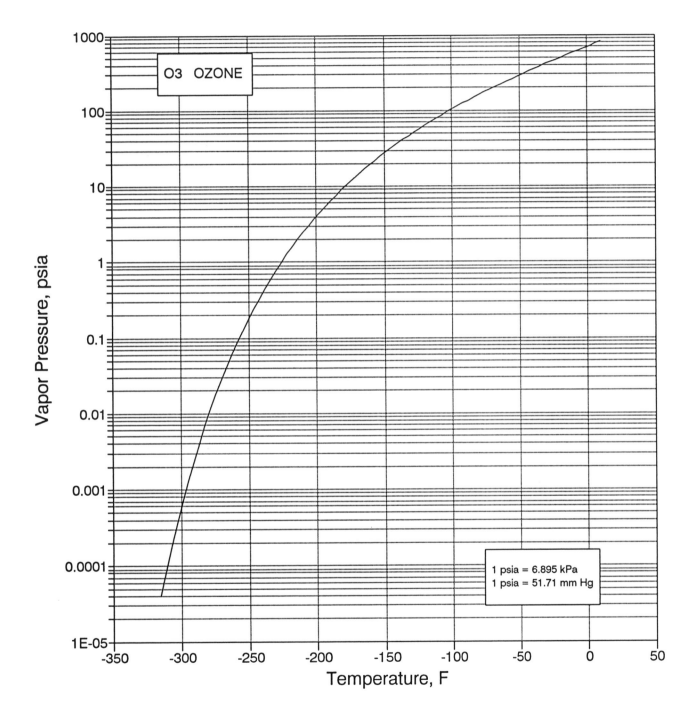

O3  OZONE

Vapor Pressure, psia

Temperature, F

1 psia = 6.895 kPa
1 psia = 51.71 mm Hg

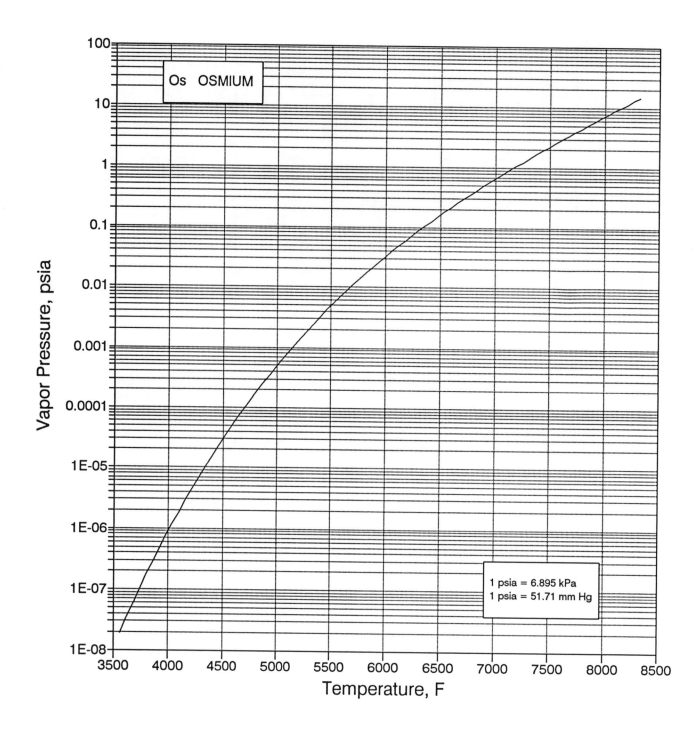

Os OSMIUM

Vapor Pressure, psia

Temperature, F

1 psia = 6.895 kPa
1 psia = 51.71 mm Hg

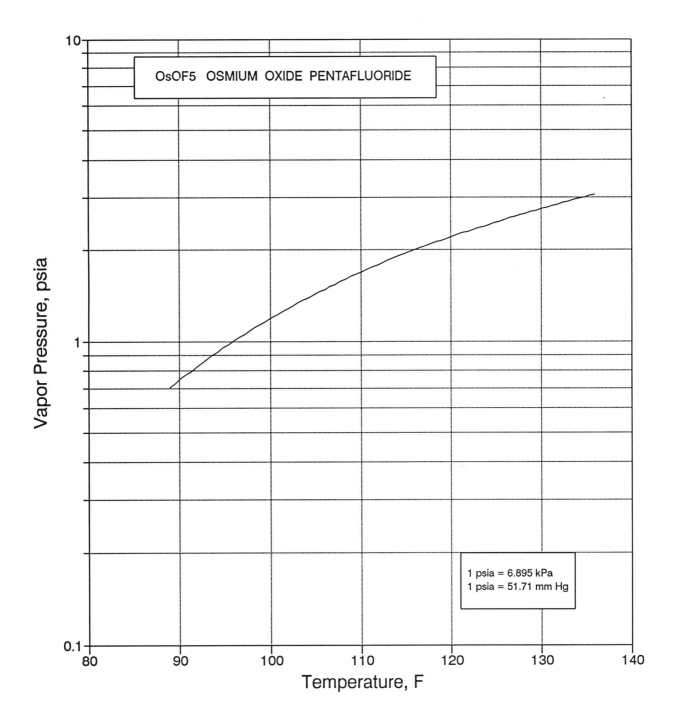

OsOF5  OSMIUM OXIDE PENTAFLUORIDE

Vapor Pressure, psia

Temperature, F

1 psia = 6.895 kPa
1 psia = 51.71 mm Hg

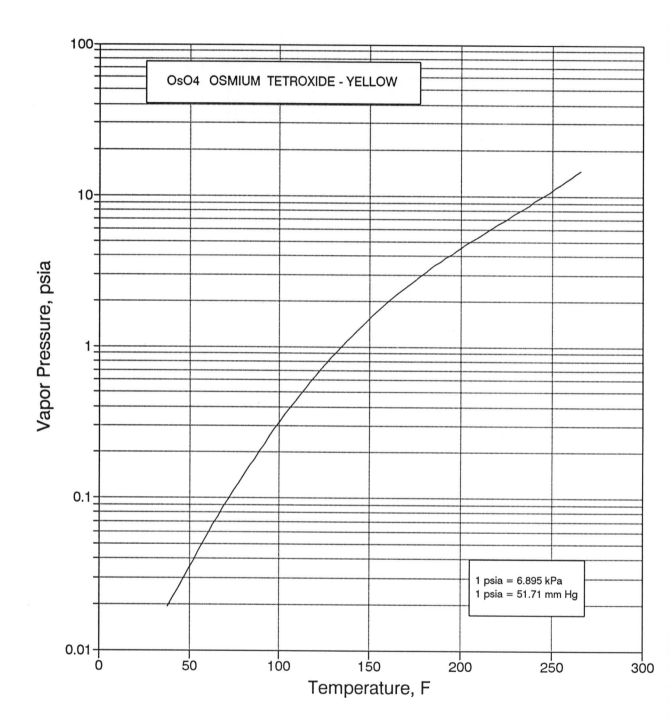

OsO4   OSMIUM  TETROXIDE - YELLOW

1 psia = 6.895 kPa
1 psia = 51.71 mm Hg

Vapor Pressure, psia

Temperature, F

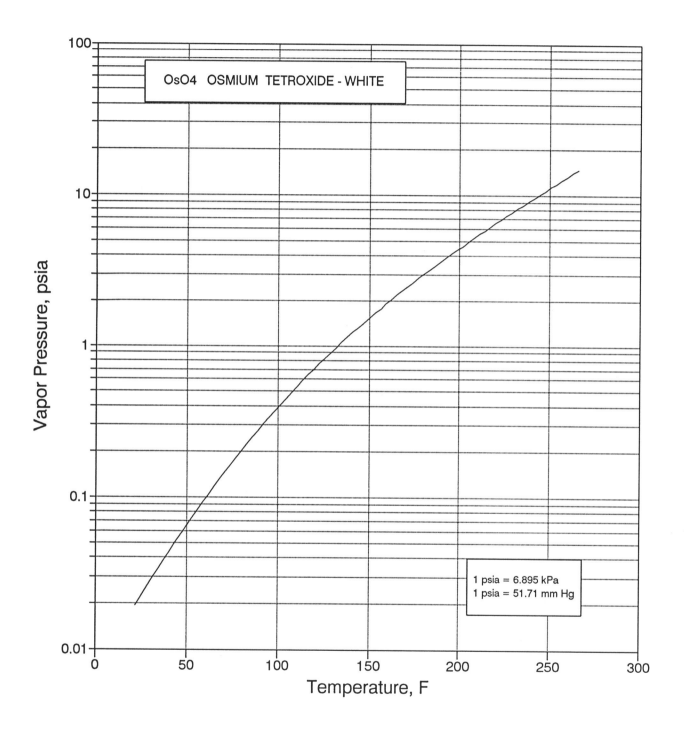

100 ─
OsO4  OSMIUM  TETROXIDE - WHITE

Vapor Pressure, psia

10 ─

1 ─

0.1 ─

1 psia = 6.895 kPa
1 psia = 51.71 mm Hg

0.01 ─

Temperature, F

0    50    100    150    200    250    300

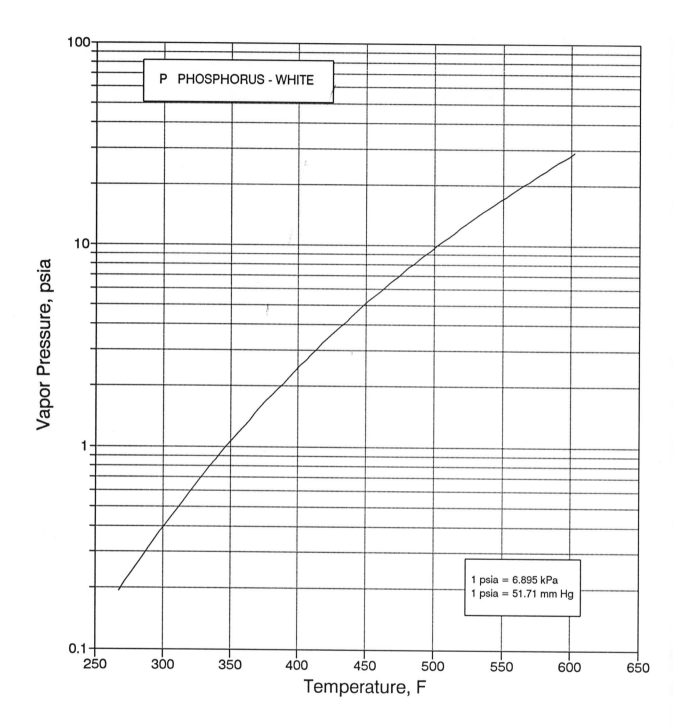

P   PHOSPHORUS - WHITE

Vapor Pressure, psia

Temperature, F

1 psia = 6.895 kPa
1 psia = 51.71 mm Hg

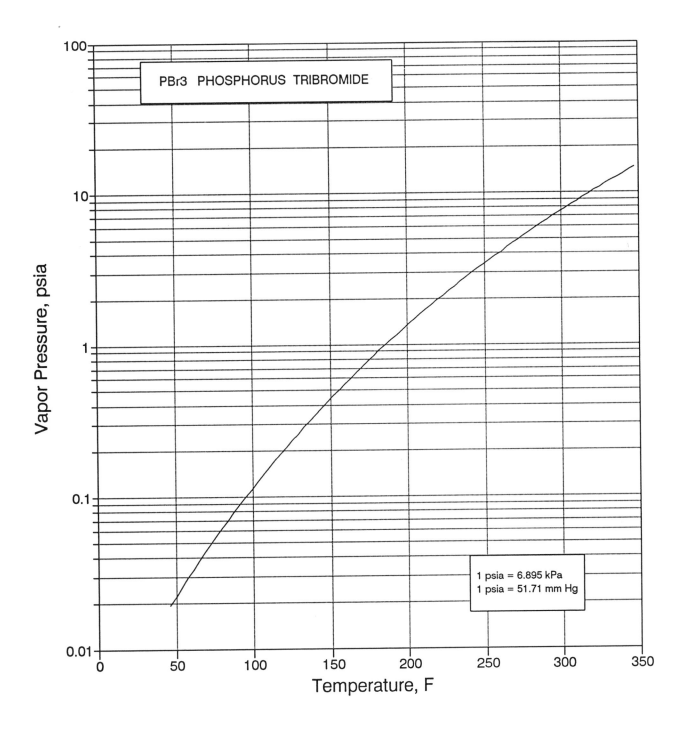

PBr3  PHOSPHORUS  TRIBROMIDE

Vapor Pressure, psia

Temperature, F

1 psia = 6.895 kPa
1 psia = 51.71 mm Hg

213

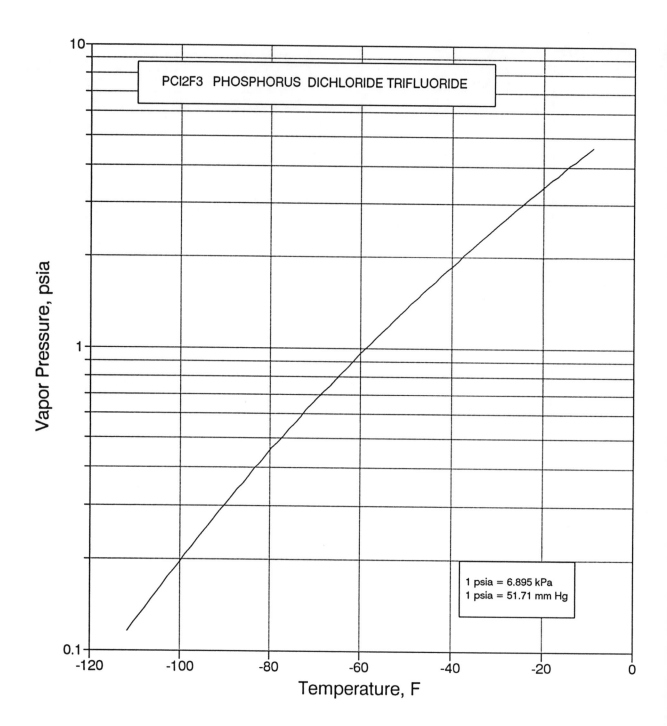

Vapor Pressure, psia

PCl2F3  PHOSPHORUS  DICHLORIDE  TRIFLUORIDE

1 psia = 6.895 kPa
1 psia = 51.71 mm Hg

Temperature, F

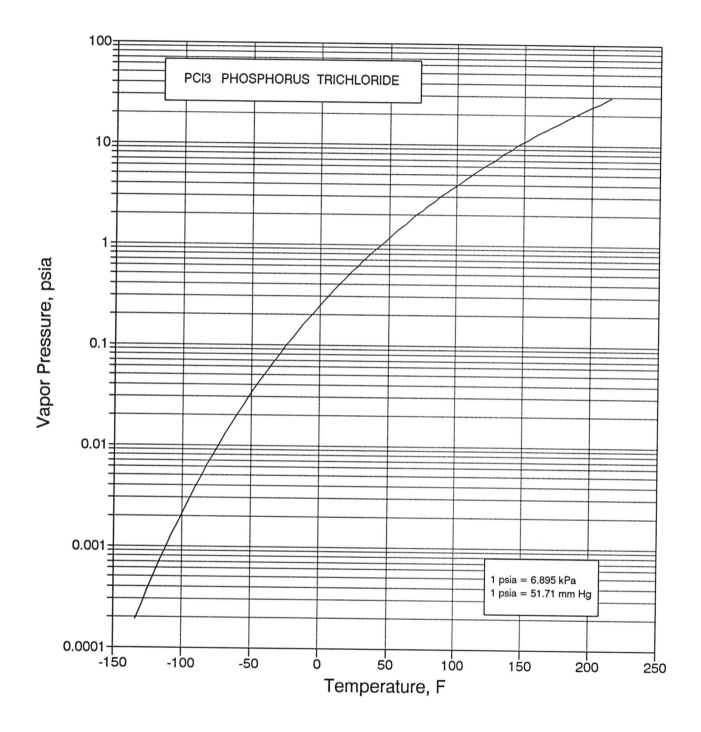

PCl3 PHOSPHORUS TRICHLORIDE

Vapor Pressure, psia

Temperature, F

1 psia = 6.895 kPa
1 psia = 51.71 mm Hg

215

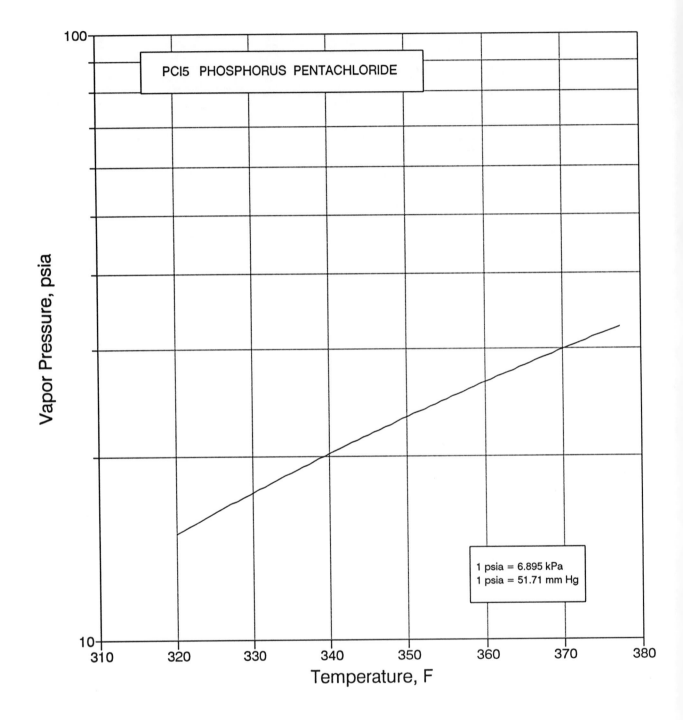

PCl5 PHOSPHORUS PENTACHLORIDE

Vapor Pressure, psia

Temperature, F

1 psia = 6.895 kPa
1 psia = 51.71 mm Hg

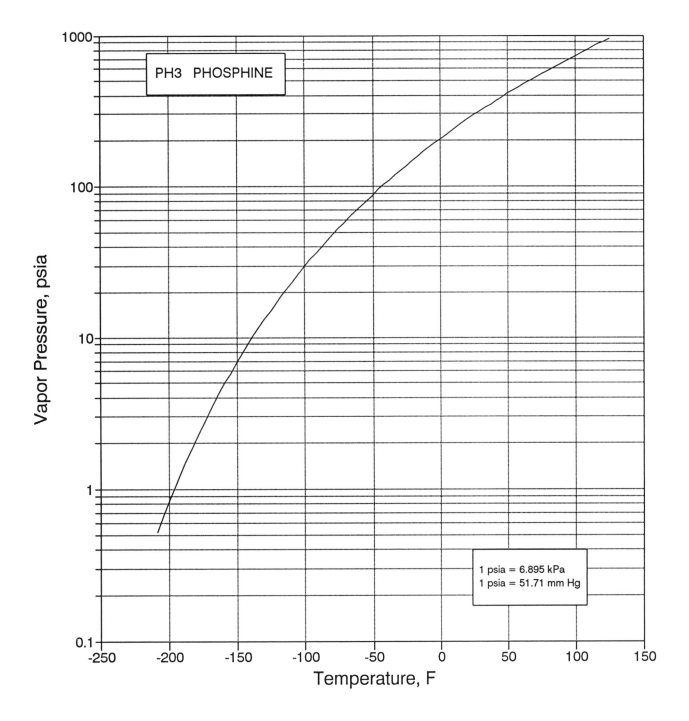

PH3 PHOSPHINE

Vapor Pressure, psia

Temperature, F

1 psia = 6.895 kPa
1 psia = 51.71 mm Hg

217

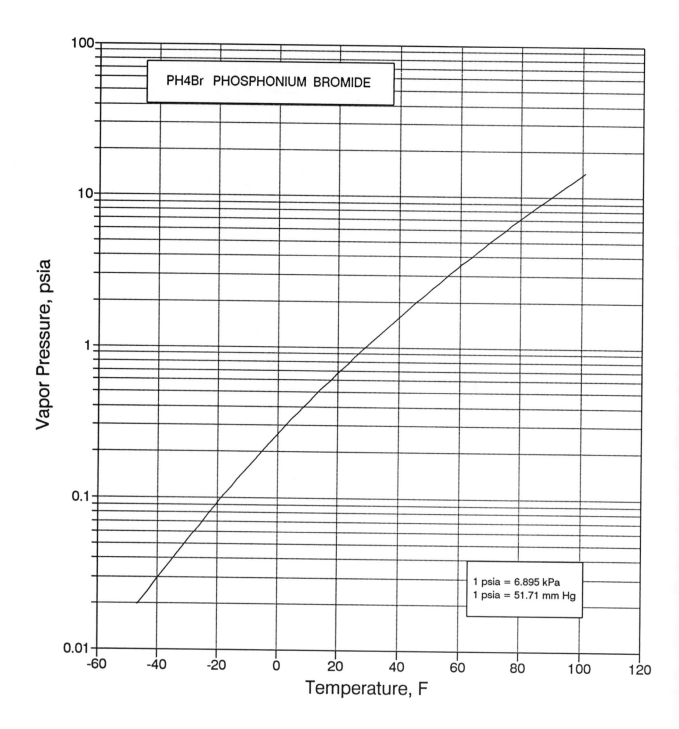

PH4Br PHOSPHONIUM BROMIDE

Vapor Pressure, psia

Temperature, F

1 psia = 6.895 kPa
1 psia = 51.71 mm Hg

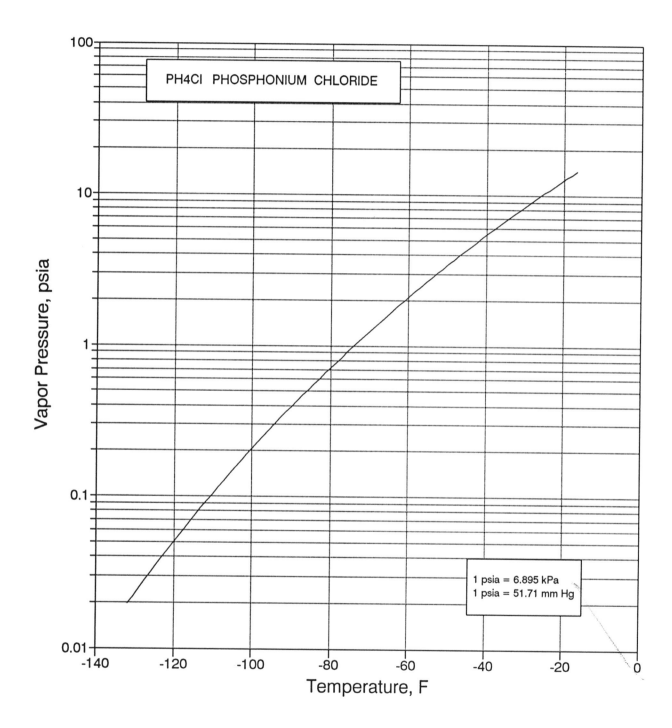

PH4Cl   PHOSPHONIUM CHLORIDE

Vapor Pressure, psia

Temperature, F

1 psia = 6.895 kPa
1 psia = 51.71 mm Hg

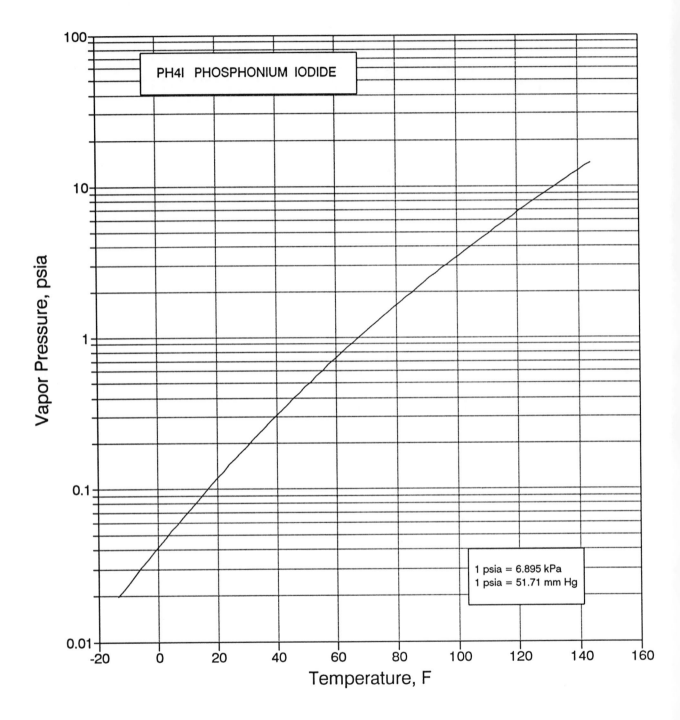

PH4I  PHOSPHONIUM IODIDE

1 psia = 6.895 kPa
1 psia = 51.71 mm Hg

Vapor Pressure, psia

Temperature, F

220

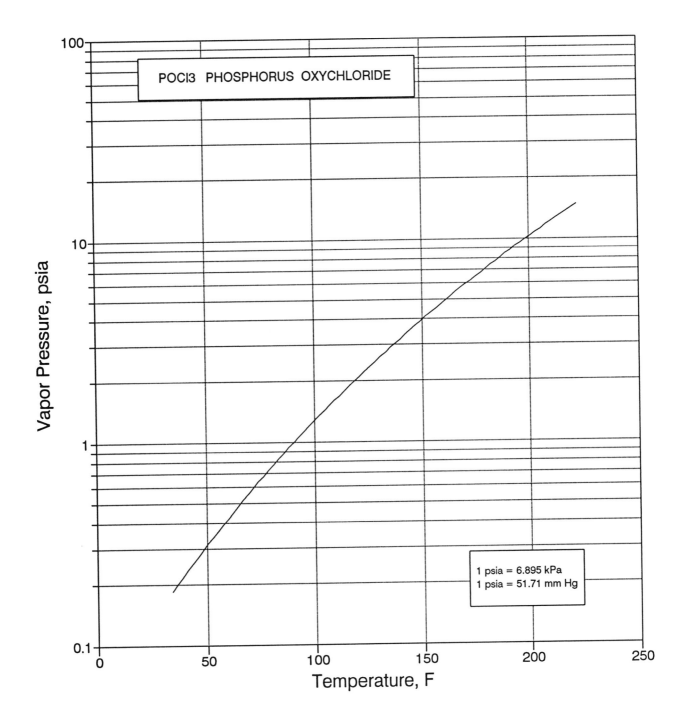

POCl3  PHOSPHORUS OXYCHLORIDE

Vapor Pressure, psia

Temperature, F

1 psia = 6.895 kPa
1 psia = 51.71 mm Hg

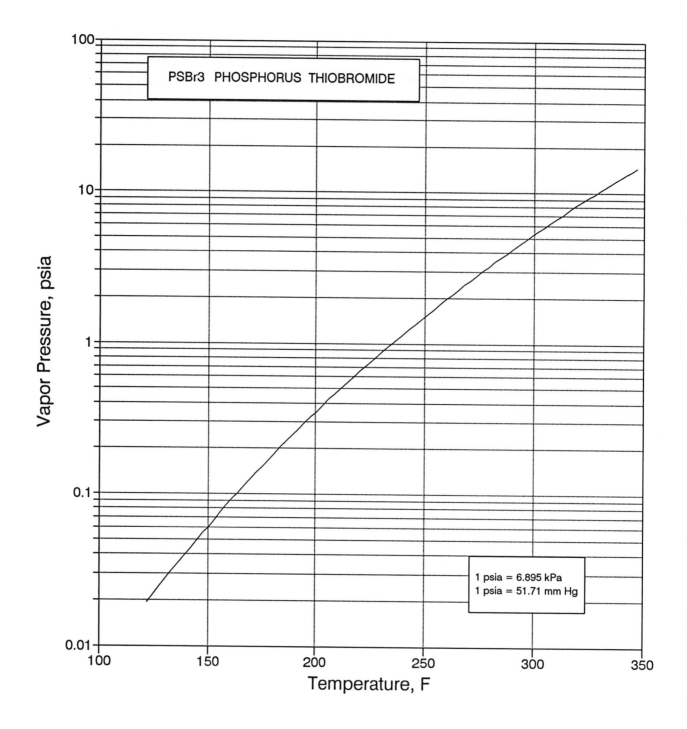

PSBr3  PHOSPHORUS THIOBROMIDE

1 psia = 6.895 kPa
1 psia = 51.71 mm Hg

Vapor Pressure, psia

Temperature, F

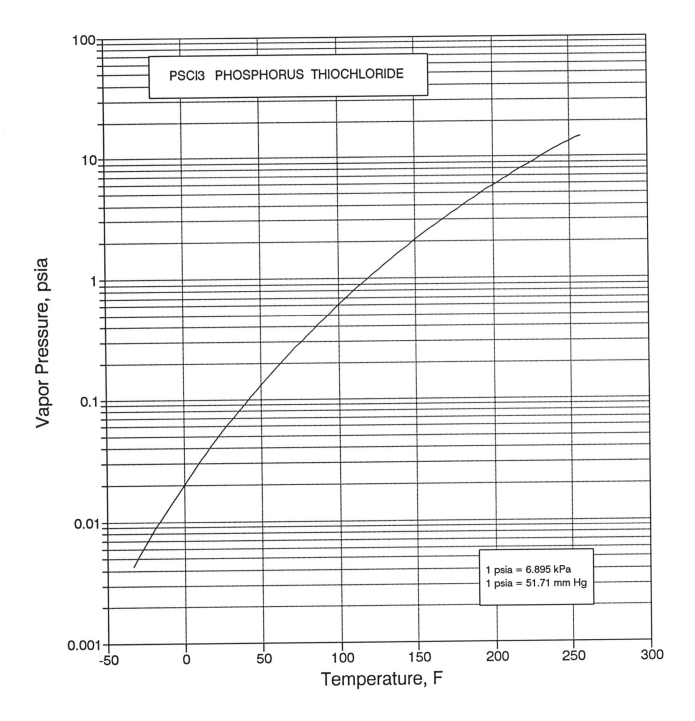

PSCl3  PHOSPHORUS  THIOCHLORIDE

1 psia = 6.895 kPa
1 psia = 51.71 mm Hg

Vapor Pressure, psia

Temperature, F

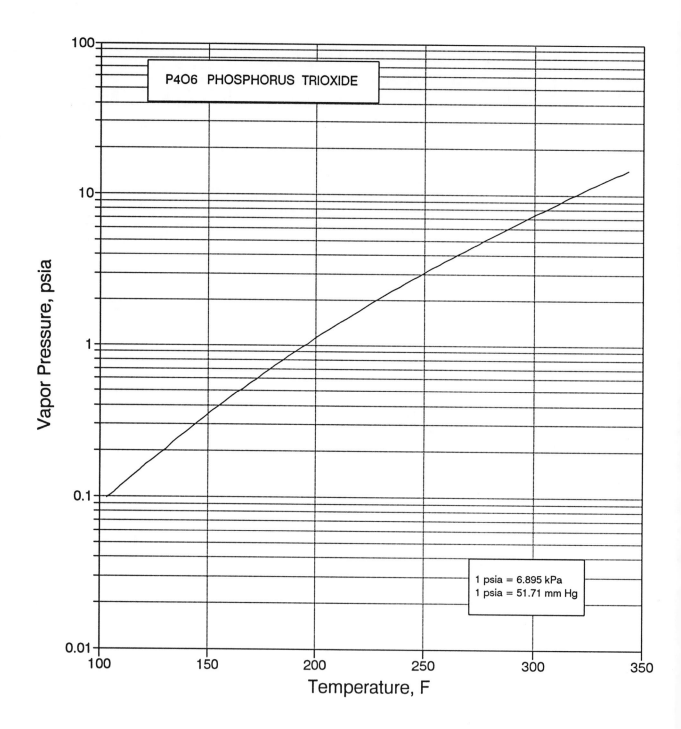

P4O6 PHOSPHORUS TRIOXIDE

1 psia = 6.895 kPa
1 psia = 51.71 mm Hg

Vapor Pressure, psia

Temperature, F

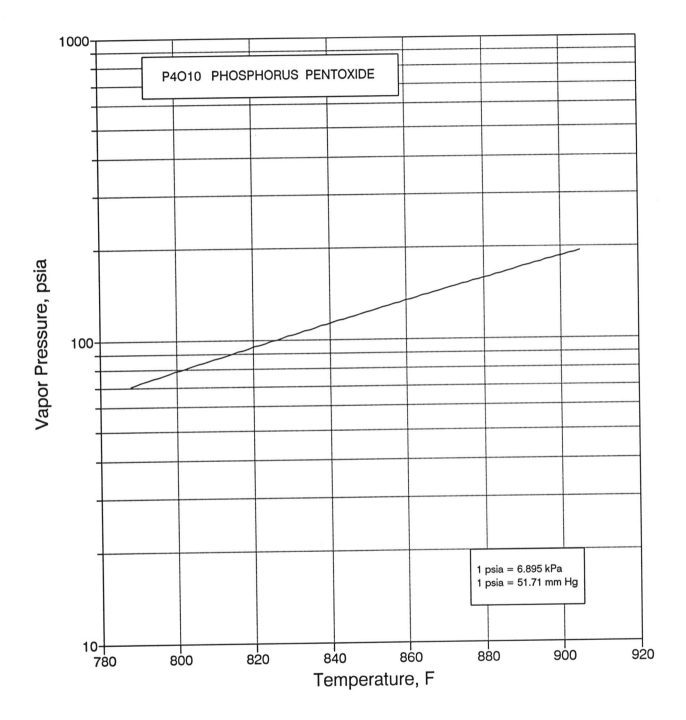

P4O10 PHOSPHORUS PENTOXIDE

Vapor Pressure, psia

Temperature, F

1 psia = 6.895 kPa
1 psia = 51.71 mm Hg

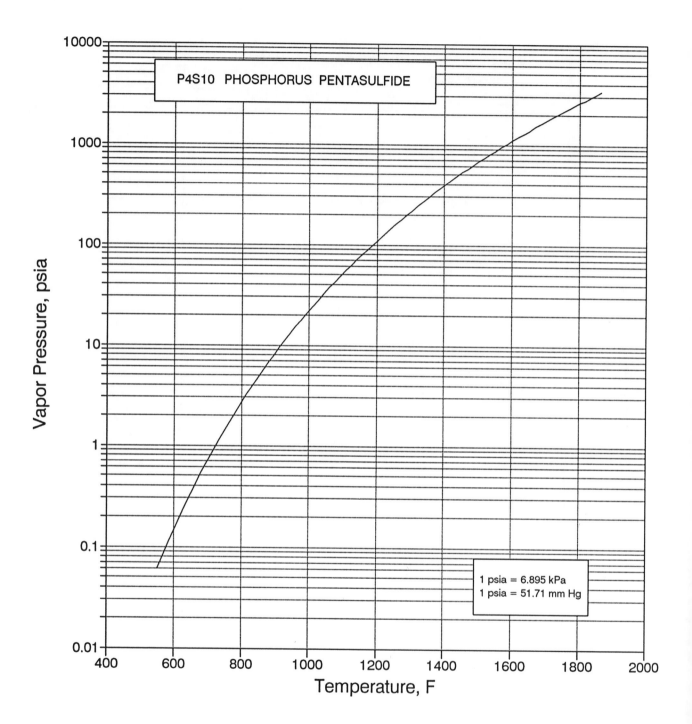

P4S10 PHOSPHORUS PENTASULFIDE

Vapor Pressure, psia

Temperature, F

1 psia = 6.895 kPa
1 psia = 51.71 mm Hg

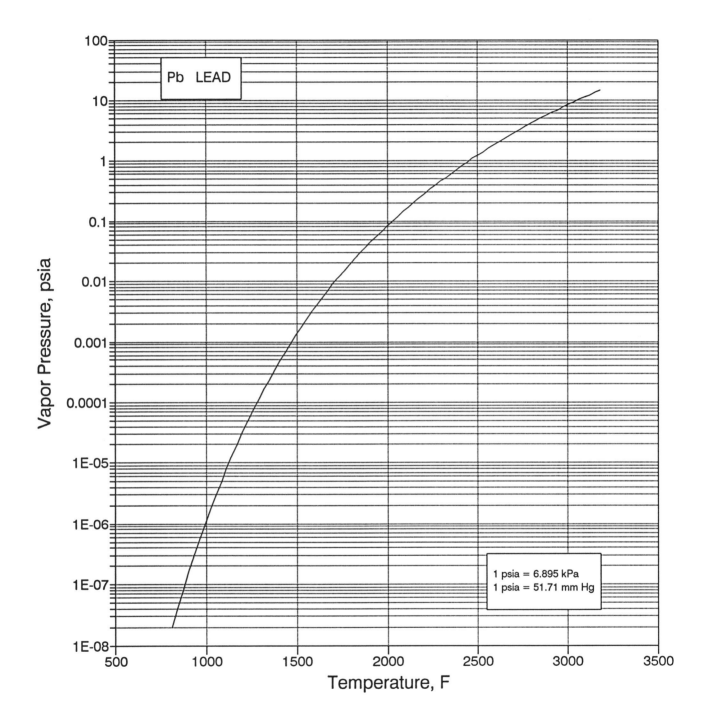

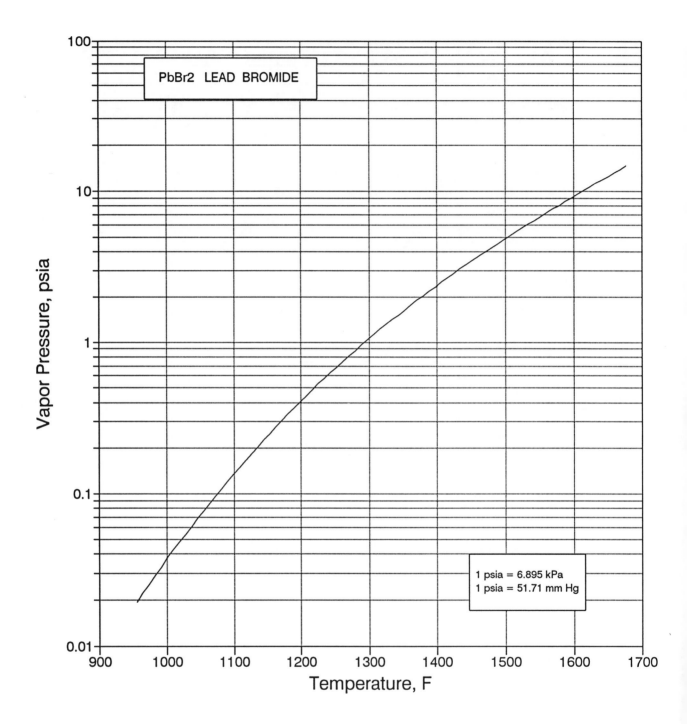

PbBr2  LEAD  BROMIDE

Vapor Pressure, psia

Temperature, F

1 psia = 6.895 kPa
1 psia = 51.71 mm Hg

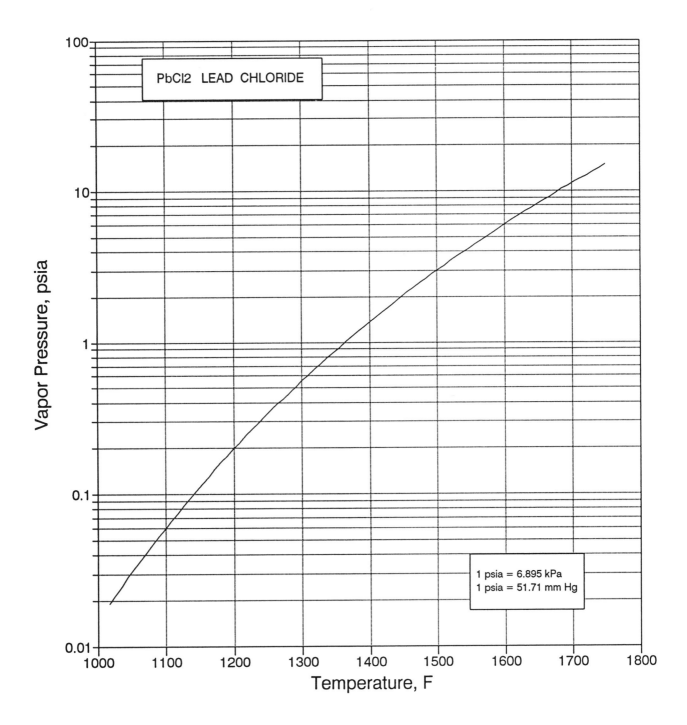

PbCl2 LEAD CHLORIDE

1 psia = 6.895 kPa
1 psia = 51.71 mm Hg

Vapor Pressure, psia

Temperature, F

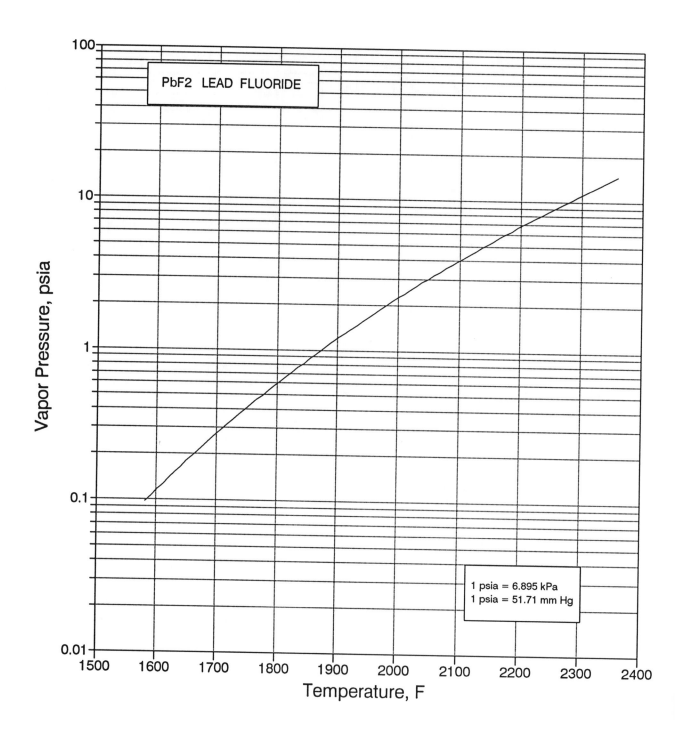

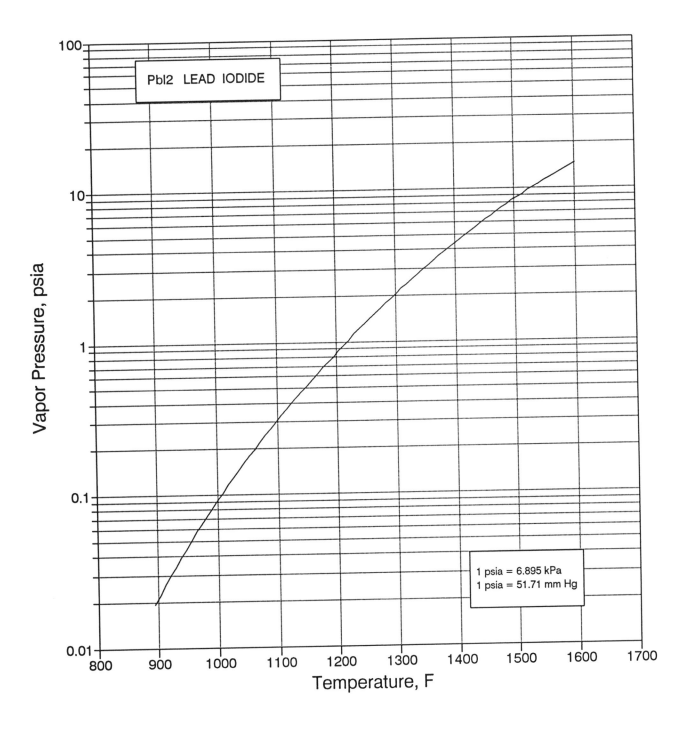

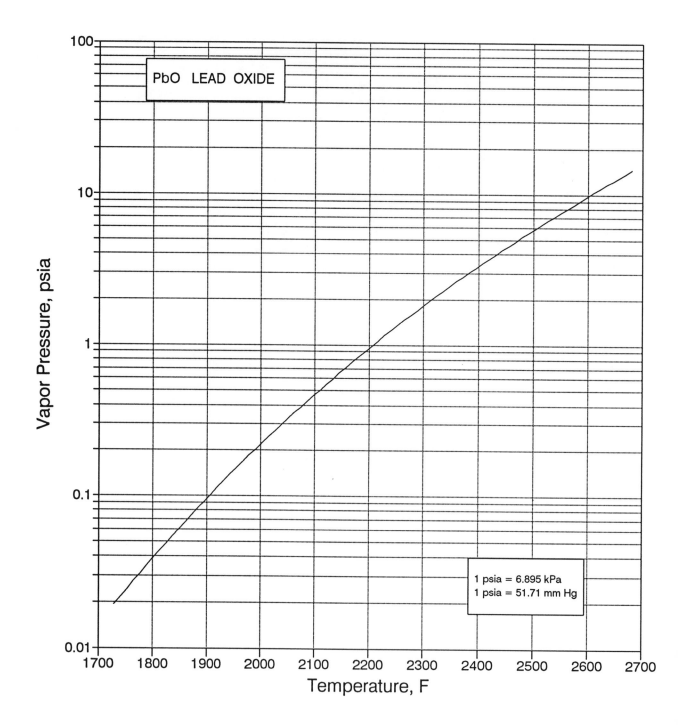

PbO LEAD OXIDE

Vapor Pressure, psia

Temperature, F

1 psia = 6.895 kPa
1 psia = 51.71 mm Hg

232

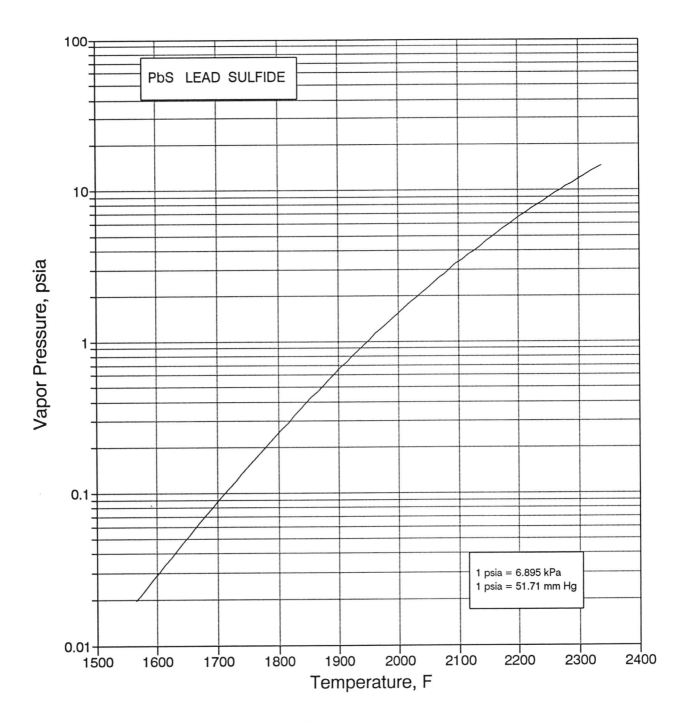

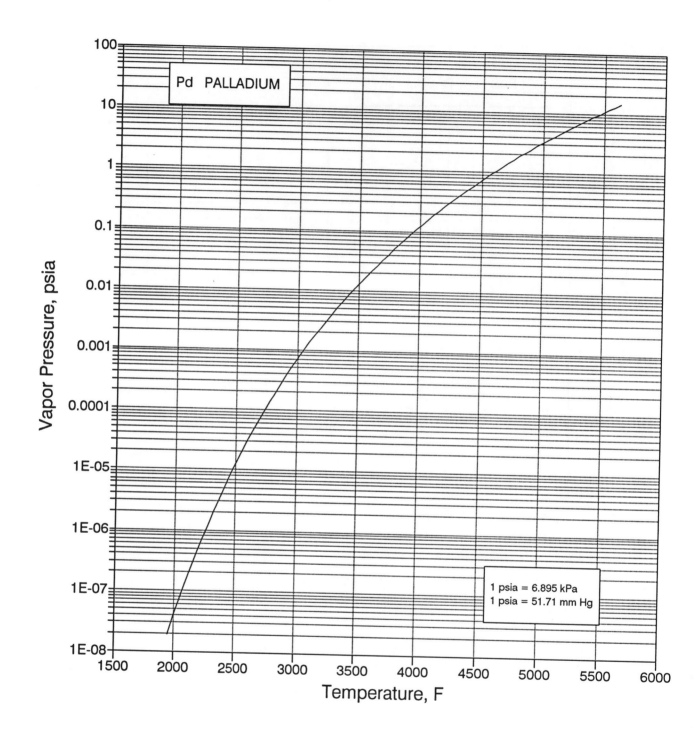

Pd   PALLADIUM

Vapor Pressure, psia

Temperature, F

1 psia = 6.895 kPa
1 psia = 51.71 mm Hg

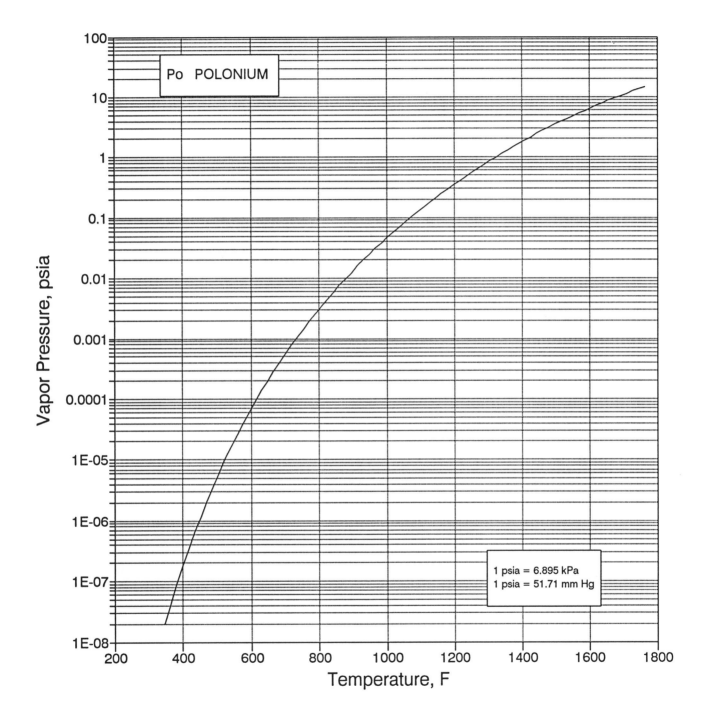

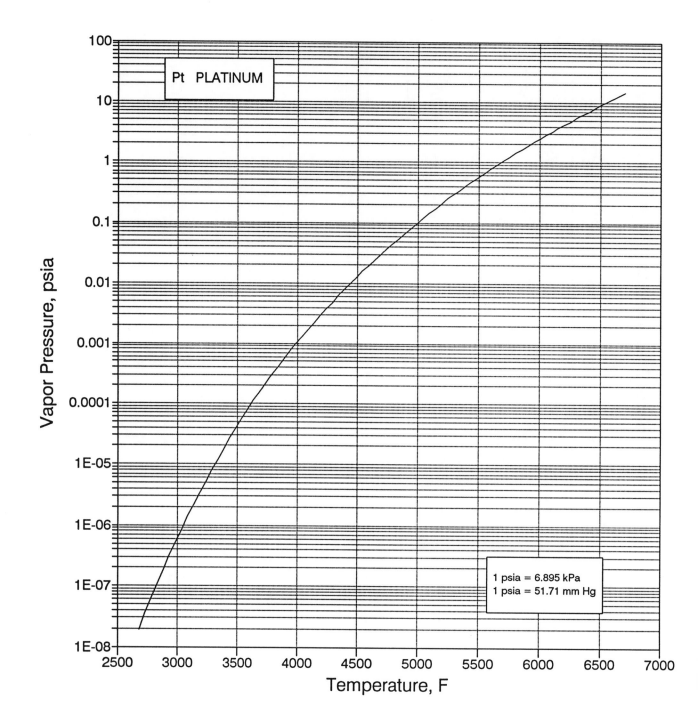

Pt  PLATINUM

1 psia = 6.895 kPa
1 psia = 51.71 mm Hg

Vapor Pressure, psia

Temperature, F

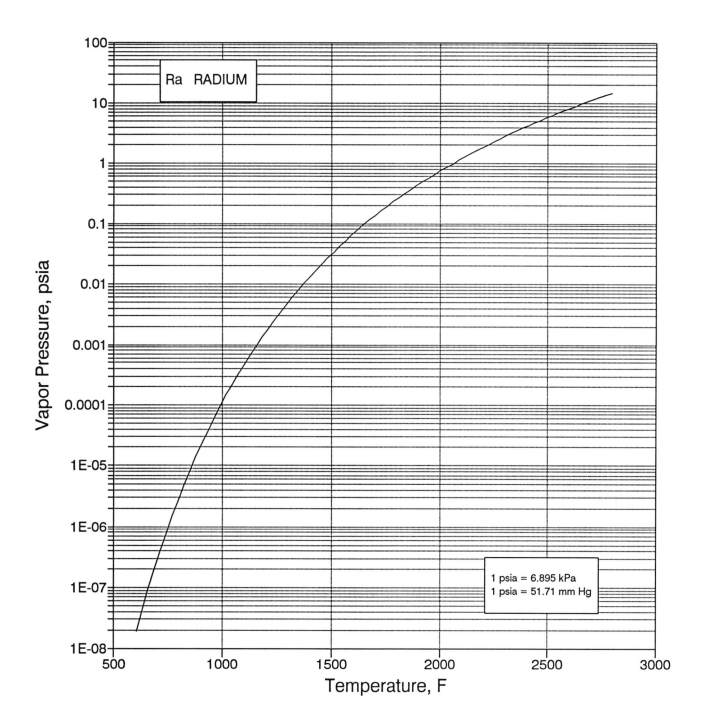

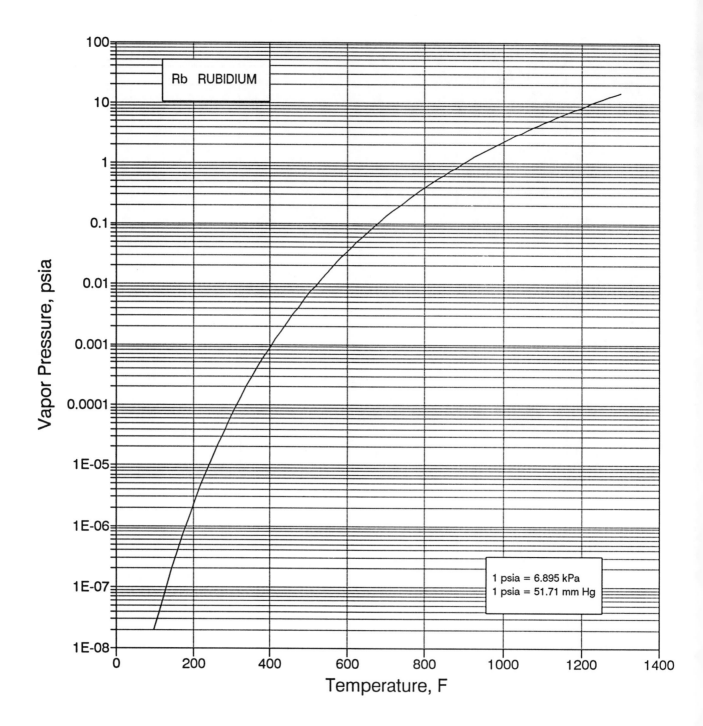

1 psia = 6.895 kPa
1 psia = 51.71 mm Hg

238

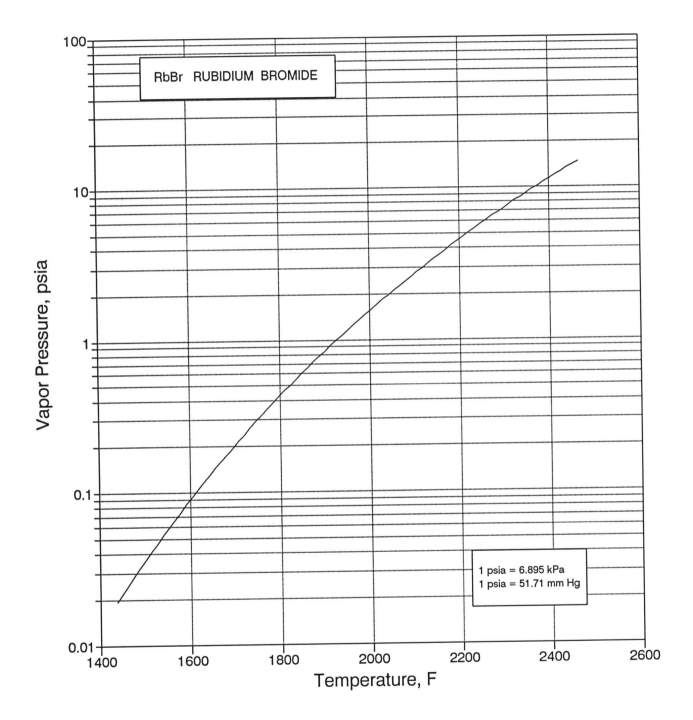

RbBr  RUBIDIUM BROMIDE

1 psia = 6.895 kPa
1 psia = 51.71 mm Hg

Vapor Pressure, psia

Temperature, F

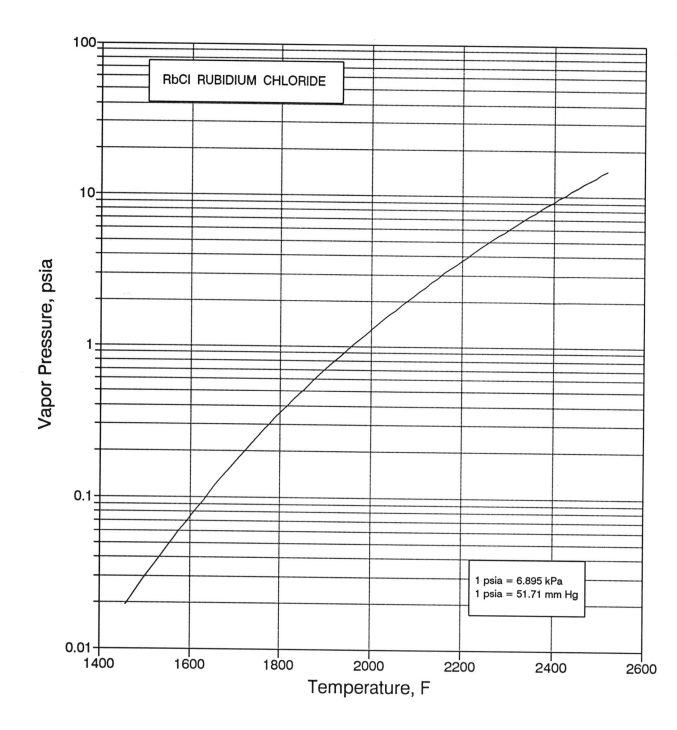

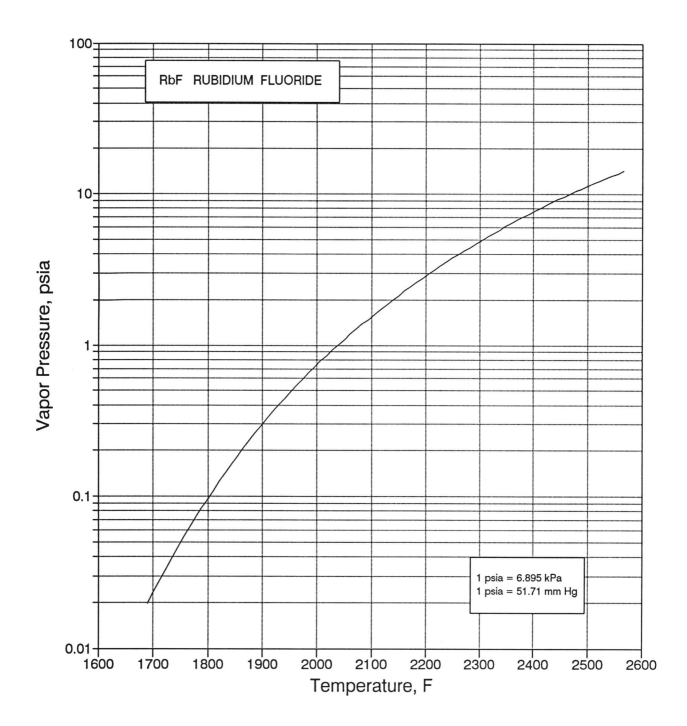

RbF  RUBIDIUM FLUORIDE

Vapor Pressure, psia

Temperature, F

1 psia = 6.895 kPa
1 psia = 51.71 mm Hg

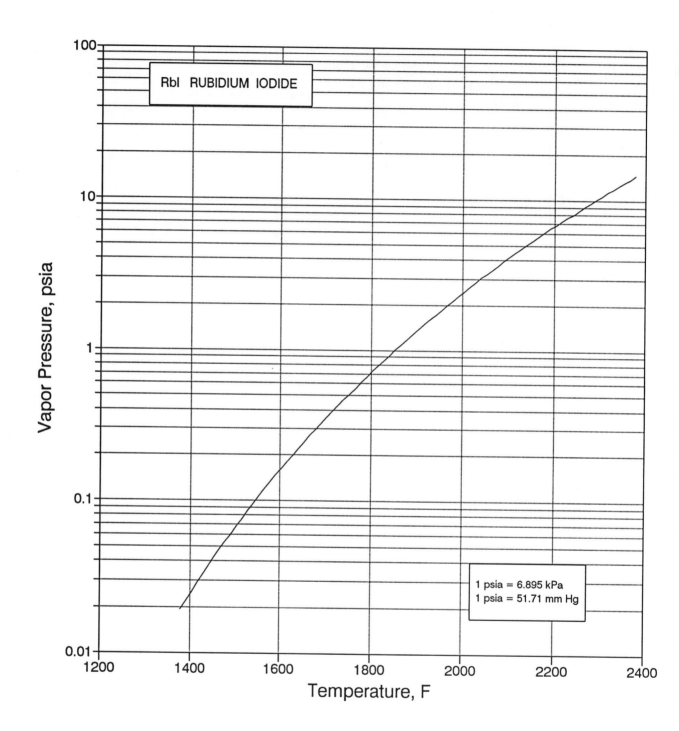

RbI  RUBIDIUM IODIDE

Vapor Pressure, psia

Temperature, F

1 psia = 6.895 kPa
1 psia = 51.71 mm Hg

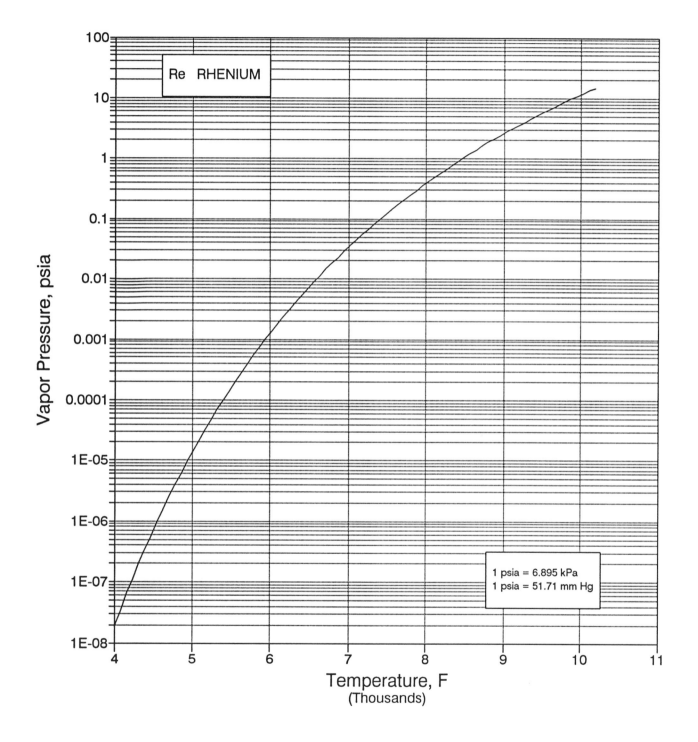

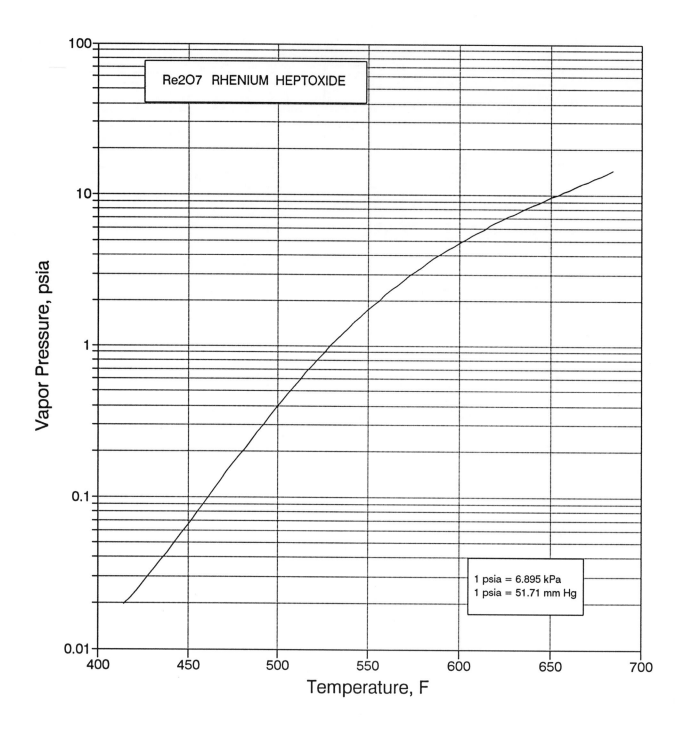

Re2O7  RHENIUM HEPTOXIDE

Vapor Pressure, psia

Temperature, F

1 psia = 6.895 kPa
1 psia = 51.71 mm Hg

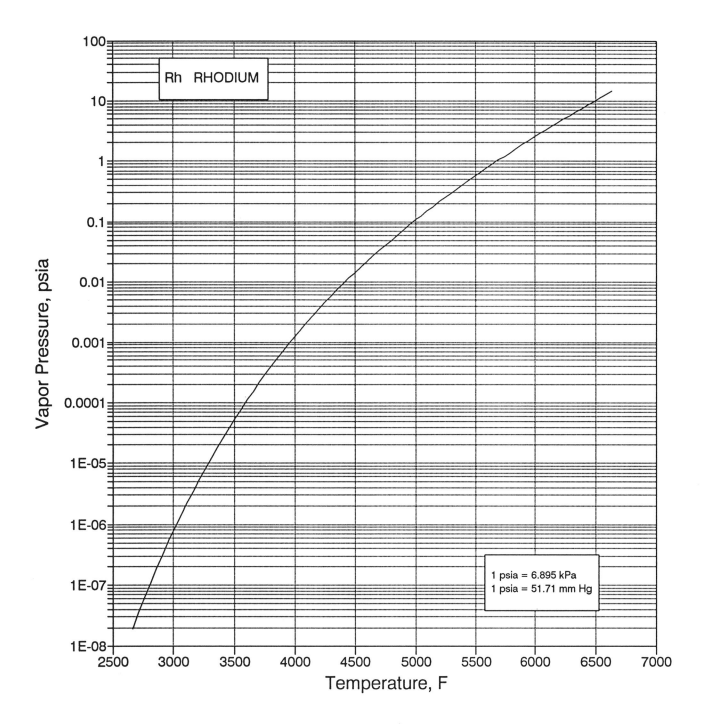

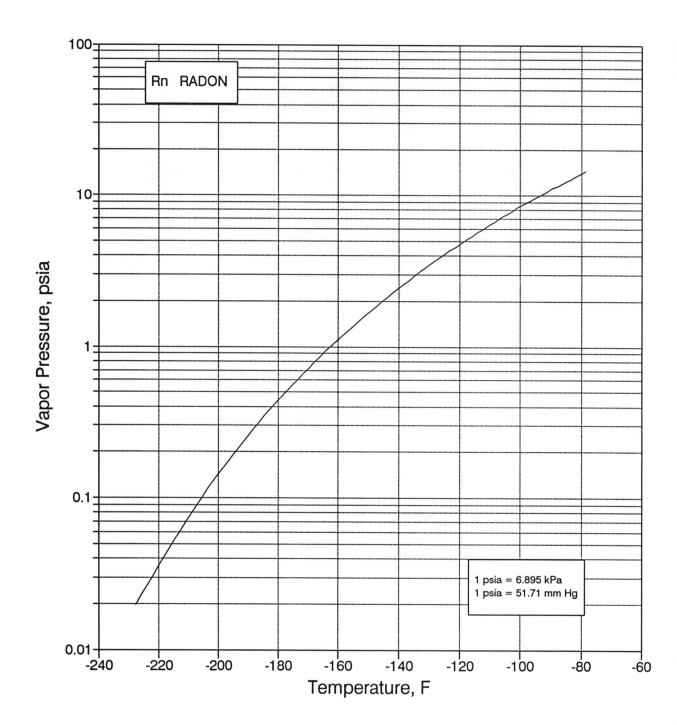

Vapor Pressure, psia

Temperature, F

Rn  RADON

1 psia = 6.895 kPa
1 psia = 51.71 mm Hg

246

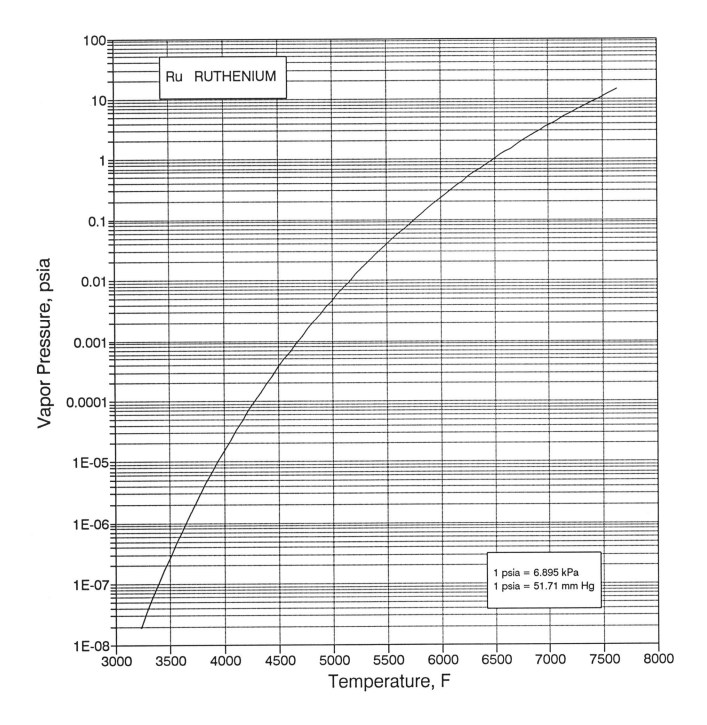

Ru RUTHENIUM

1 psia = 6.895 kPa
1 psia = 51.71 mm Hg

Vapor Pressure, psia

Temperature, F

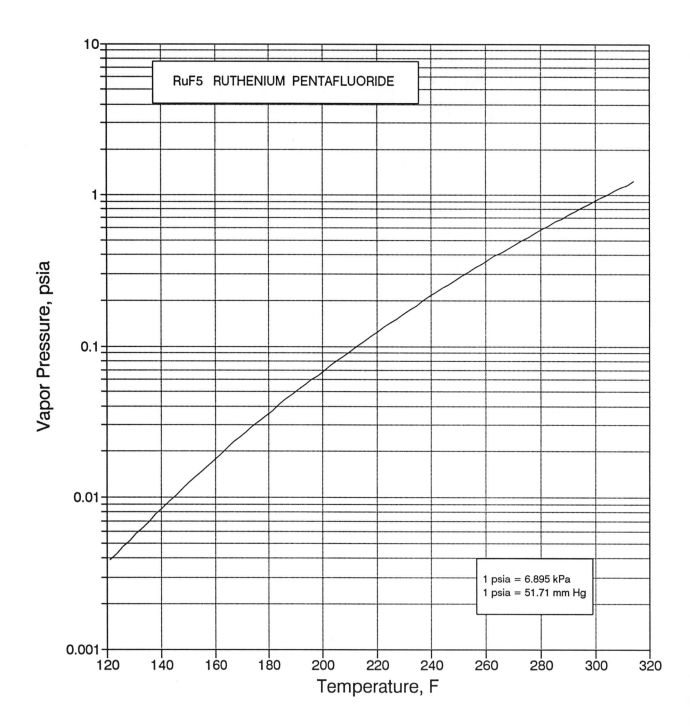

RuF5 RUTHENIUM PENTAFLUORIDE

Vapor Pressure, psia

Temperature, F

1 psia = 6.895 kPa
1 psia = 51.71 mm Hg

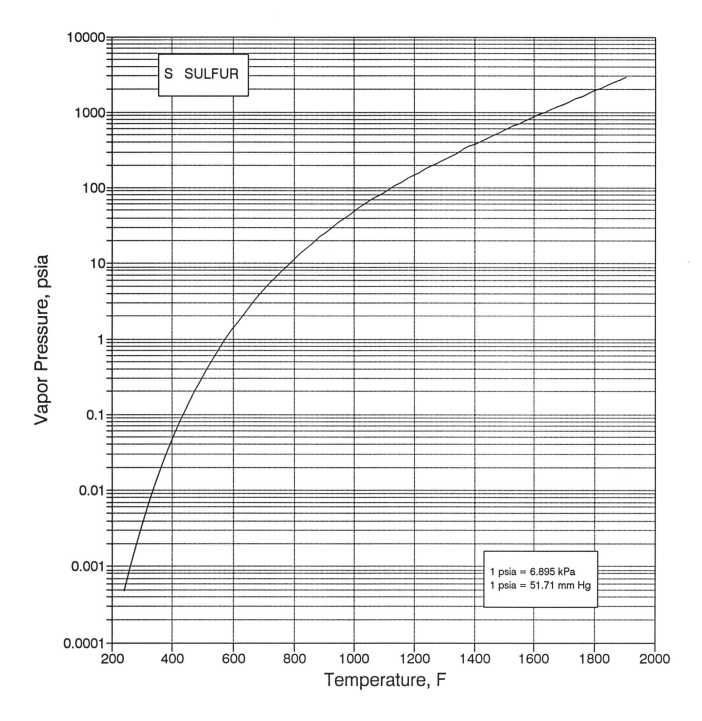

S SULFUR

Vapor Pressure, psia

Temperature, F

1 psia = 6.895 kPa
1 psia = 51.71 mm Hg

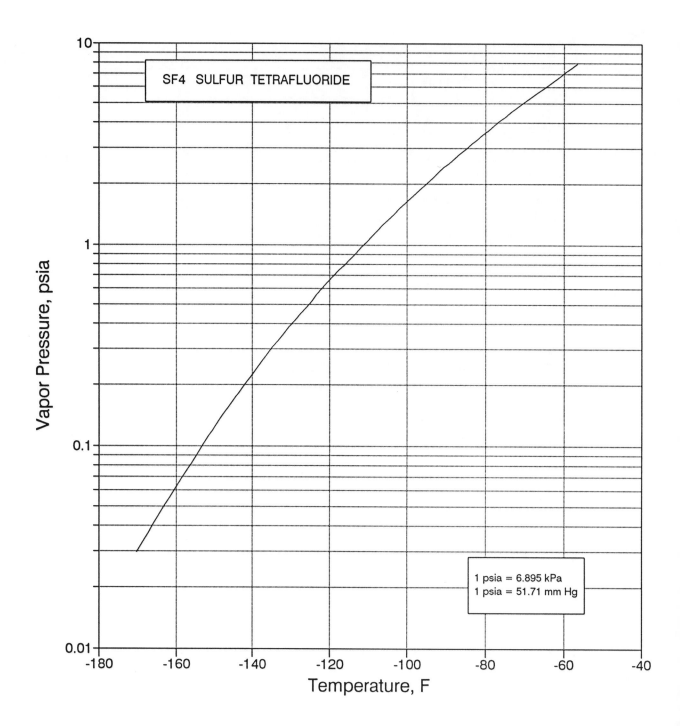

SF4  SULFUR TETRAFLUORIDE

Vapor Pressure, psia

Temperature, F

1 psia = 6.895 kPa
1 psia = 51.71 mm Hg

250

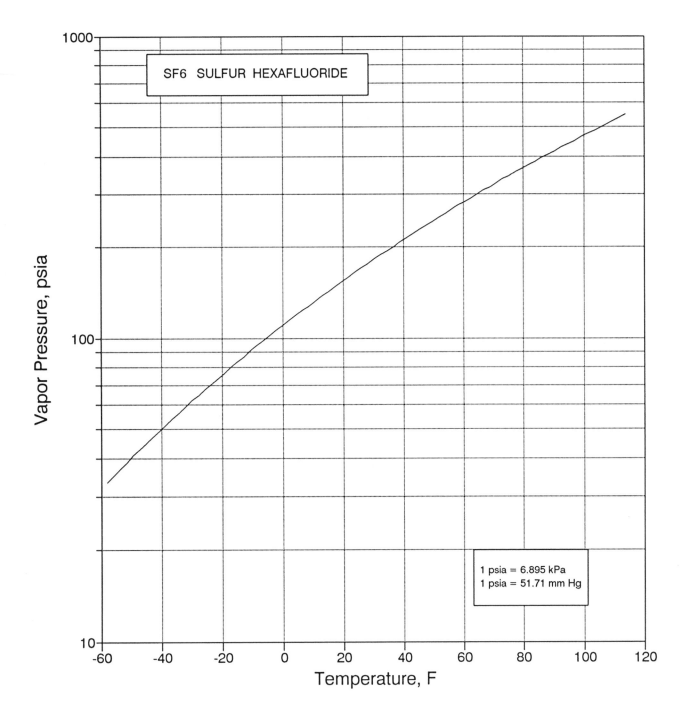

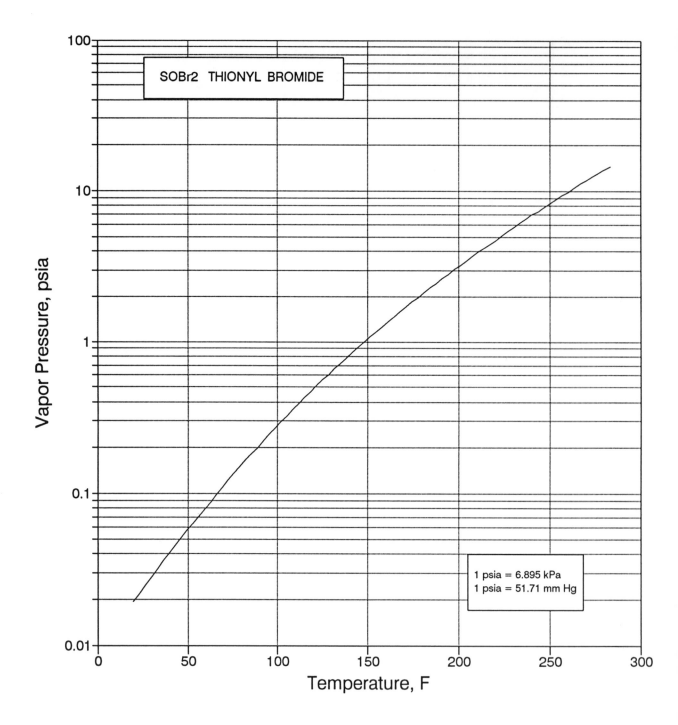

SOBr2 THIONYL BROMIDE

Vapor Pressure, psia

Temperature, F

1 psia = 6.895 kPa
1 psia = 51.71 mm Hg

252

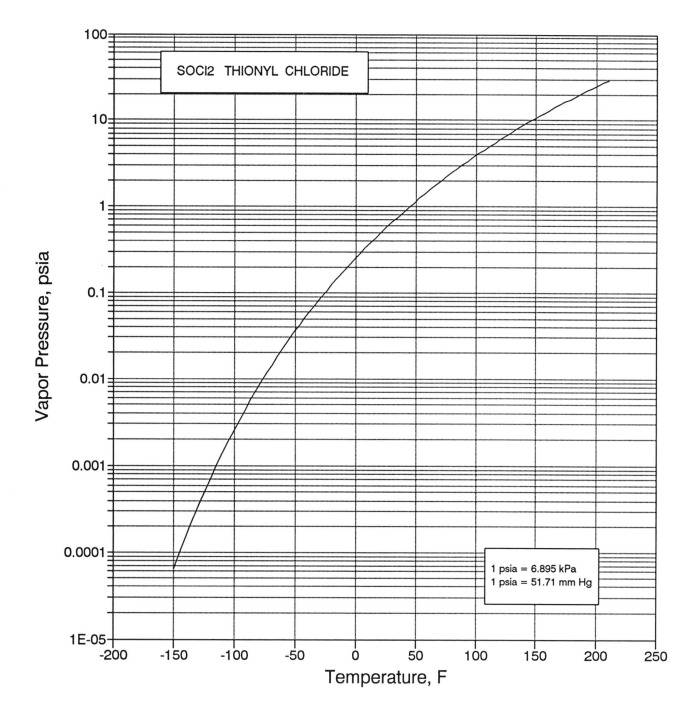

SOCl2   THIONYL CHLORIDE

1 psia = 6.895 kPa
1 psia = 51.71 mm Hg

Vapor Pressure, psia

Temperature, F

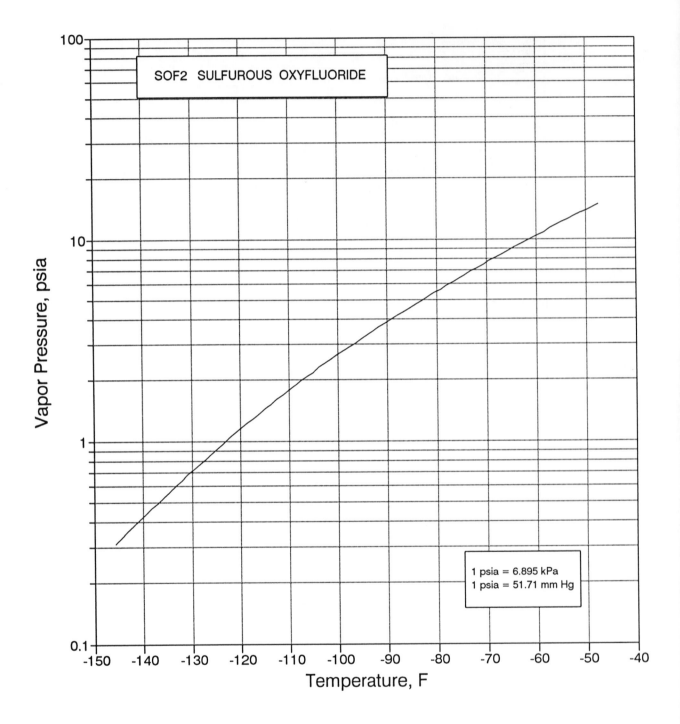

SOF2 SULFUROUS OXYFLUORIDE

Vapor Pressure, psia

Temperature, F

1 psia = 6.895 kPa
1 psia = 51.71 mm Hg

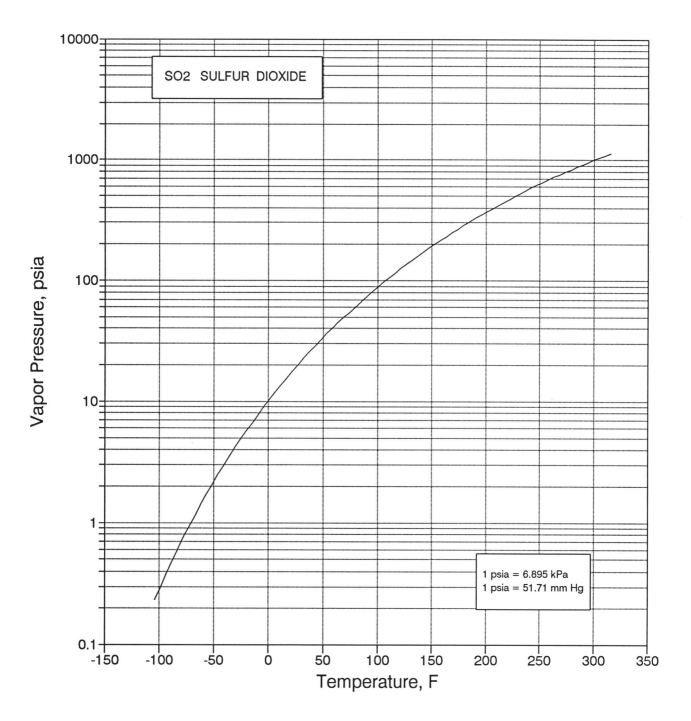

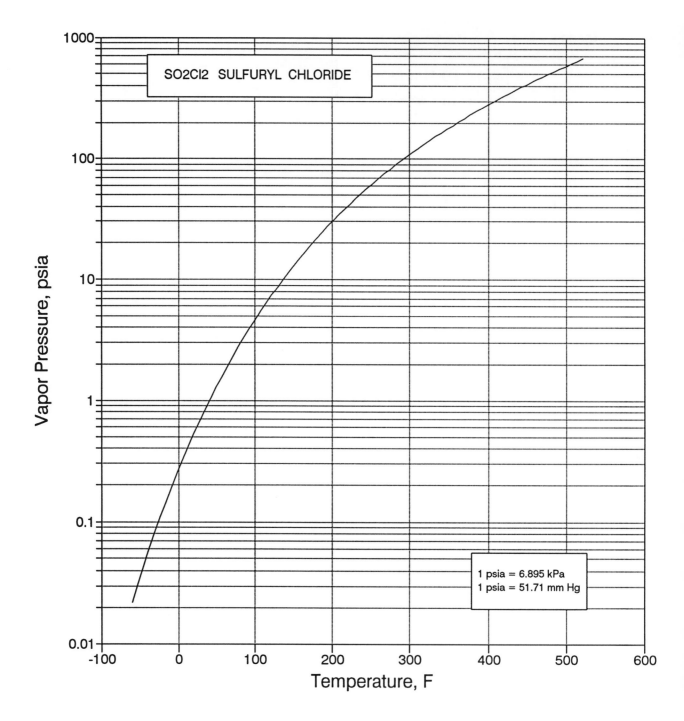

SO2Cl2 SULFURYL CHLORIDE

Vapor Pressure, psia

Temperature, F

1 psia = 6.895 kPa
1 psia = 51.71 mm Hg

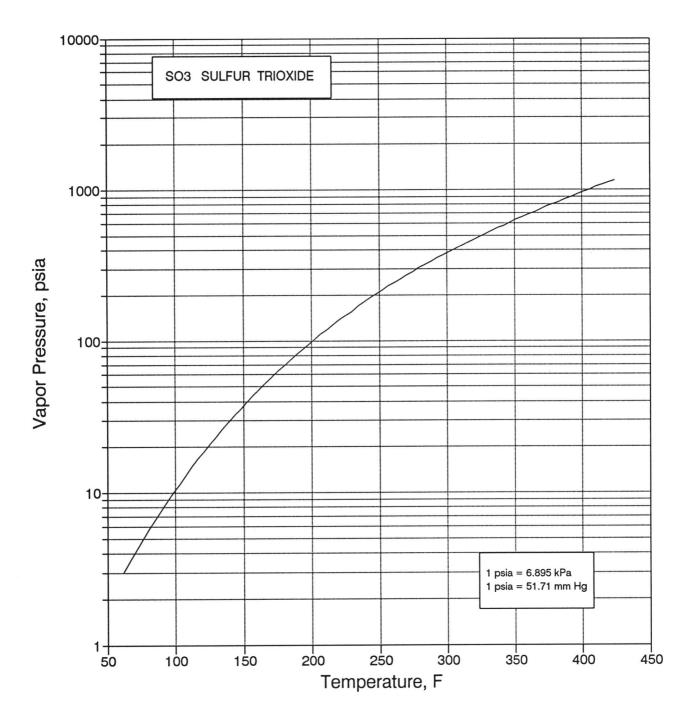

SO3 SULFUR TRIOXIDE

Vapor Pressure, psia

Temperature, F

1 psia = 6.895 kPa
1 psia = 51.71 mm Hg

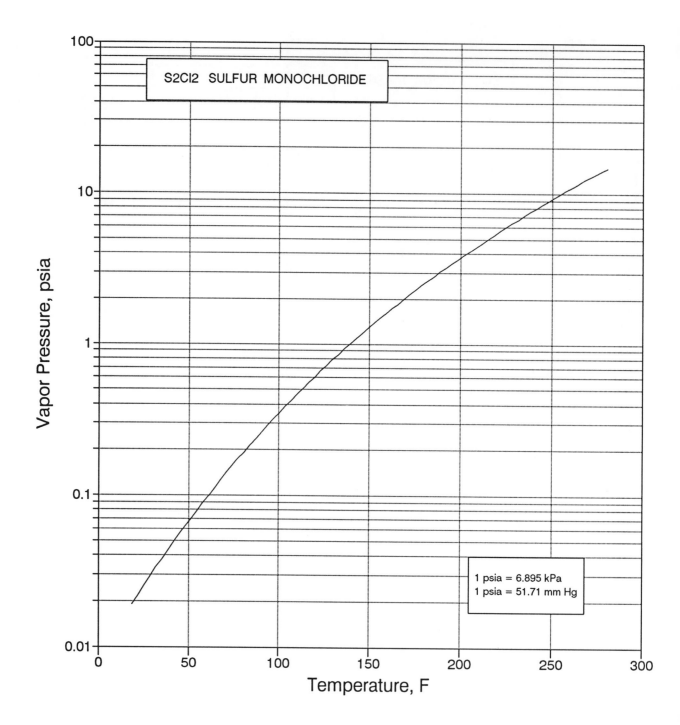

Vapor Pressure, psia

Temperature, F

S2Cl2  SULFUR MONOCHLORIDE

1 psia = 6.895 kPa
1 psia = 51.71 mm Hg

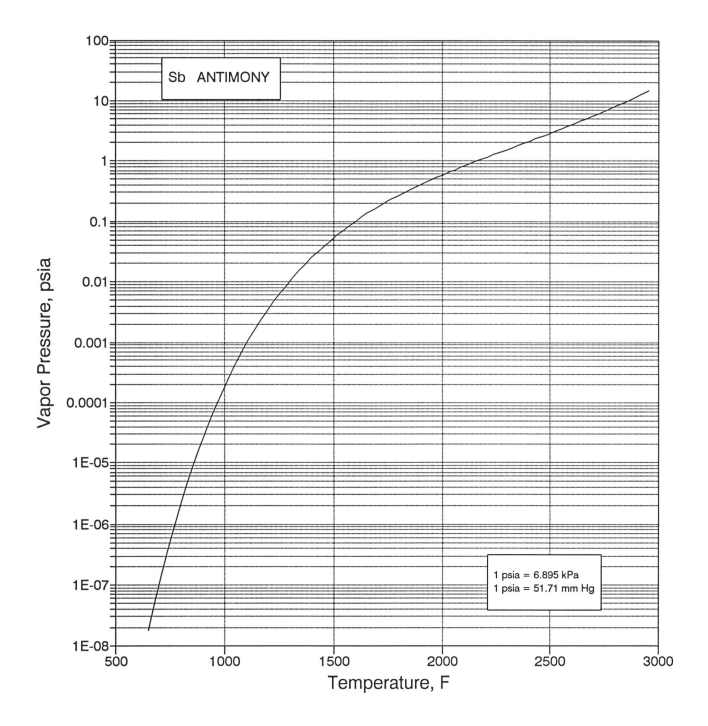

Vapor Pressure, psia vs Temperature, F

Sb  ANTIMONY

1 psia = 6.895 kPa
1 psia = 51.71 mm Hg

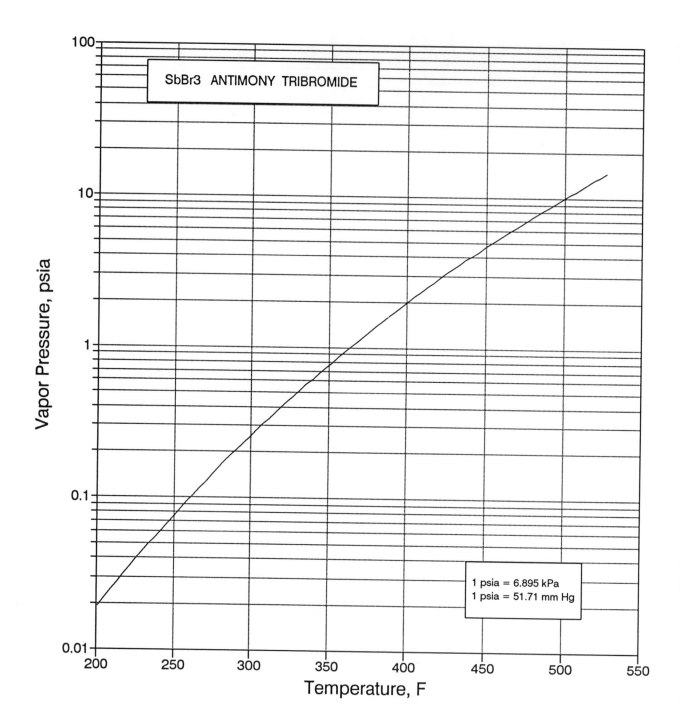

SbBr3  ANTIMONY  TRIBROMIDE

Vapor Pressure, psia

Temperature, F

1 psia = 6.895 kPa
1 psia = 51.71 mm Hg

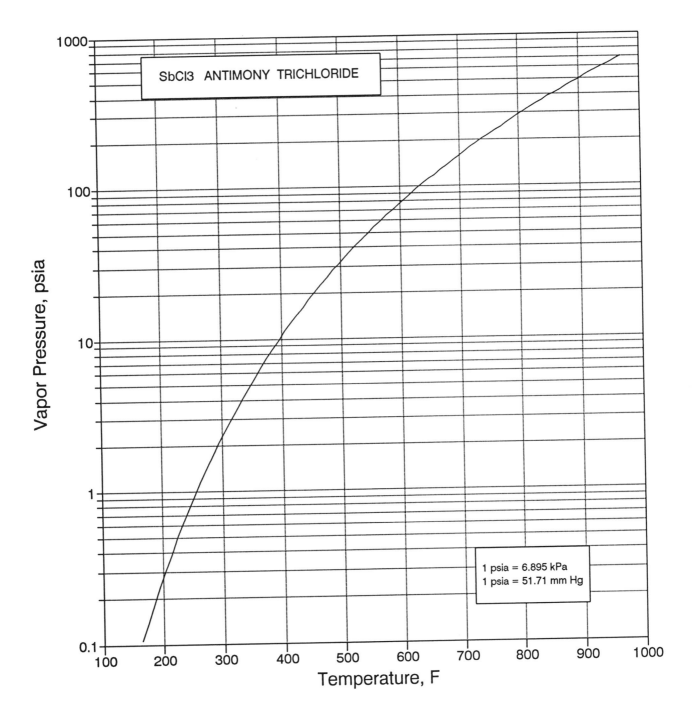

SbCl3 ANTIMONY TRICHLORIDE

Vapor Pressure, psia

Temperature, F

1 psia = 6.895 kPa
1 psia = 51.71 mm Hg

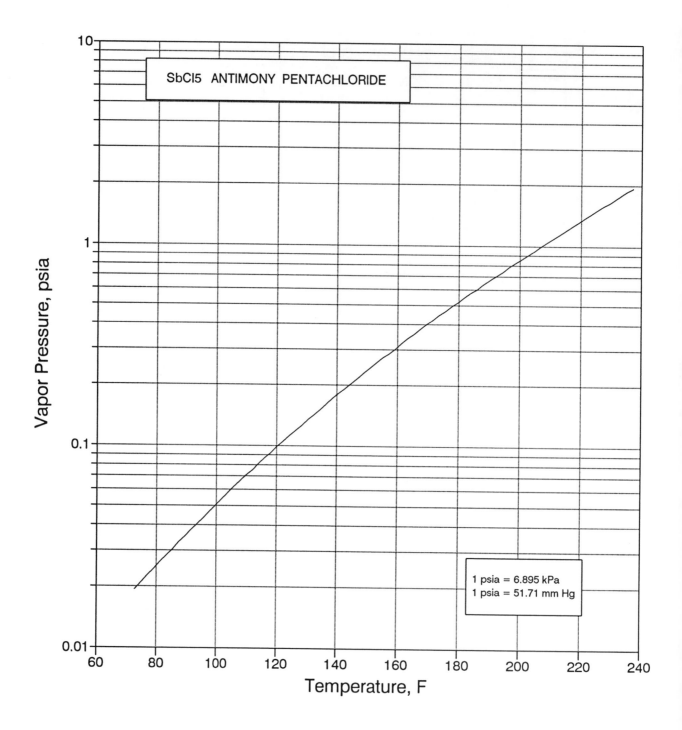

SbCl5  ANTIMONY  PENTACHLORIDE

Vapor Pressure, psia

Temperature, F

1 psia = 6.895 kPa
1 psia = 51.71 mm Hg

262

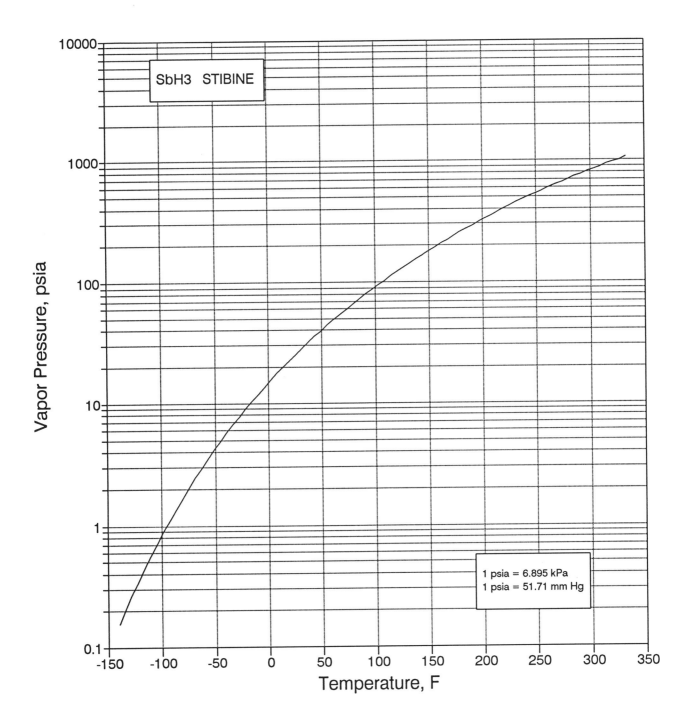

SbH3  STIBINE

Vapor Pressure, psia

Temperature, F

1 psia = 6.895 kPa
1 psia = 51.71 mm Hg

263

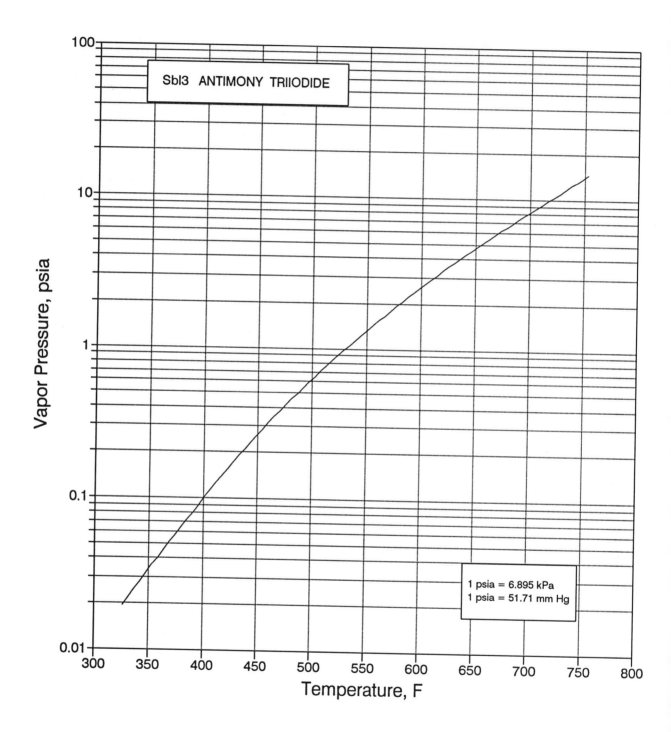

Vapor Pressure, psia — Temperature, F

SbI3  ANTIMONY TRIIODIDE

1 psia = 6.895 kPa
1 psia = 51.71 mm Hg

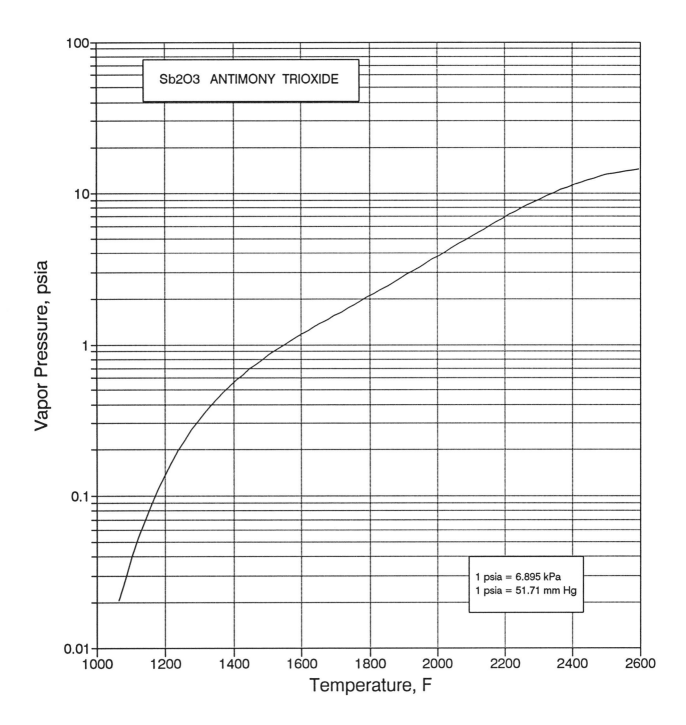

1 psia = 6.895 kPa
1 psia = 51.71 mm Hg

Sb2O3  ANTIMONY TRIOXIDE

Vapor Pressure, psia

Temperature, F

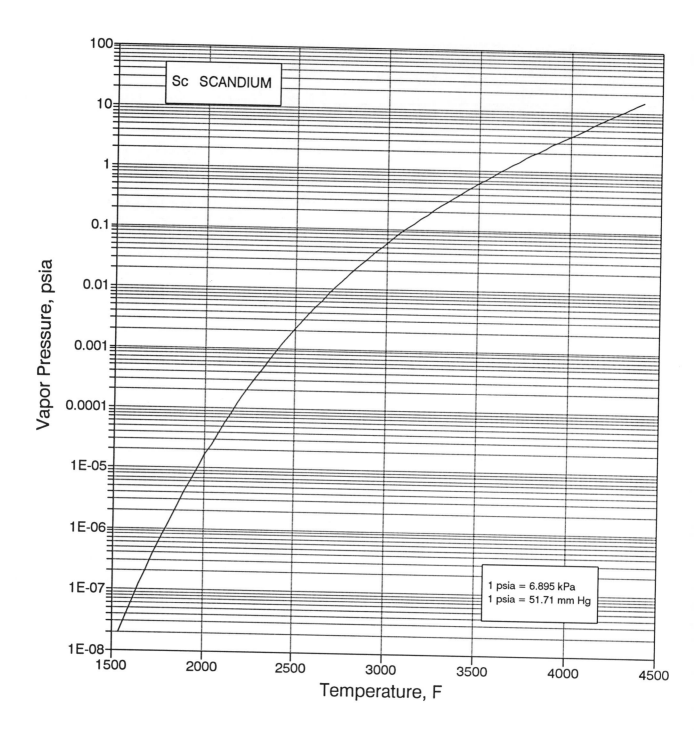

Sc  SCANDIUM

1 psia = 6.895 kPa
1 psia = 51.71 mm Hg

Vapor Pressure, psia

Temperature, F

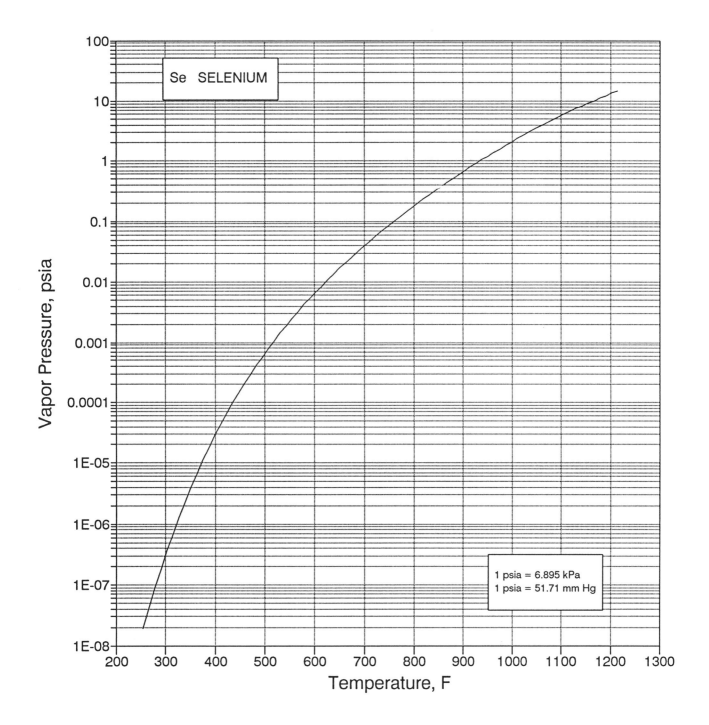

Vapor Pressure, psia

Se   SELENIUM

1 psia = 6.895 kPa
1 psia = 51.71 mm Hg

Temperature, F

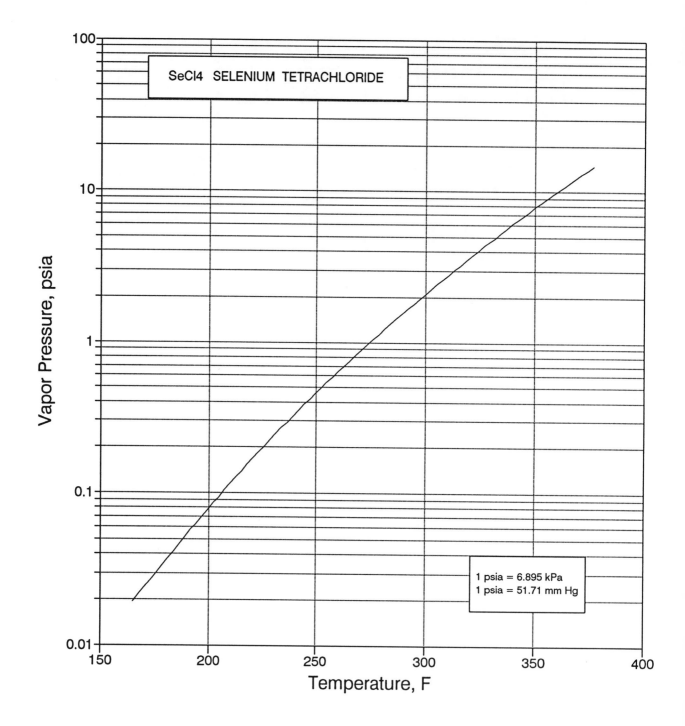

SeCl4  SELENIUM TETRACHLORIDE

1 psia = 6.895 kPa
1 psia = 51.71 mm Hg

Vapor Pressure, psia

Temperature, F

268

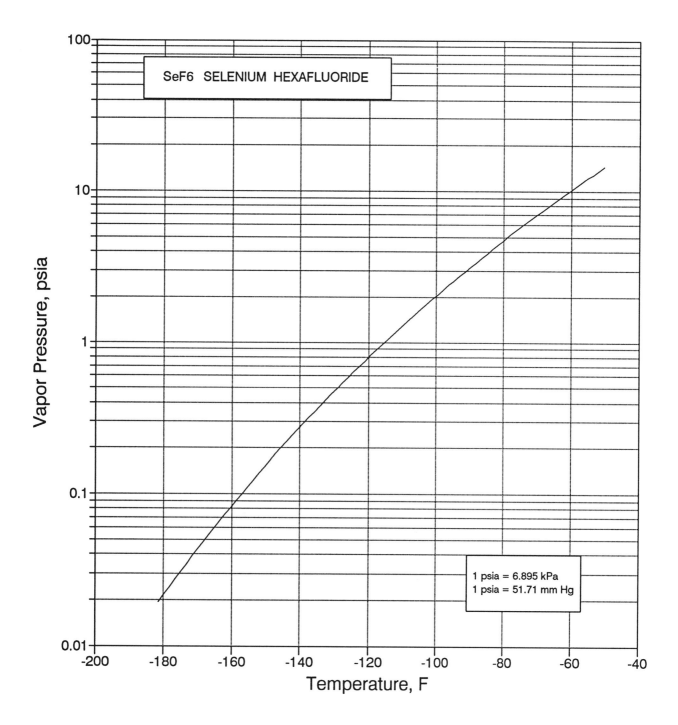

SeF6  SELENIUM  HEXAFLUORIDE

1 psia = 6.895 kPa
1 psia = 51.71 mm Hg

Vapor Pressure, psia

Temperature, F

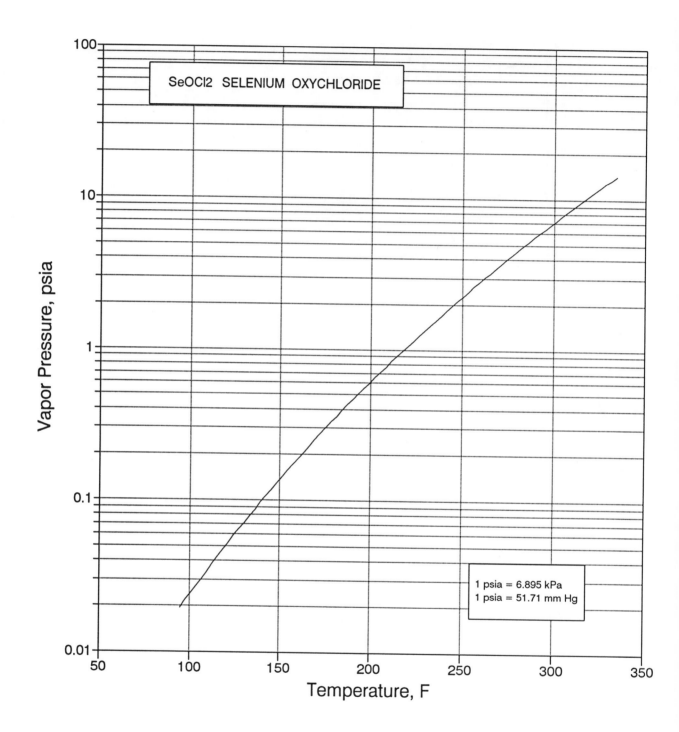

SeOCl2 SELENIUM OXYCHLORIDE

Vapor Pressure, psia

Temperature, F

1 psia = 6.895 kPa
1 psia = 51.71 mm Hg

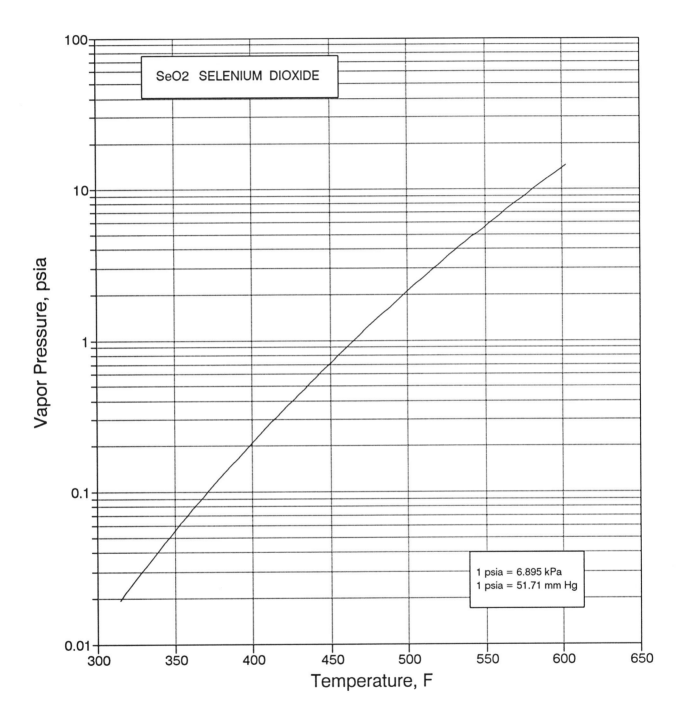

SeO2 SELENIUM DIOXIDE

Vapor Pressure, psia

Temperature, F

1 psia = 6.895 kPa
1 psia = 51.71 mm Hg

271

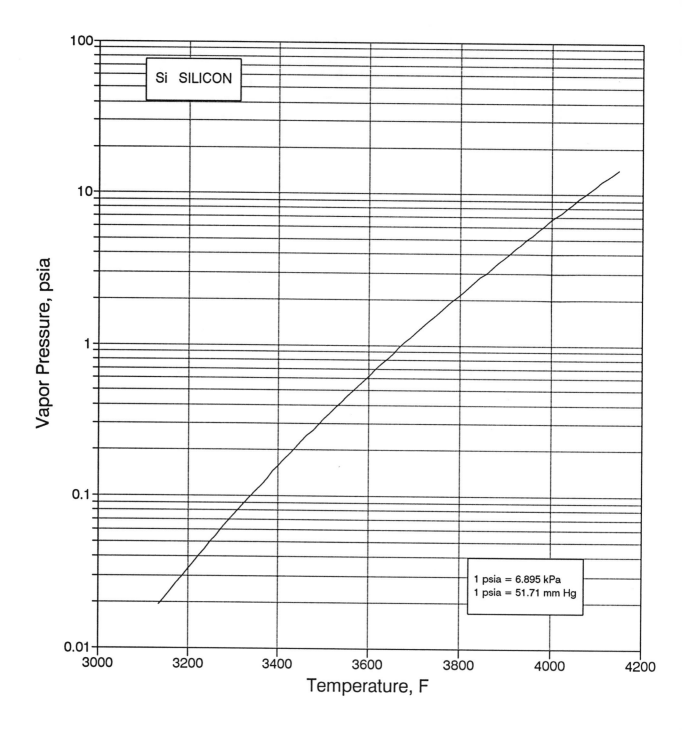

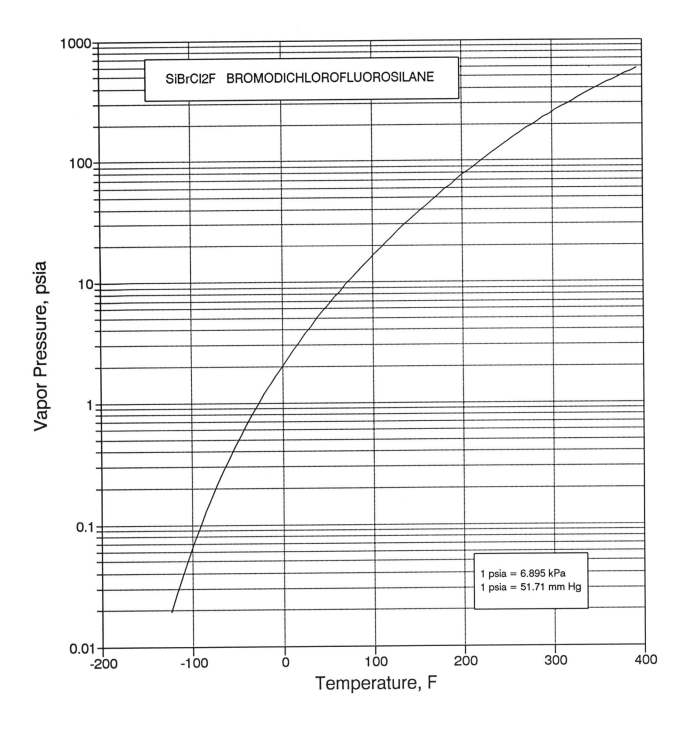

SiBrCl2F  BROMODICHLOROFLUOROSILANE

Vapor Pressure, psia

Temperature, F

1 psia = 6.895 kPa
1 psia = 51.71 mm Hg

273

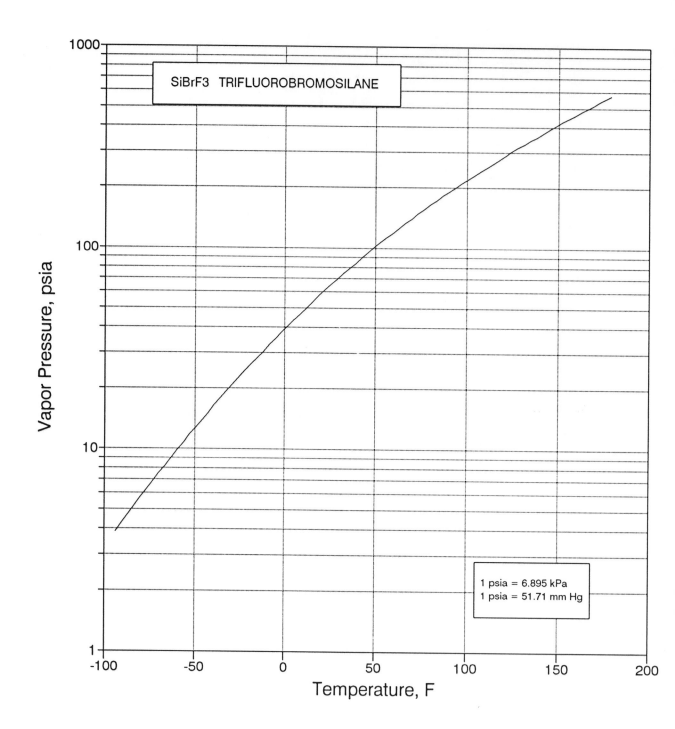

SiBrF3   TRIFLUOROBROMOSILANE

Vapor Pressure, psia

Temperature, F

1 psia = 6.895 kPa
1 psia = 51.71 mm Hg

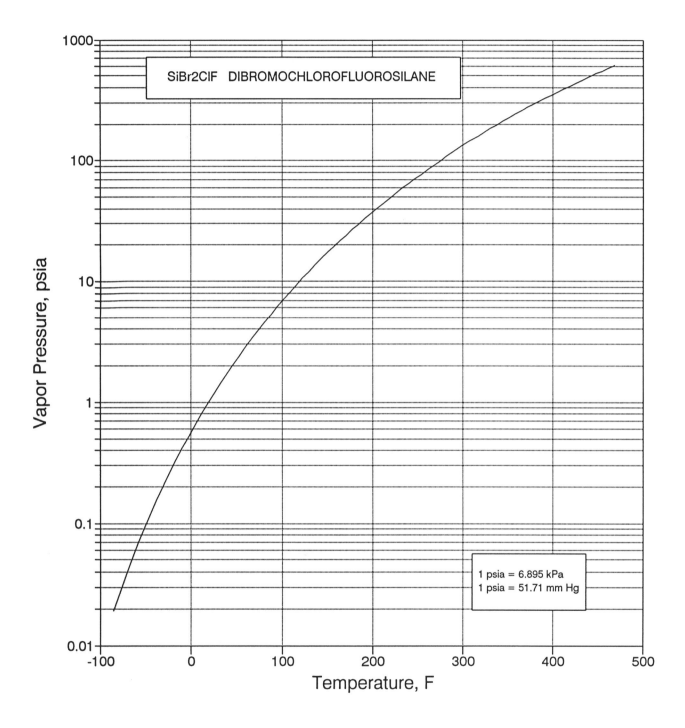

SiBr2ClF   DIBROMOCHLOROFLUOROSILANE

Vapor Pressure, psia

Temperature, F

1 psia = 6.895 kPa
1 psia = 51.71 mm Hg

275

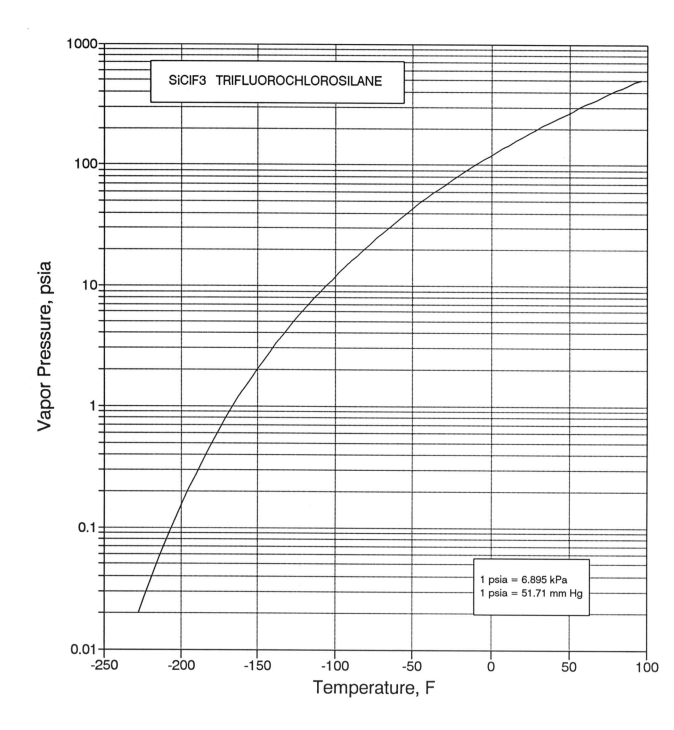

SiCIF3   TRIFLUOROCHLOROSILANE

Vapor Pressure, psia

Temperature, F

1 psia = 6.895 kPa
1 psia = 51.71 mm Hg

276

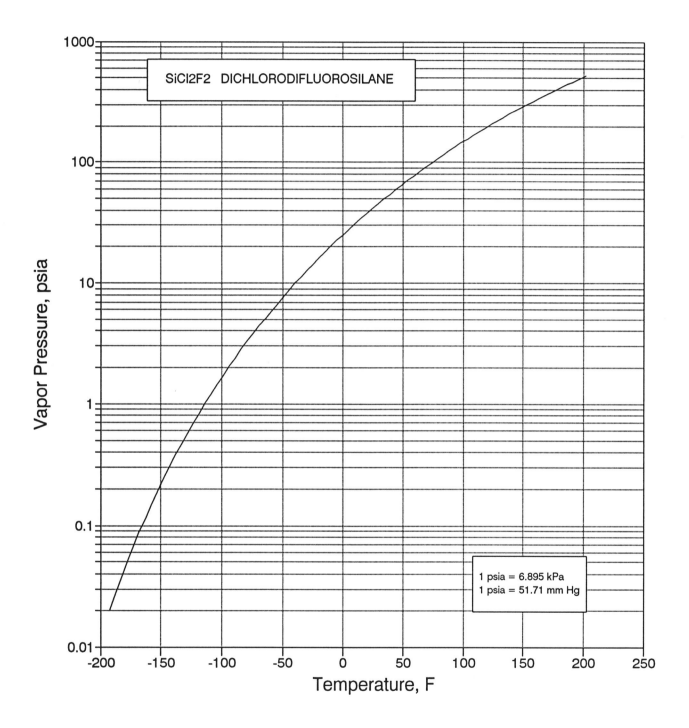

SiCl2F2 DICHLORODIFLUOROSILANE

Vapor Pressure, psia

Temperature, F

1 psia = 6.895 kPa
1 psia = 51.71 mm Hg

277

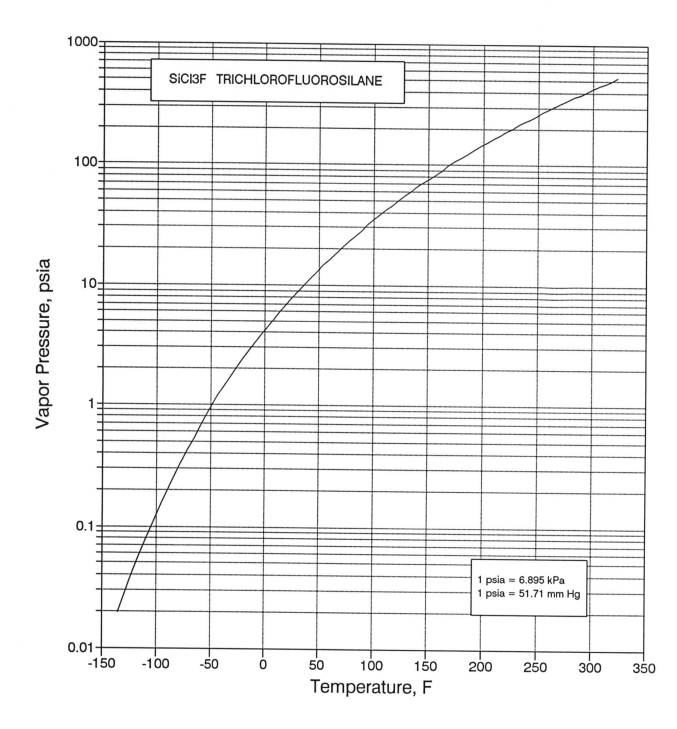

SiCl3F  TRICHLOROFLUOROSILANE

1 psia = 6.895 kPa
1 psia = 51.71 mm Hg

Vapor Pressure, psia

Temperature, F

278

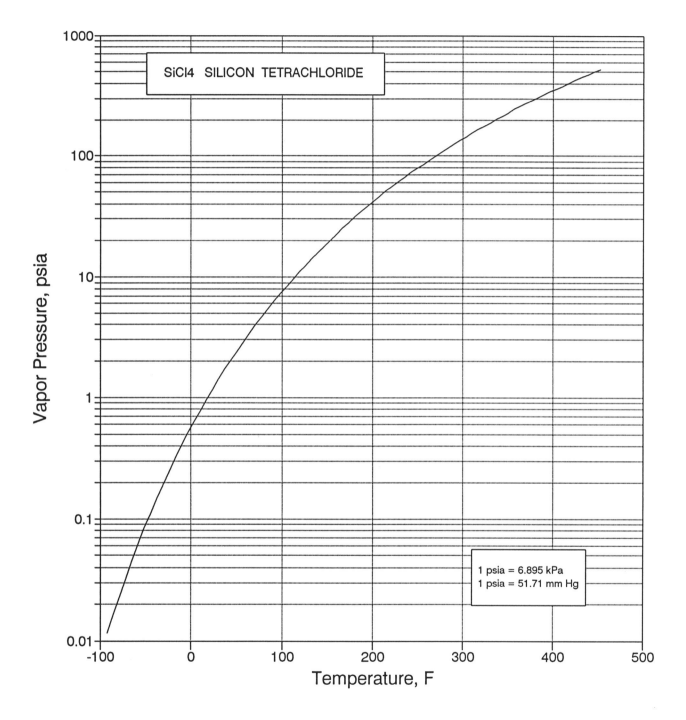

SiCl4  SILICON TETRACHLORIDE

1 psia = 6.895 kPa
1 psia = 51.71 mm Hg

Vapor Pressure, psia

Temperature, F

279

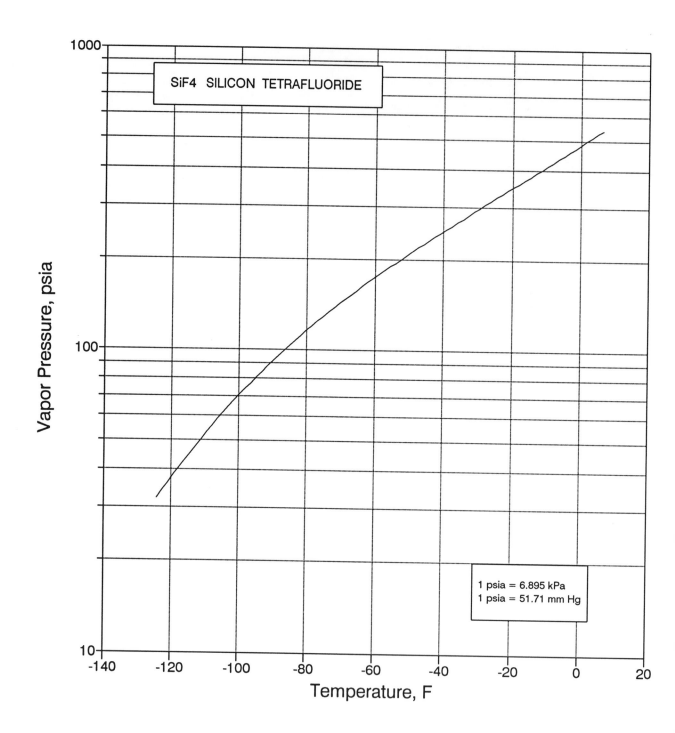

SiF4  SILICON TETRAFLUORIDE

Vapor Pressure, psia

Temperature, F

1 psia = 6.895 kPa
1 psia = 51.71 mm Hg

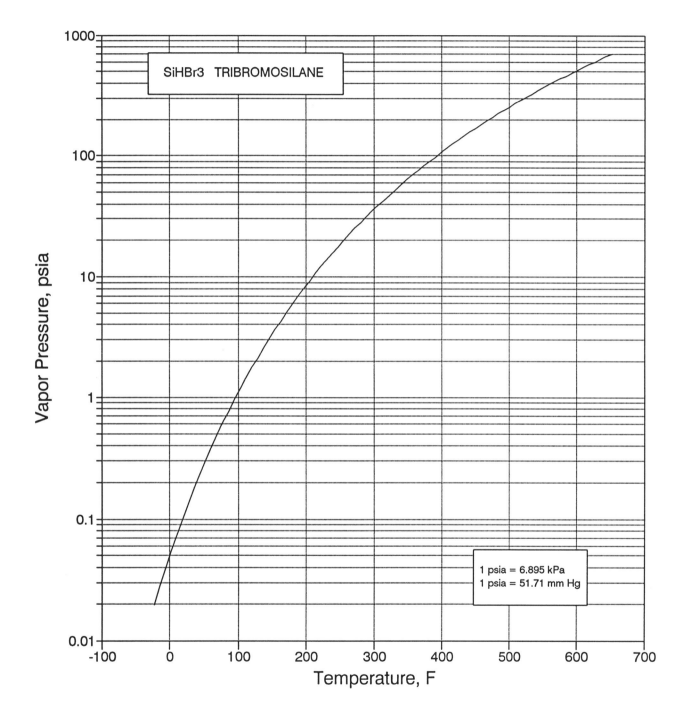

SiHBr3   TRIBROMOSILANE

Vapor Pressure, psia

Temperature, F

1 psia = 6.895 kPa
1 psia = 51.71 mm Hg

281

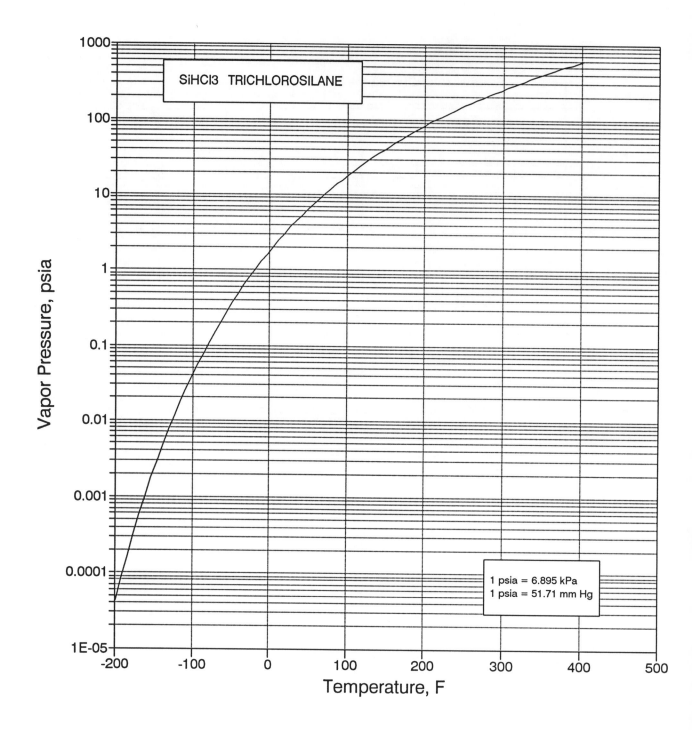

SiHCl3  TRICHLOROSILANE

Vapor Pressure, psia

Temperature, F

1 psia = 6.895 kPa
1 psia = 51.71 mm Hg

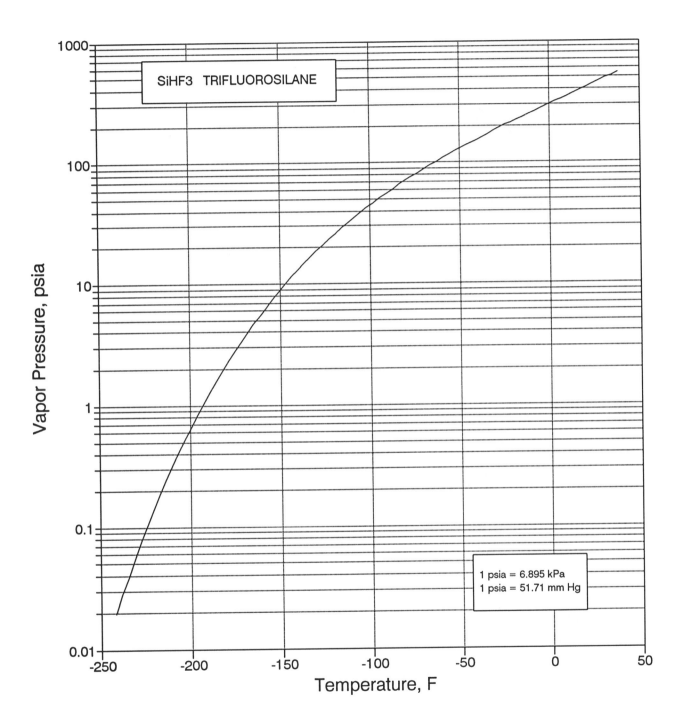

Vapor Pressure, psia

SiHF3  TRIFLUOROSILANE

1 psia = 6.895 kPa
1 psia = 51.71 mm Hg

Temperature, F

283

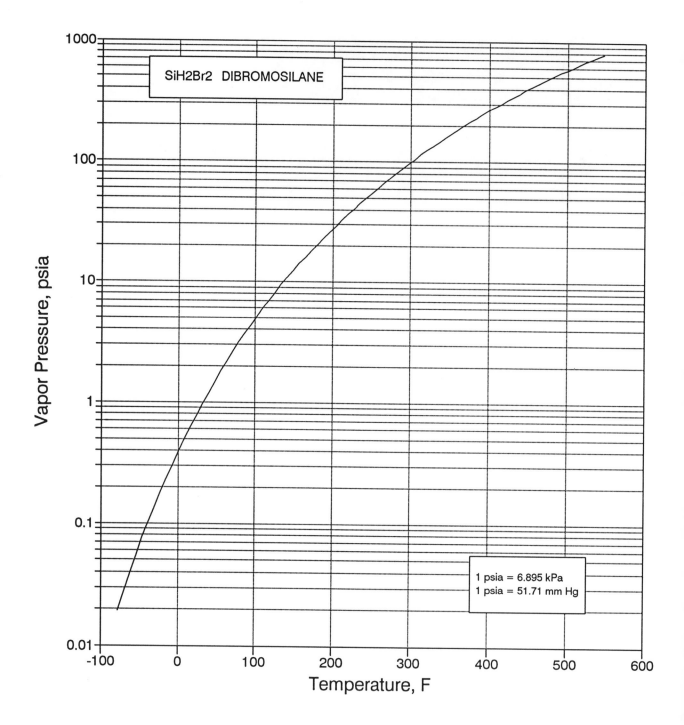

SiH2Br2 DIBROMOSILANE

Vapor Pressure, psia

Temperature, F

1 psia = 6.895 kPa
1 psia = 51.71 mm Hg

284

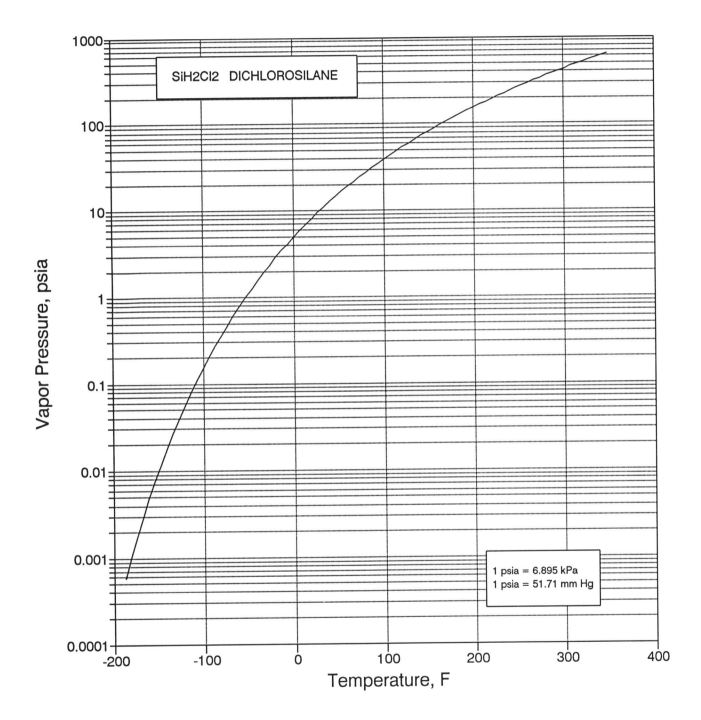

SiH2Cl2  DICHLOROSILANE

Vapor Pressure, psia

Temperature, F

1 psia = 6.895 kPa
1 psia = 51.71 mm Hg

285

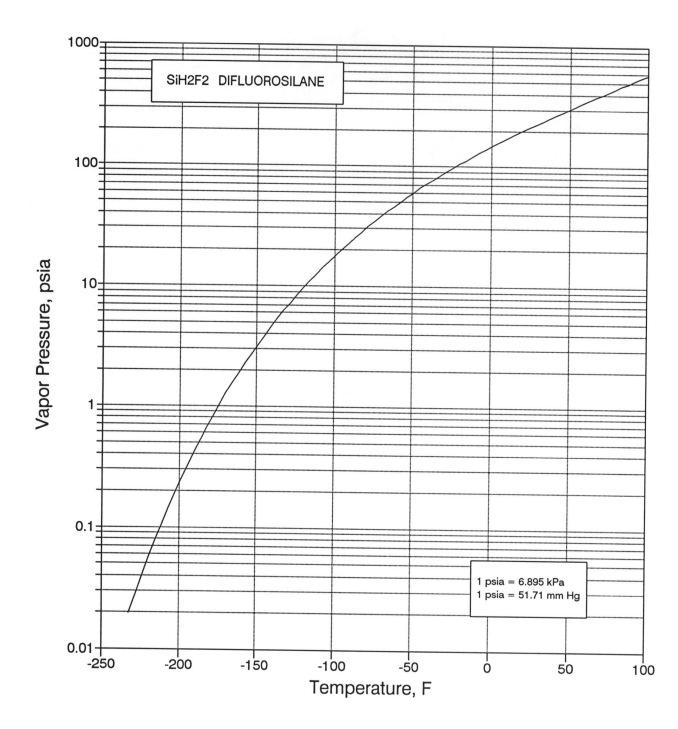

SiH2F2  DIFLUOROSILANE

Vapor Pressure, psia

Temperature, F

1 psia = 6.895 kPa
1 psia = 51.71 mm Hg

286

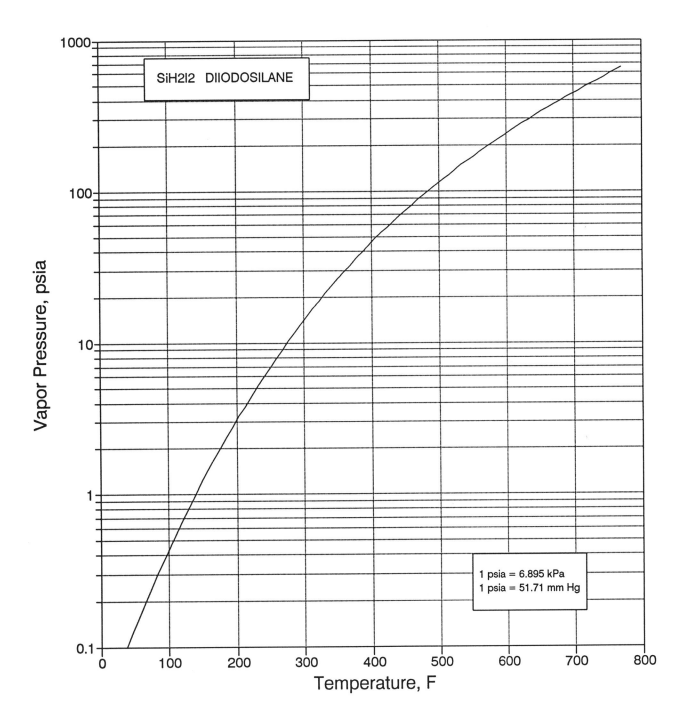

SiH2I2   DIIODOSILANE

Vapor Pressure, psia

Temperature, F

1 psia = 6.895 kPa
1 psia = 51.71 mm Hg

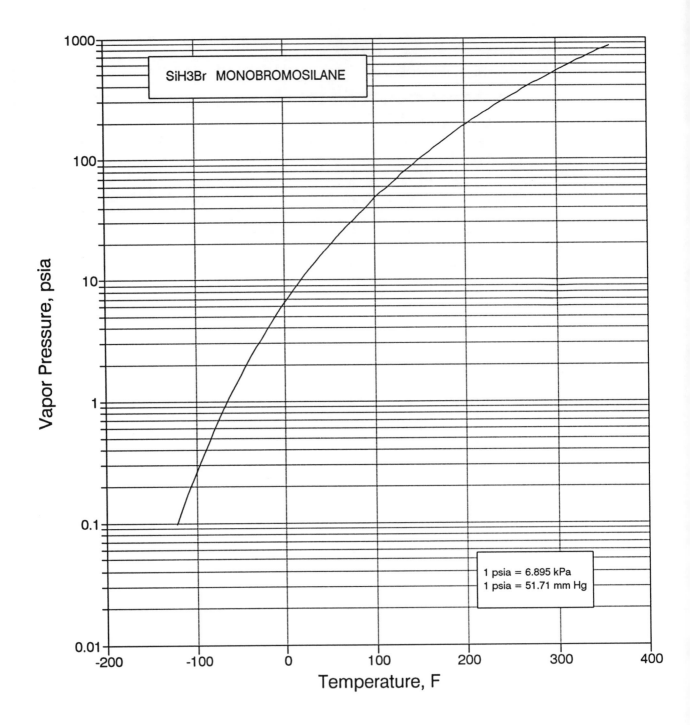

SiH3Br  MONOBROMOSILANE

Vapor Pressure, psia

Temperature, F

1 psia = 6.895 kPa
1 psia = 51.71 mm Hg

288

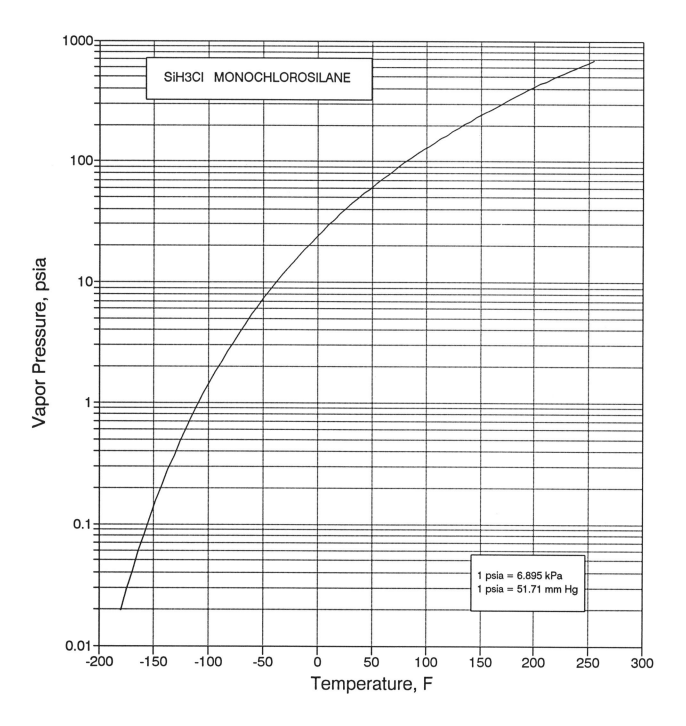

1 psia = 6.895 kPa
1 psia = 51.71 mm Hg

SiH3Cl MONOCHLOROSILANE

Vapor Pressure, psia

Temperature, F

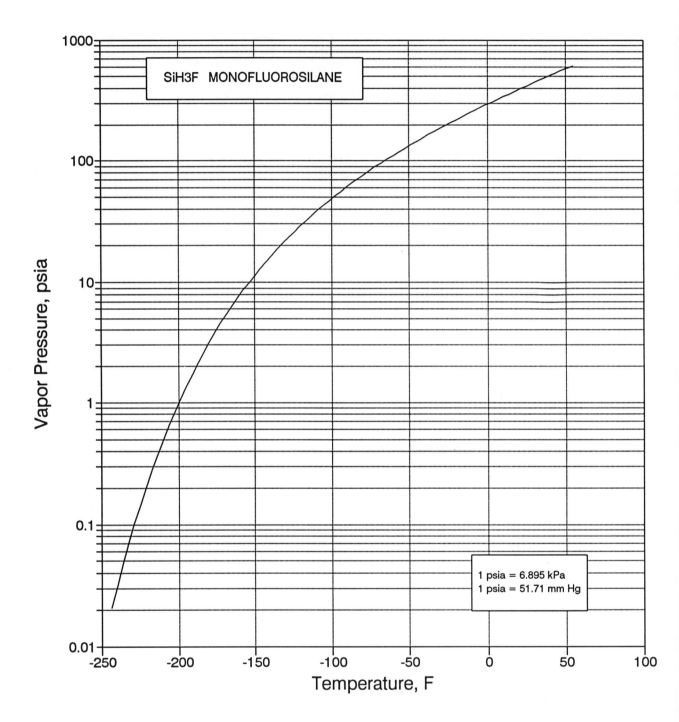

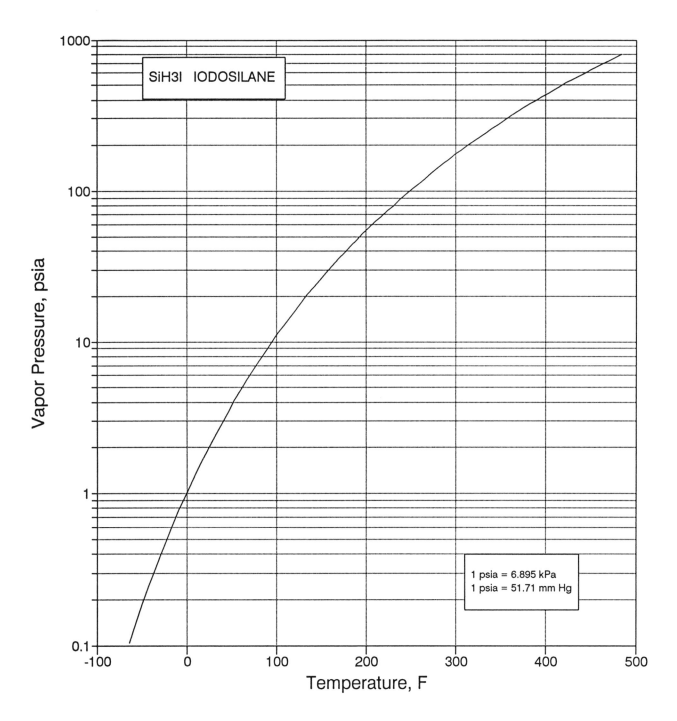

SiH3I   IODOSILANE

Vapor Pressure, psia

Temperature, F

1 psia = 6.895 kPa
1 psia = 51.71 mm Hg

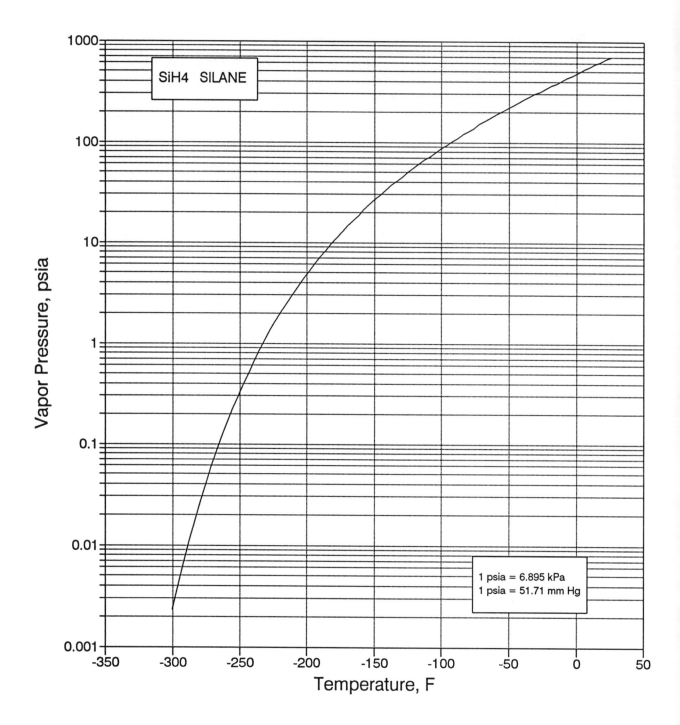

SiH4 SILANE

Vapor Pressure, psia

Temperature, F

1 psia = 6.895 kPa
1 psia = 51.71 mm Hg

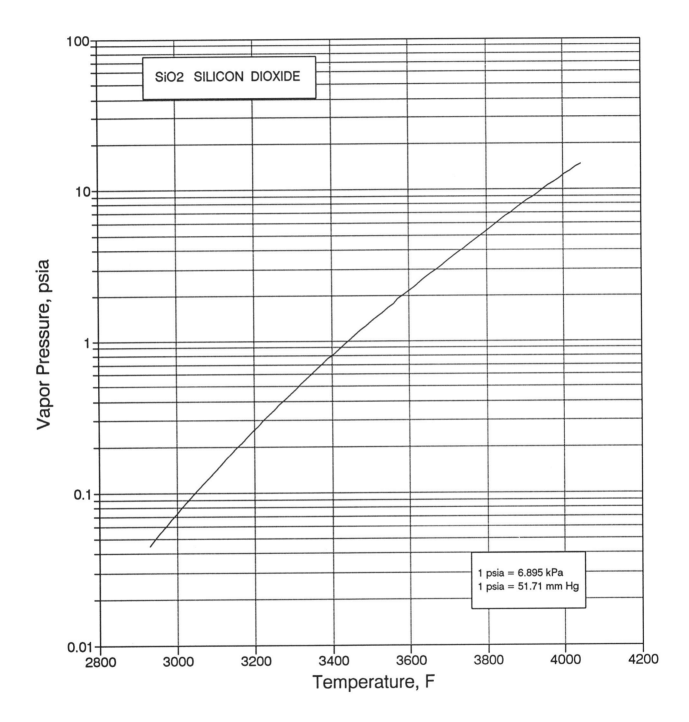

SiO2 SILICON DIOXIDE

Vapor Pressure, psia

Temperature, F

1 psia = 6.895 kPa
1 psia = 51.71 mm Hg

293

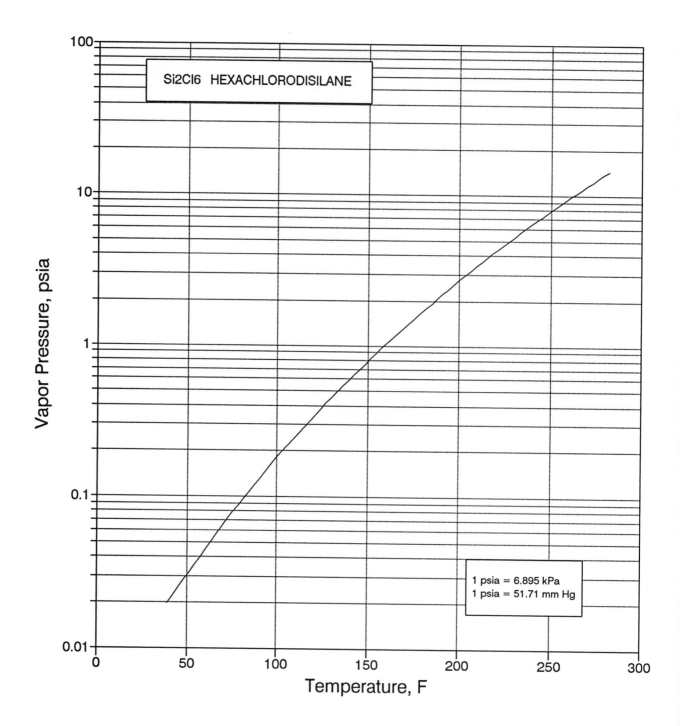

Si2Cl6  HEXACHLORODISILANE

Vapor Pressure, psia

Temperature, F

1 psia = 6.895 kPa
1 psia = 51.71 mm Hg

294

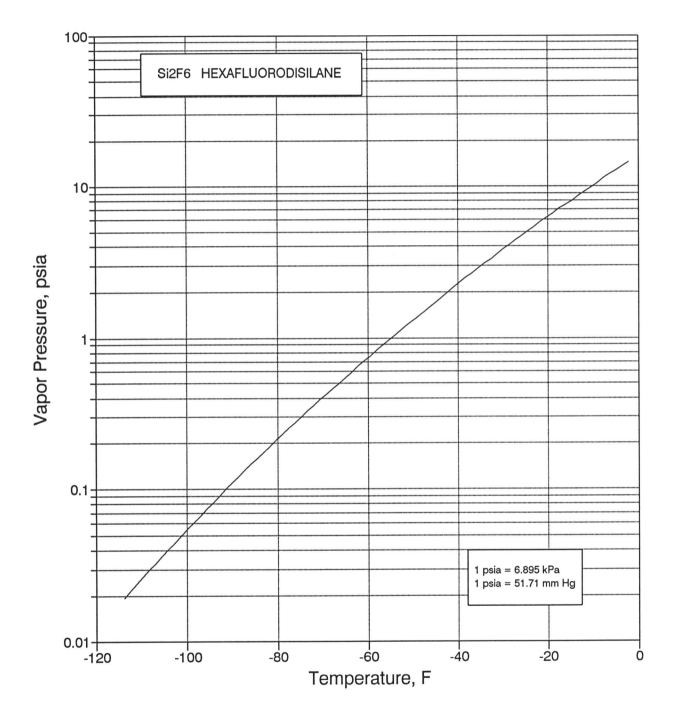

Si2F6  HEXAFLUORODISILANE

Vapor Pressure, psia

Temperature, F

1 psia = 6.895 kPa
1 psia = 51.71 mm Hg

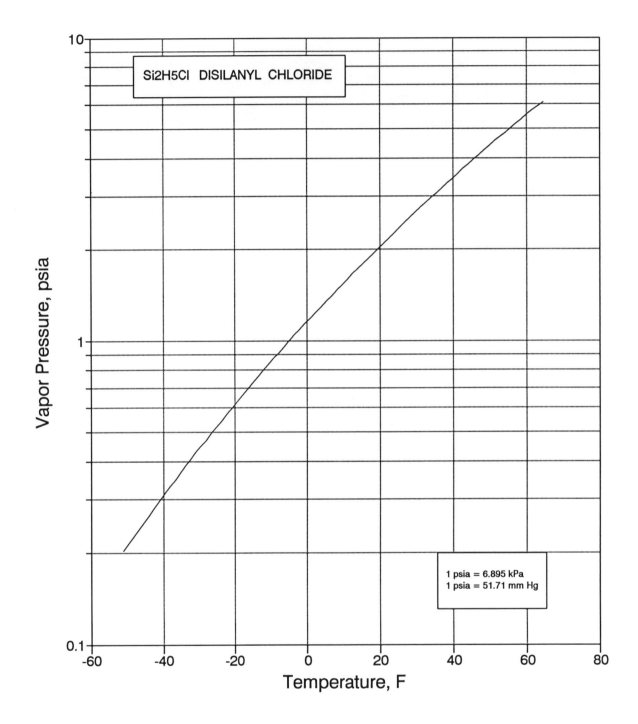

Si2H5Cl  DISILANYL CHLORIDE

1 psia = 6.895 kPa
1 psia = 51.71 mm Hg

Vapor Pressure, psia

Temperature, F

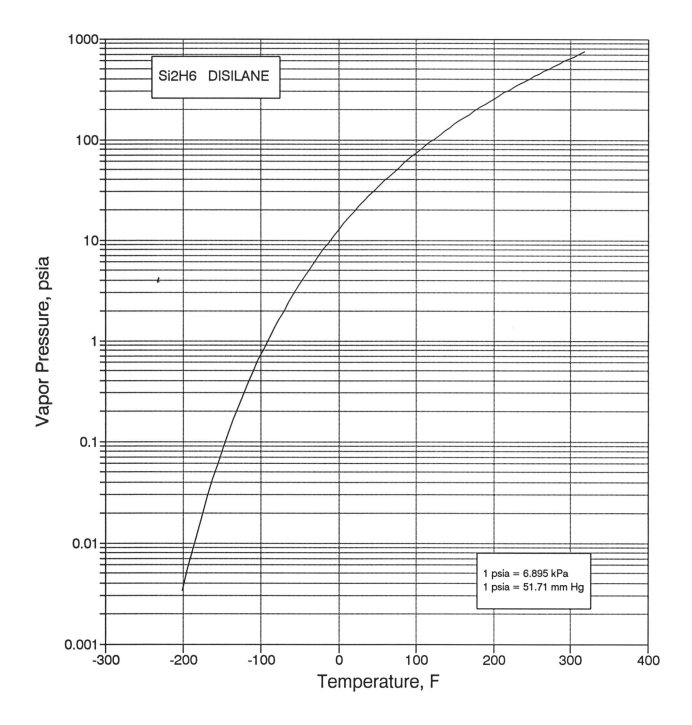

Si2H6   DISILANE

Vapor Pressure, psia

1 psia = 6.895 kPa
1 psia = 51.71 mm Hg

Temperature, F

297

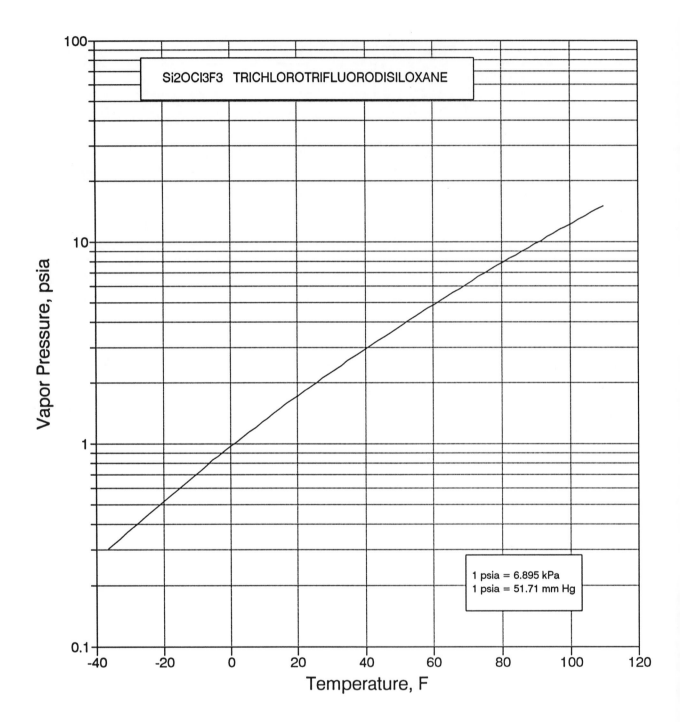

Si2OCl3F3   TRICHLOROTRIFLUORODISILOXANE

Vapor Pressure, psia

Temperature, F

1 psia = 6.895 kPa
1 psia = 51.71 mm Hg

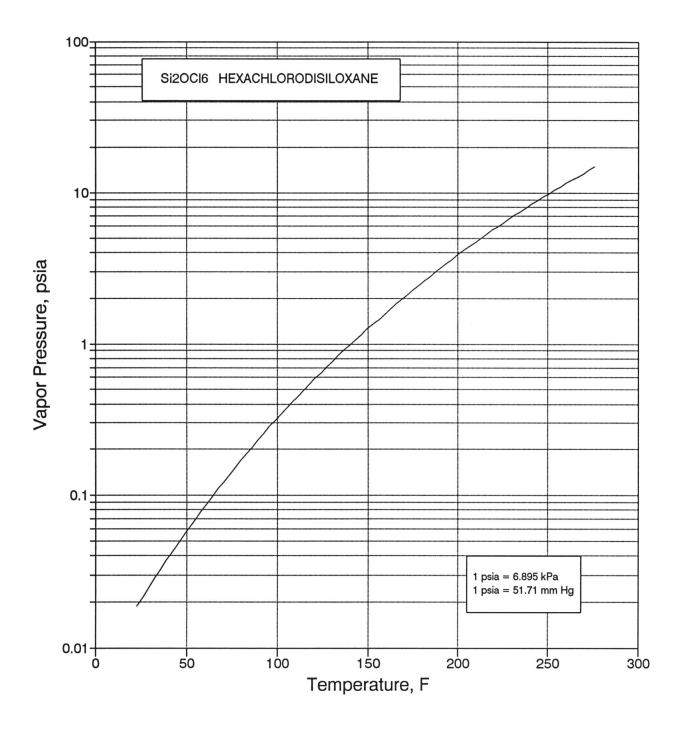

Si2OCl6   HEXACHLORODISILOXANE

1 psia = 6.895 kPa
1 psia = 51.71 mm Hg

Vapor Pressure, psia

Temperature, F

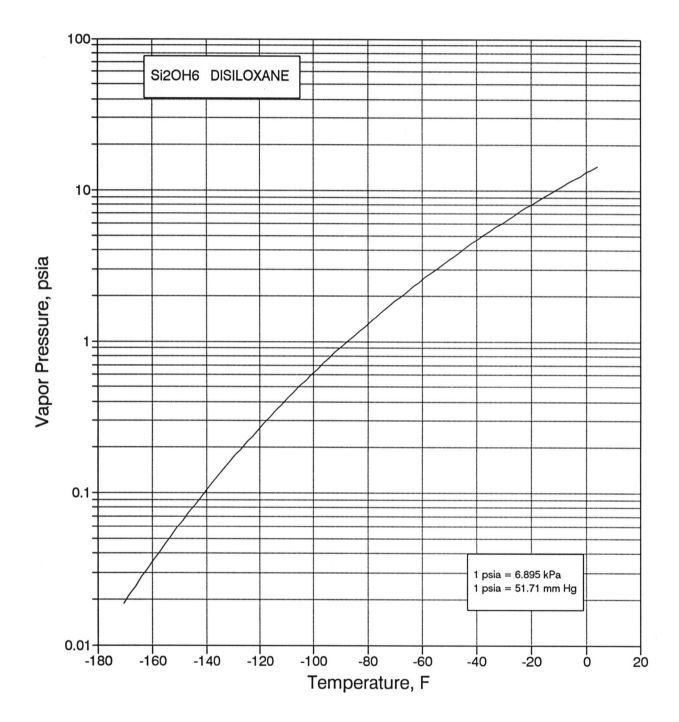

Si2OH6  DISILOXANE

1 psia = 6.895 kPa
1 psia = 51.71 mm Hg

Vapor Pressure, psia

Temperature, F

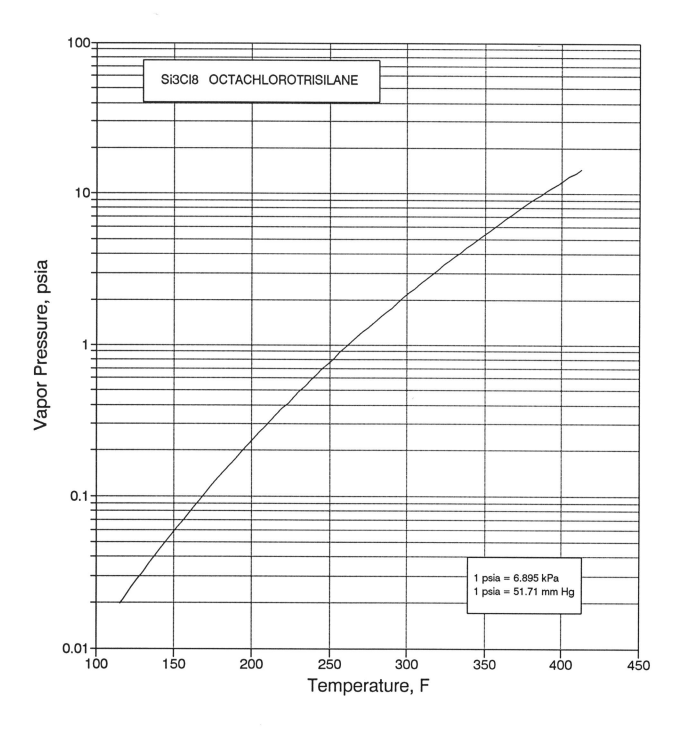

Si3Cl8  OCTACHLOROTRISILANE

Vapor Pressure, psia

Temperature, F

1 psia = 6.895 kPa
1 psia = 51.71 mm Hg

301

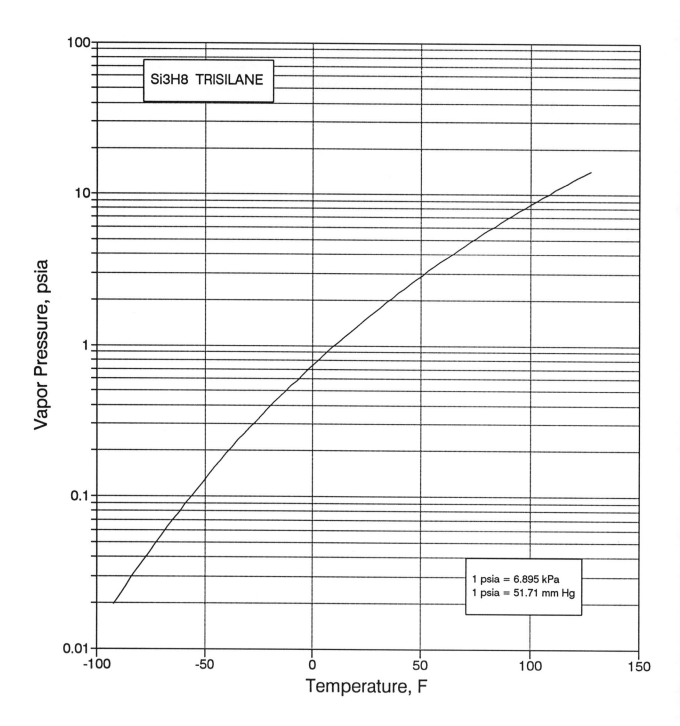

Si3H8 TRISILANE

Vapor Pressure, psia

Temperature, F

1 psia = 6.895 kPa
1 psia = 51.71 mm Hg

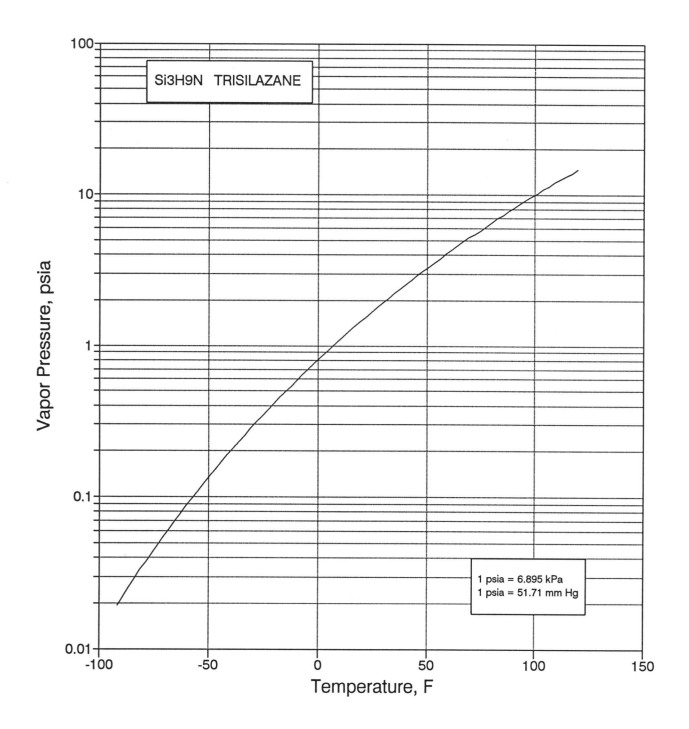

Si3H9N  TRISILAZANE

Vapor Pressure, psia

Temperature, F

1 psia = 6.895 kPa
1 psia = 51.71 mm Hg

303

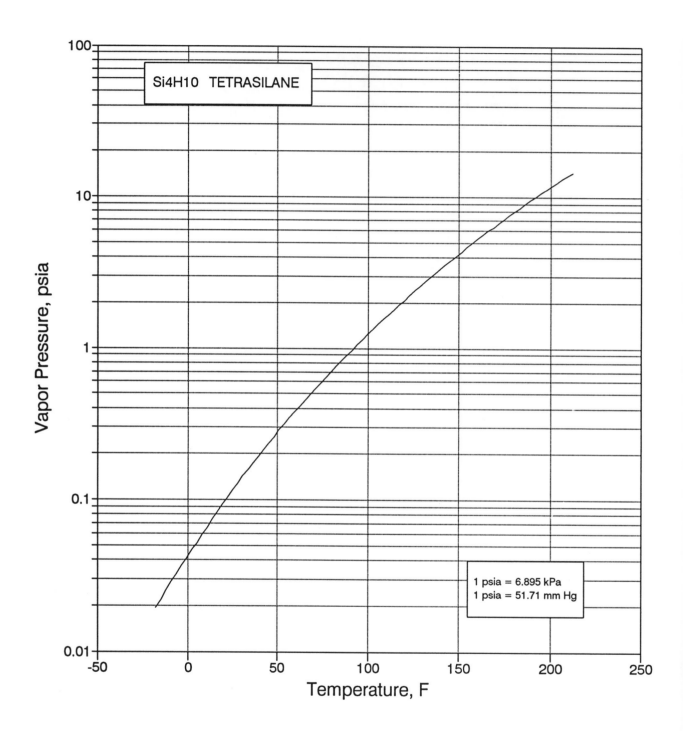

Si4H10  TETRASILANE

Vapor Pressure, psia

Temperature, F

1 psia = 6.895 kPa
1 psia = 51.71 mm Hg

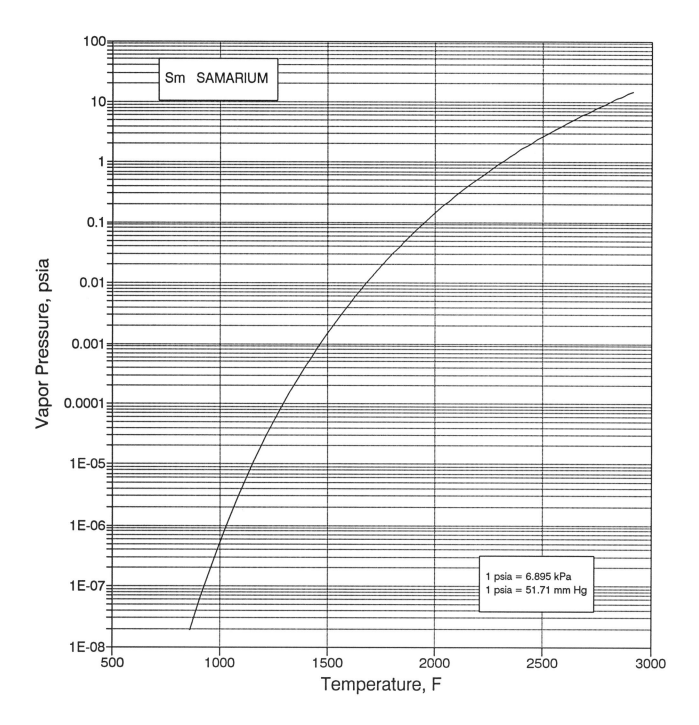

Sm SAMARIUM

Vapor Pressure, psia

Temperature, F

1 psia = 6.895 kPa
1 psia = 51.71 mm Hg

305

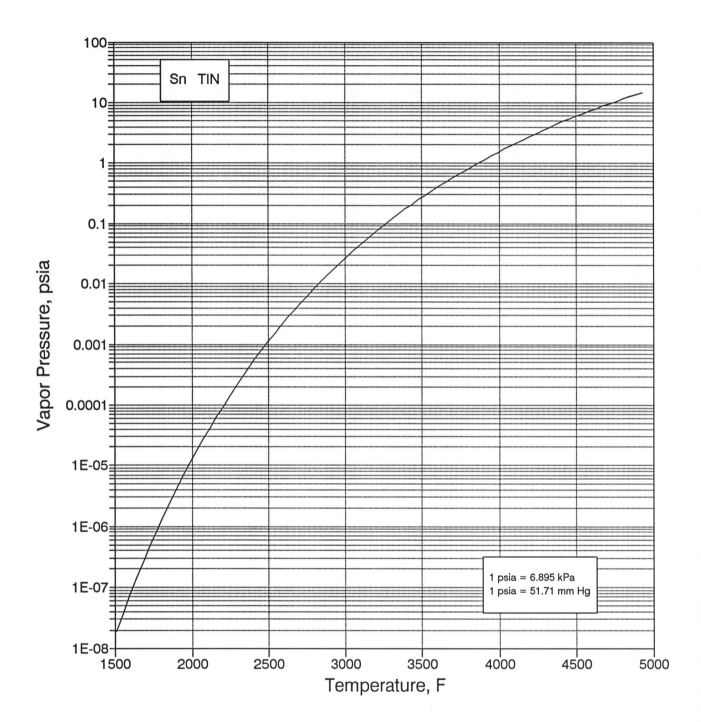

Sn   TIN

1 psia = 6.895 kPa
1 psia = 51.71 mm Hg

Vapor Pressure, psia

Temperature, F

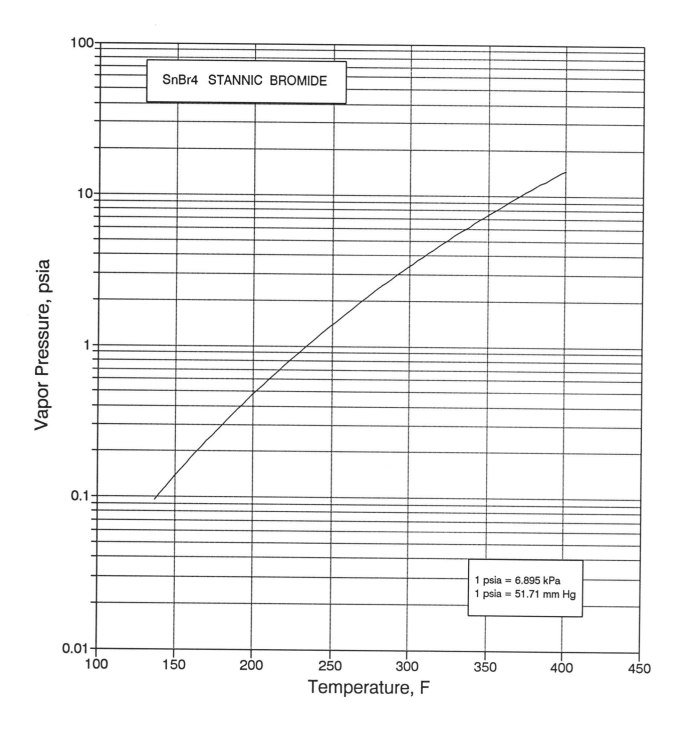

SnBr4  STANNIC BROMIDE

1 psia = 6.895 kPa
1 psia = 51.71 mm Hg

Vapor Pressure, psia

Temperature, F

307

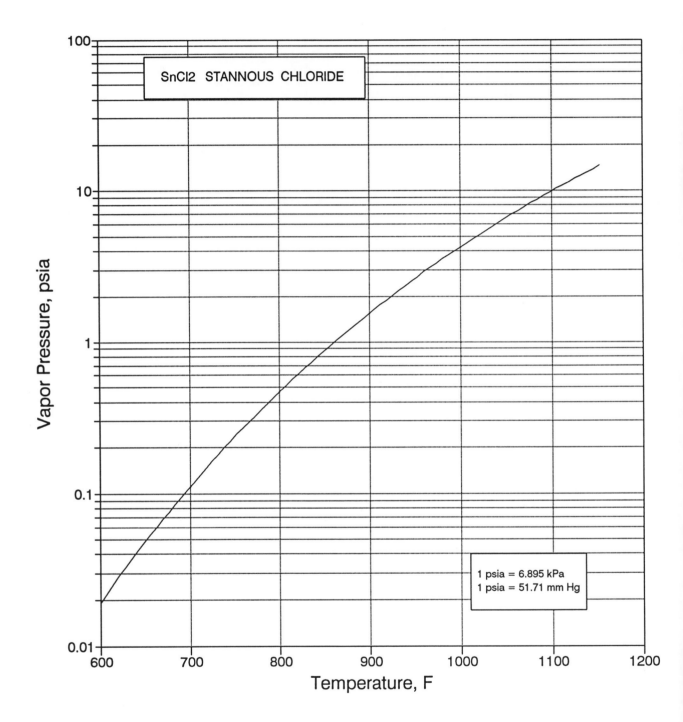

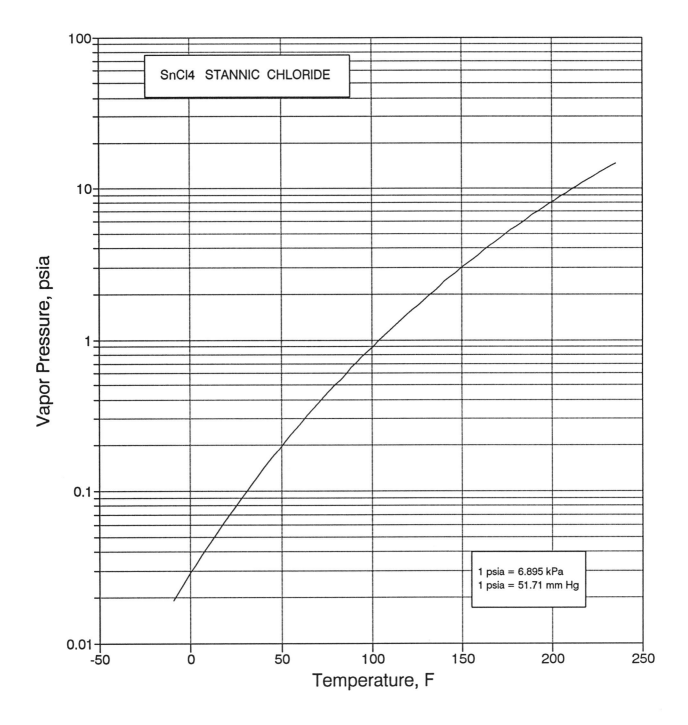

SnCl4 STANNIC CHLORIDE

Vapor Pressure, psia

Temperature, F

1 psia = 6.895 kPa
1 psia = 51.71 mm Hg

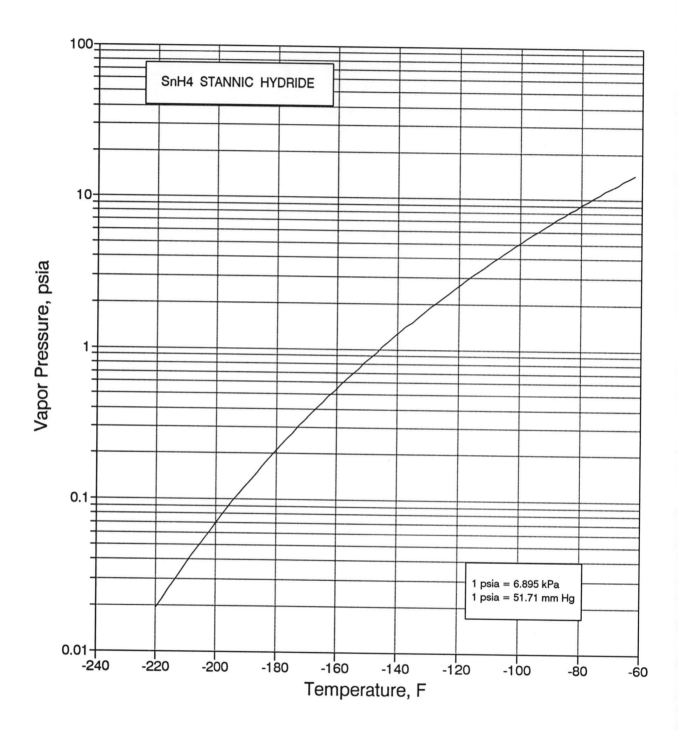

SnH4 STANNIC HYDRIDE

Vapor Pressure, psia

Temperature, F

1 psia = 6.895 kPa
1 psia = 51.71 mm Hg

310

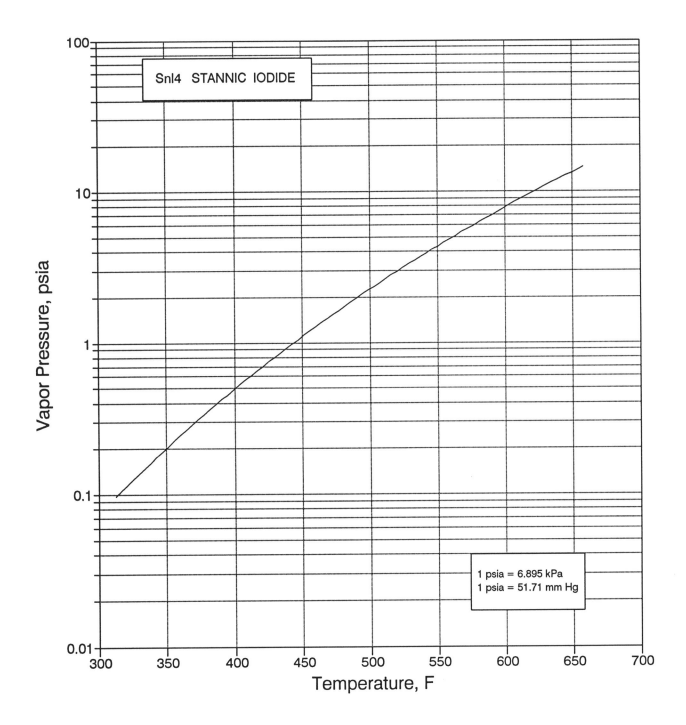

Vapor Pressure, psia

Temperature, F

SnI4  STANNIC IODIDE

1 psia = 6.895 kPa
1 psia = 51.71 mm Hg

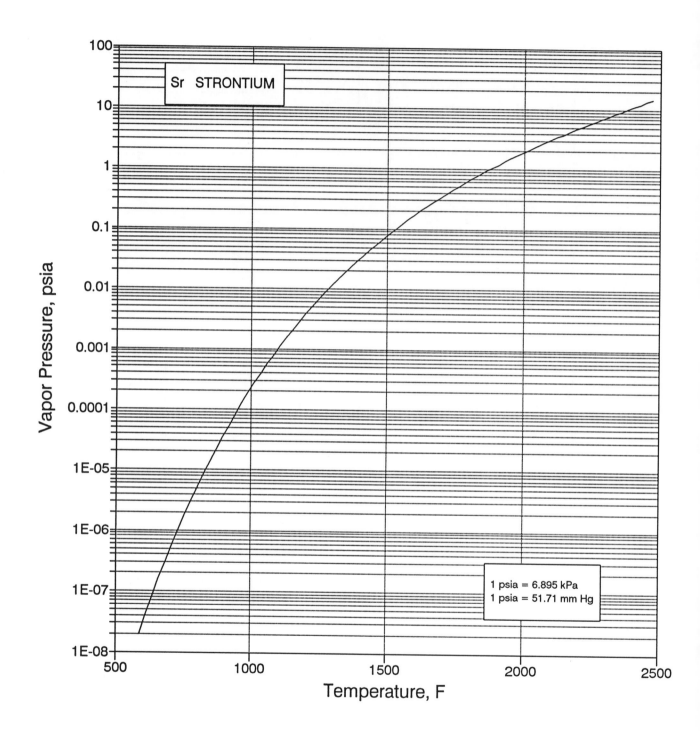

Sr STRONTIUM

Vapor Pressure, psia

Temperature, F

1 psia = 6.895 kPa
1 psia = 51.71 mm Hg

312

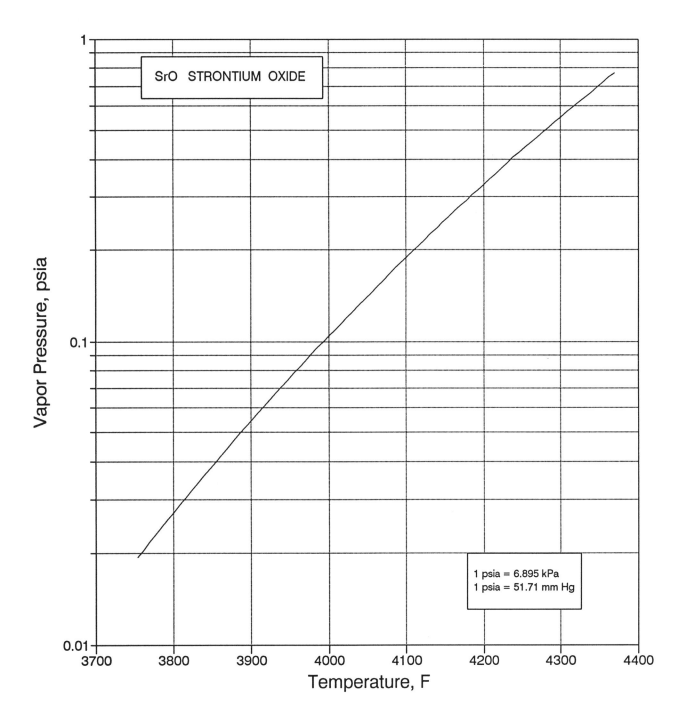

SrO  STRONTIUM OXIDE

1 psia = 6.895 kPa
1 psia = 51.71 mm Hg

Vapor Pressure, psia

Temperature, F

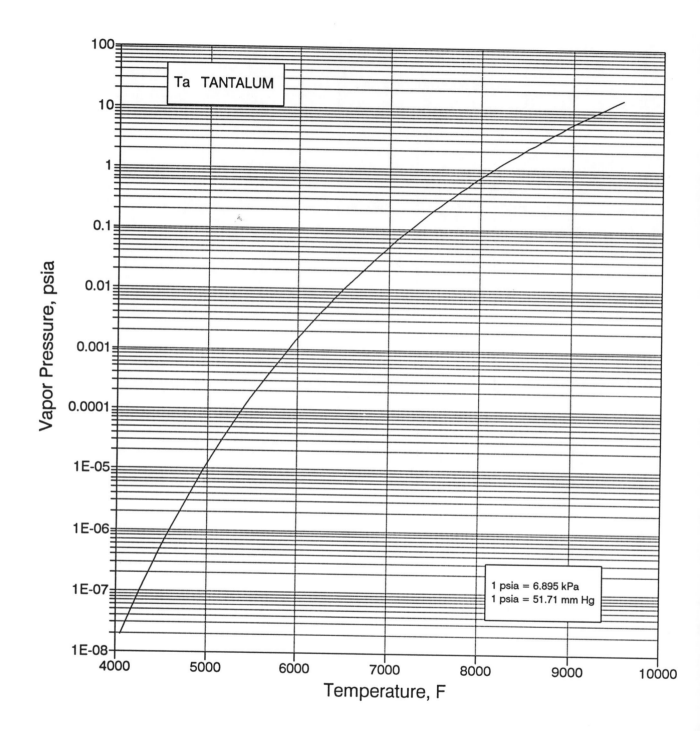

Ta TANTALUM

Vapor Pressure, psia

Temperature, F

1 psia = 6.895 kPa
1 psia = 51.71 mm Hg

314

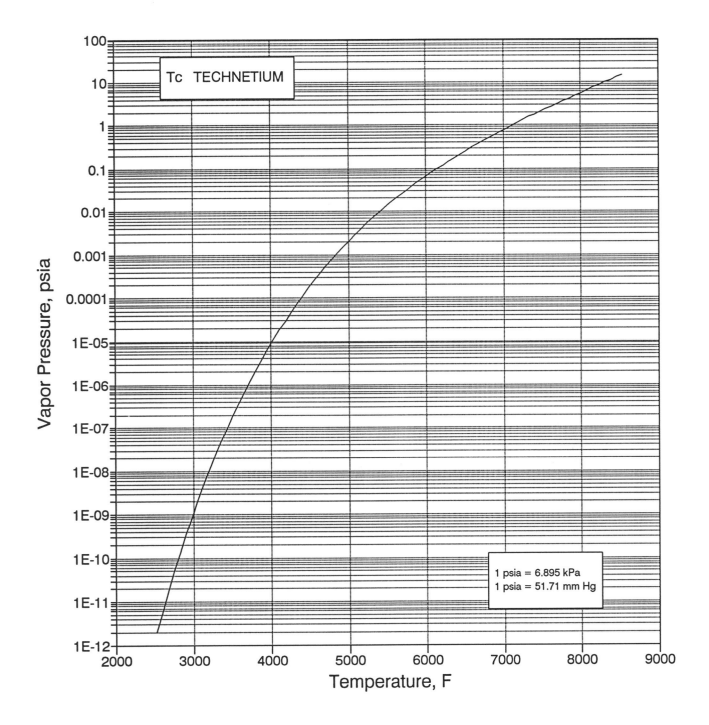

Tc TECHNETIUM

1 psia = 6.895 kPa
1 psia = 51.71 mm Hg

Vapor Pressure, psia

Temperature, F

315

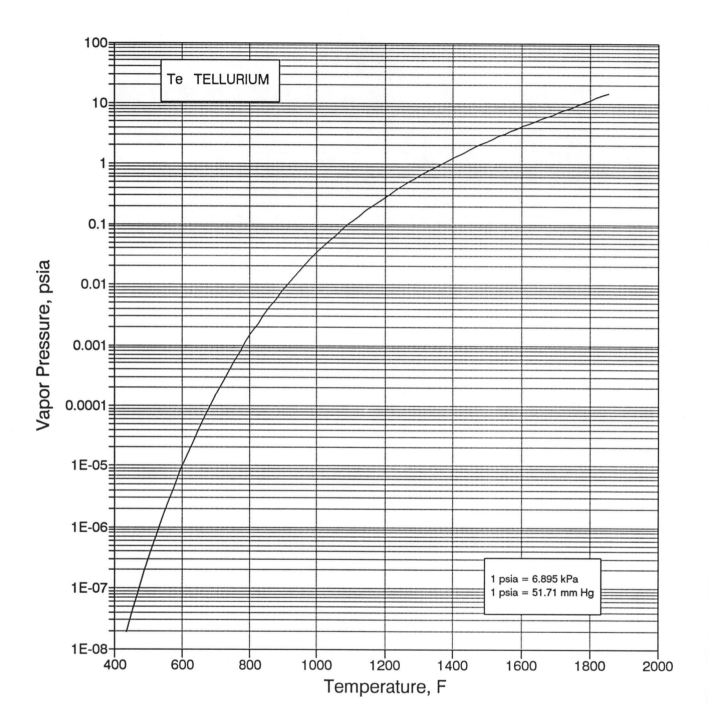

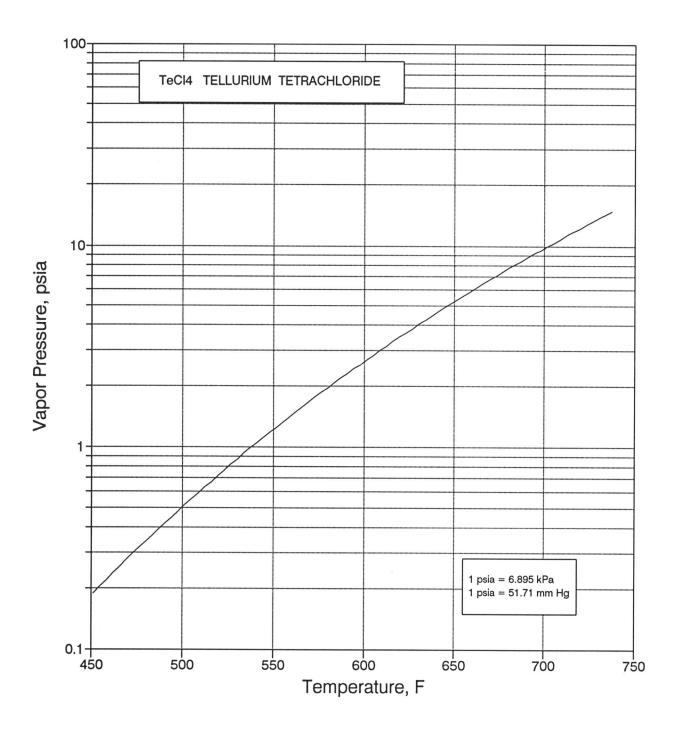

TeCl4  TELLURIUM TETRACHLORIDE

Vapor Pressure, psia

Temperature, F

1 psia = 6.895 kPa
1 psia = 51.71 mm Hg

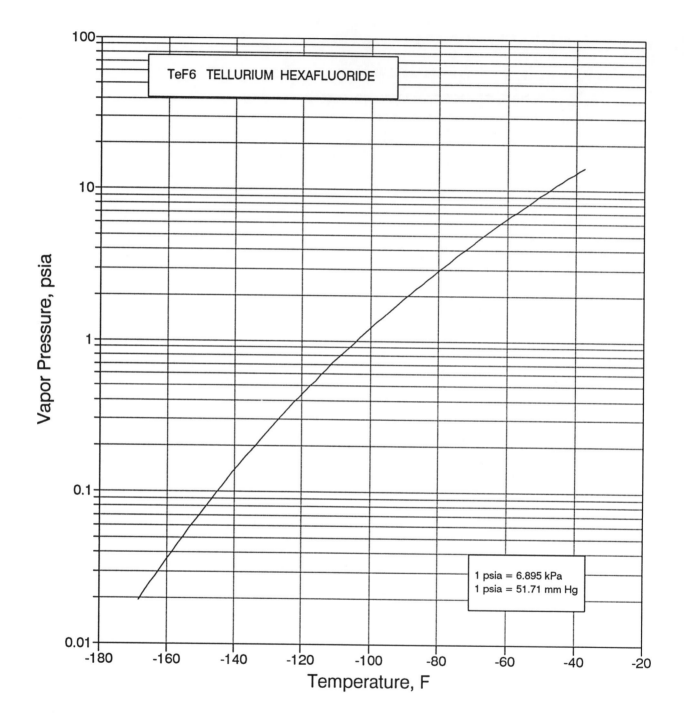

TeF6   TELLURIUM HEXAFLUORIDE

1 psia = 6.895 kPa
1 psia = 51.71 mm Hg

318

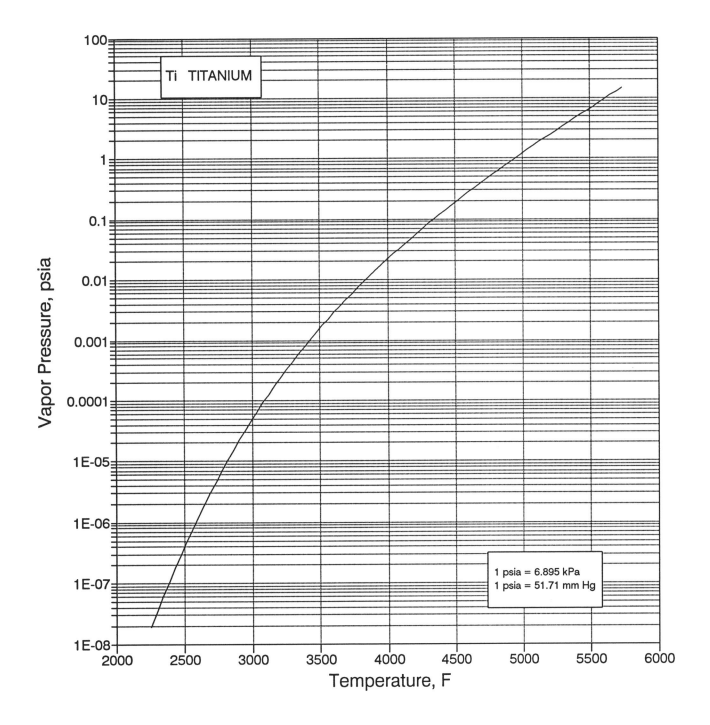

Ti TITANIUM

Vapor Pressure, psia

Temperature, F

1 psia = 6.895 kPa
1 psia = 51.71 mm Hg

319

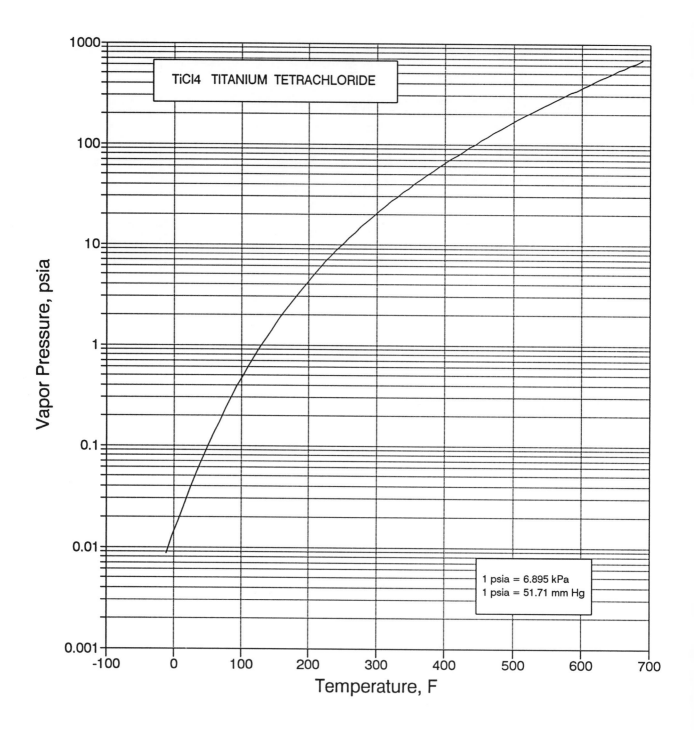

TiCl4  TITANIUM TETRACHLORIDE

Vapor Pressure, psia

Temperature, F

1 psia = 6.895 kPa
1 psia = 51.71 mm Hg

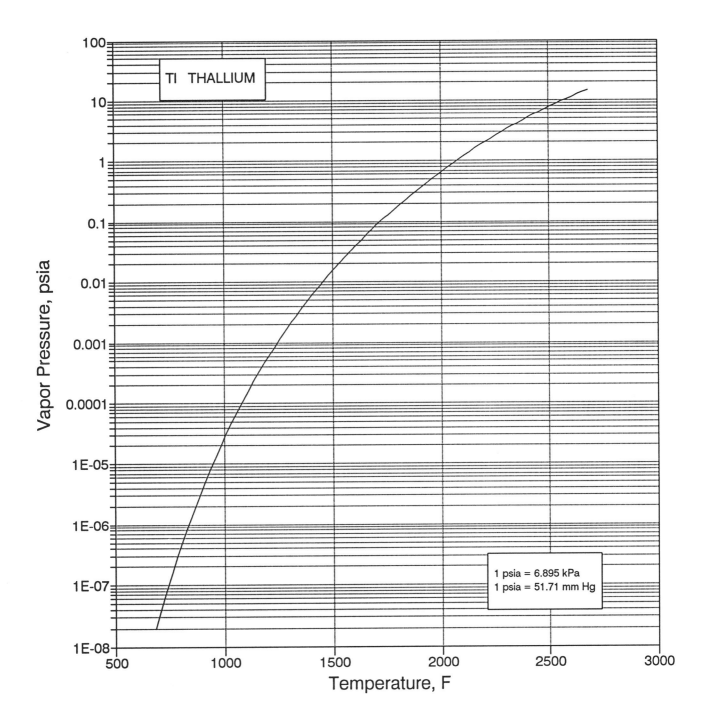

TI  THALLIUM

Vapor Pressure, psia

Temperature, F

1 psia = 6.895 kPa
1 psia = 51.71 mm Hg

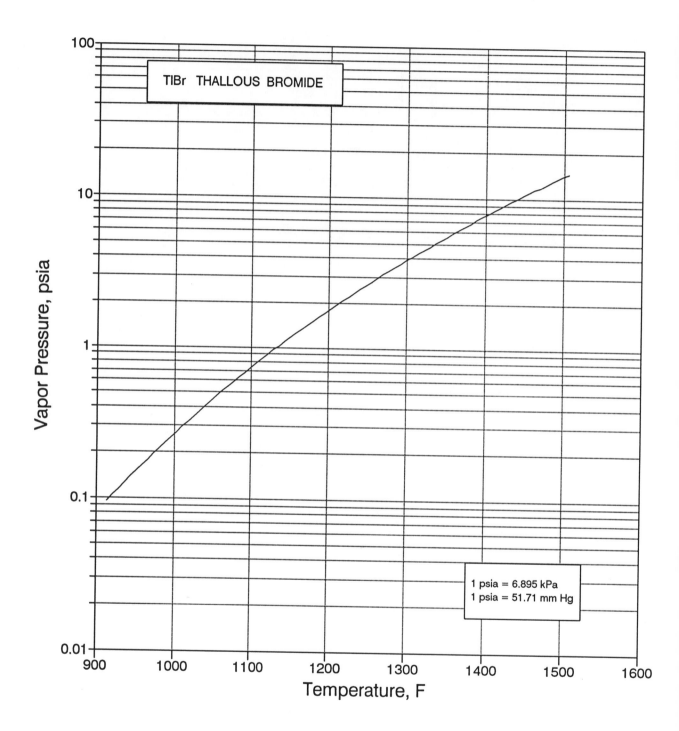

TlBr THALLOUS BROMIDE

Vapor Pressure, psia

Temperature, F

1 psia = 6.895 kPa
1 psia = 51.71 mm Hg

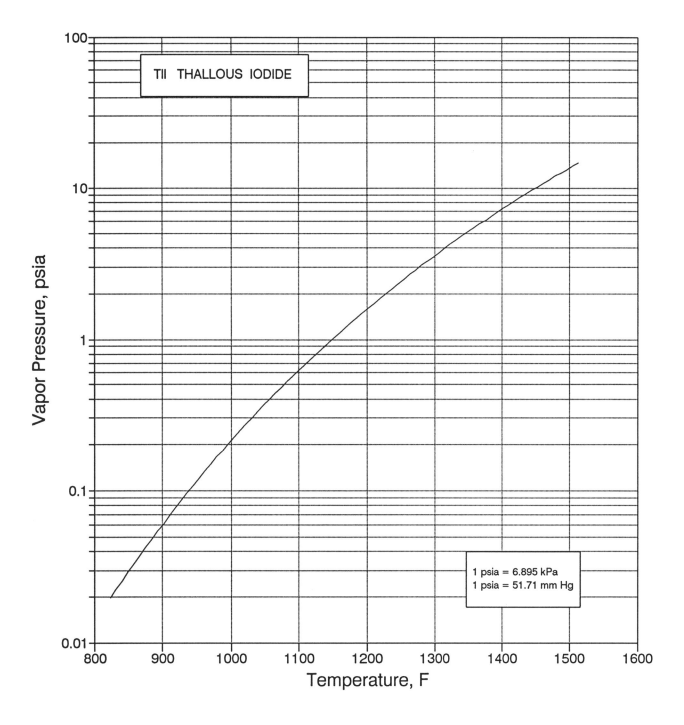

TlI THALLOUS IODIDE

Vapor Pressure, psia

Temperature, F

1 psia = 6.895 kPa
1 psia = 51.71 mm Hg

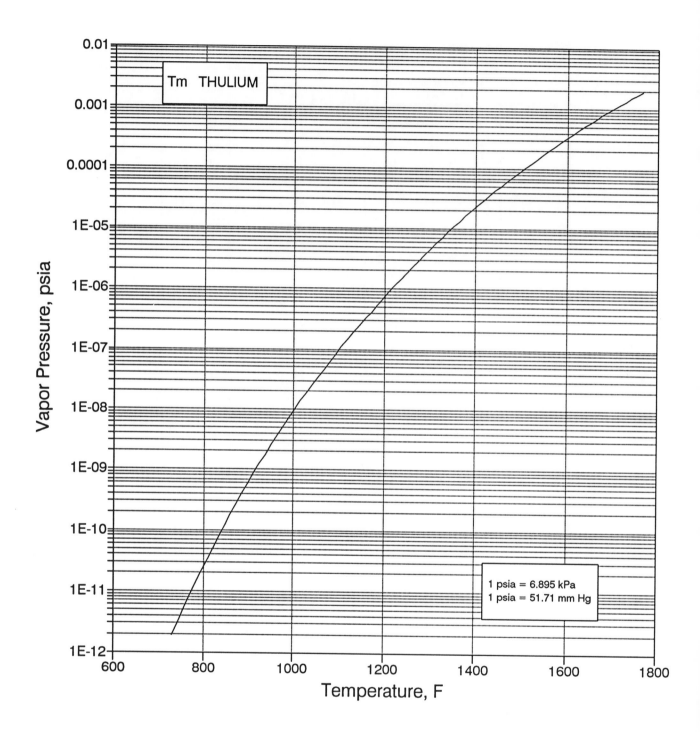

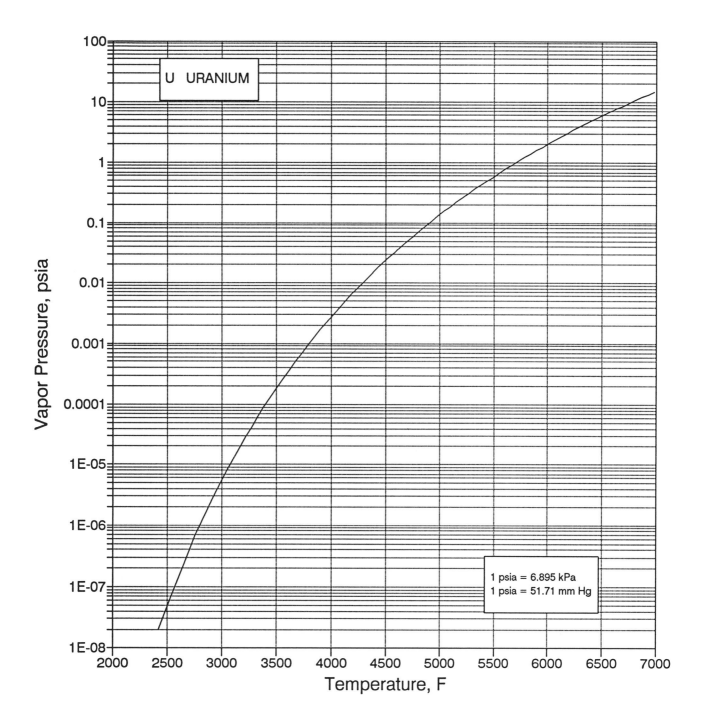

The chart shows Vapor Pressure (psia) versus Temperature (F) for U URANIUM.

U URANIUM

1 psia = 6.895 kPa
1 psia = 51.71 mm Hg

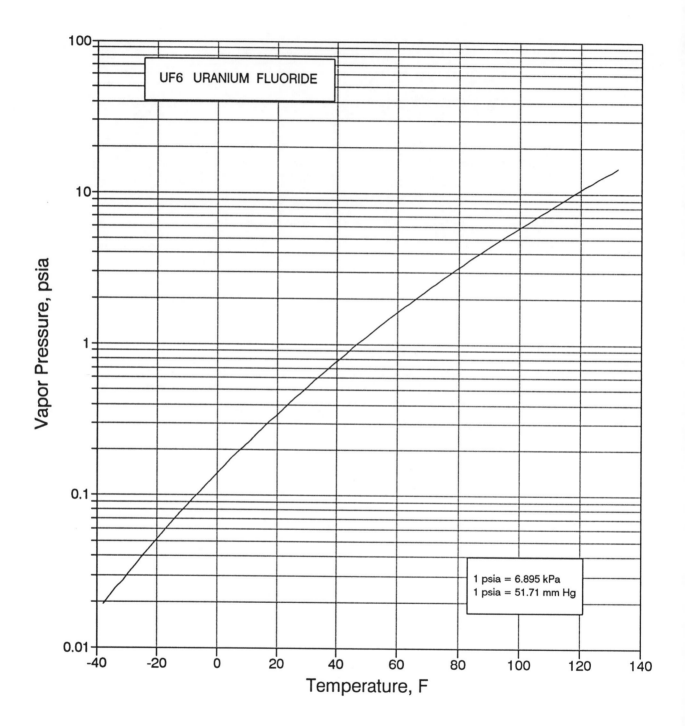

UF6 URANIUM FLUORIDE

1 psia = 6.895 kPa
1 psia = 51.71 mm Hg

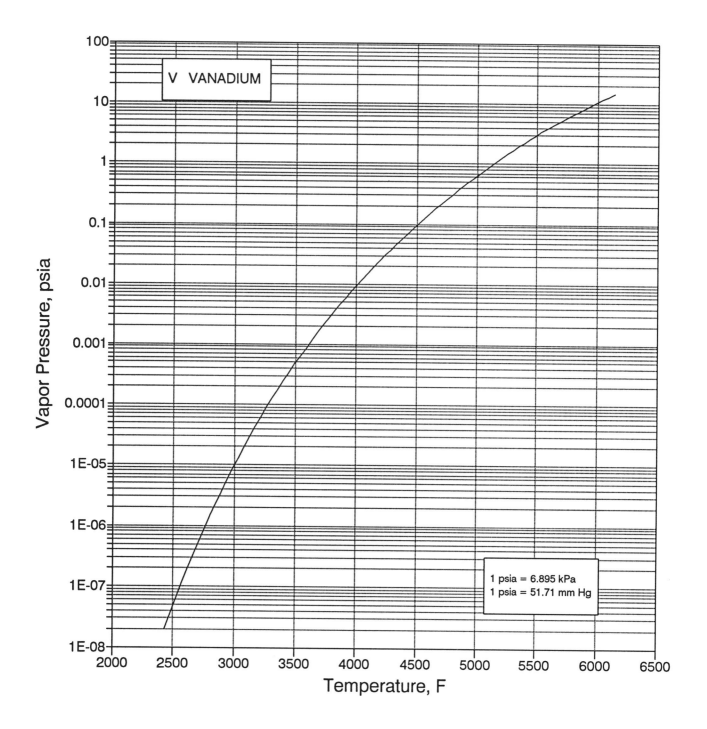

The chart shows Vapor Pressure, psia (y-axis, logarithmic from 1E-08 to 100) versus Temperature, F (x-axis, from 2000 to 6500) for V VANADIUM.

1 psia = 6.895 kPa
1 psia = 51.71 mm Hg

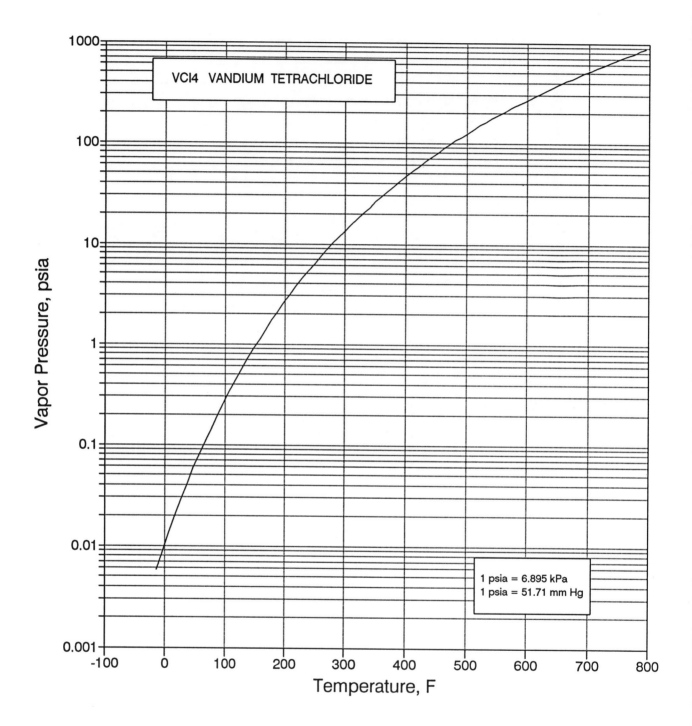

VCl4 VANDIUM TETRACHLORIDE

1 psia = 6.895 kPa
1 psia = 51.71 mm Hg

Vapor Pressure, psia

Temperature, F

328

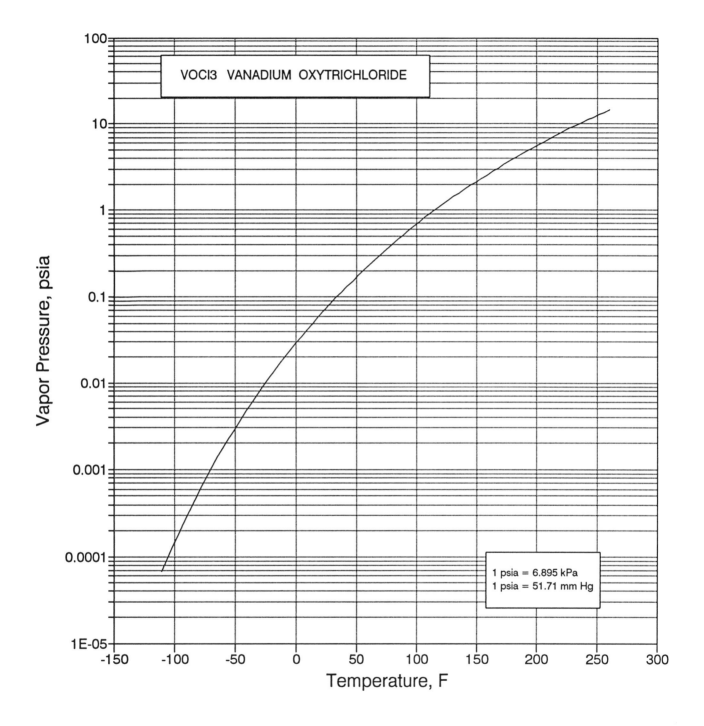

VOCl3  VANADIUM OXYTRICHLORIDE

Vapor Pressure, psia

Temperature, F

1 psia = 6.895 kPa
1 psia = 51.71 mm Hg

329

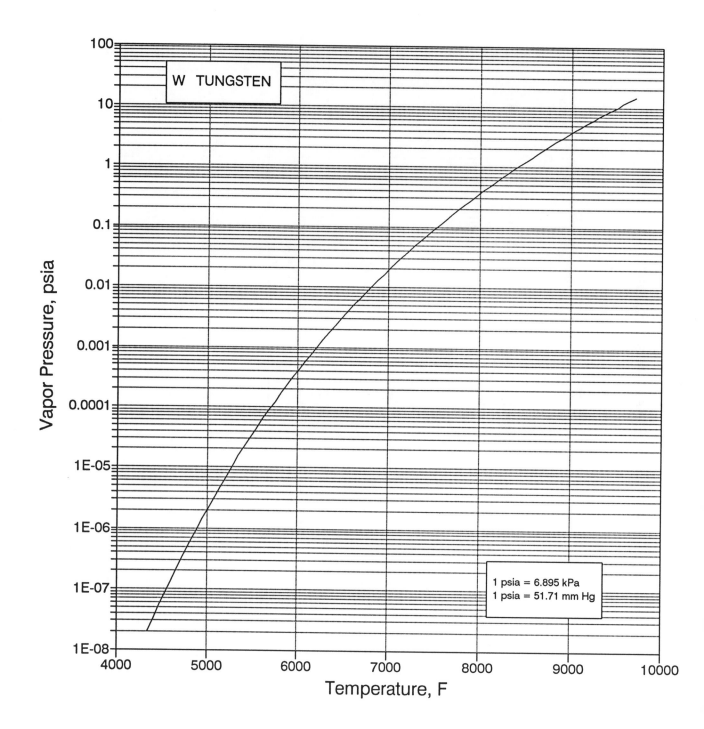

W TUNGSTEN

Vapor Pressure, psia

Temperature, F

1 psia = 6.895 kPa
1 psia = 51.71 mm Hg

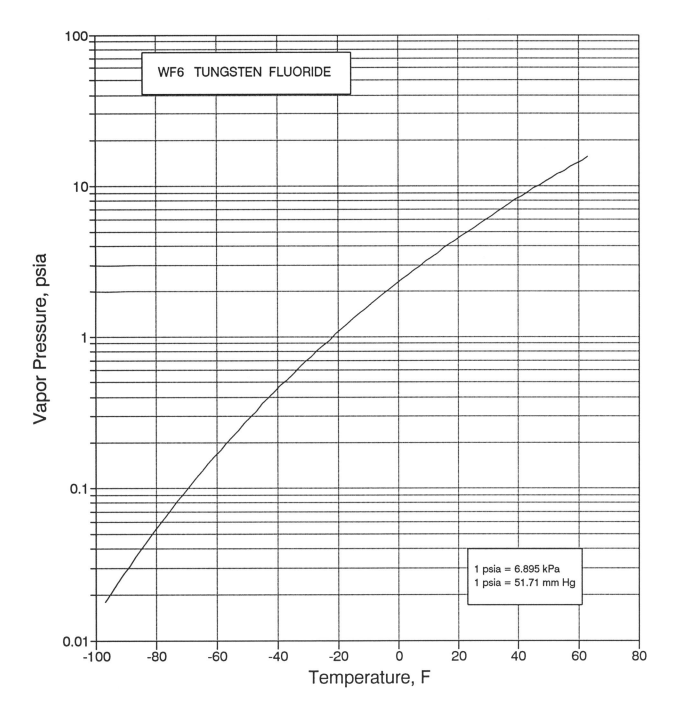

WF6  TUNGSTEN FLUORIDE

Vapor Pressure, psia

Temperature, F

1 psia = 6.895 kPa
1 psia = 51.71 mm Hg

331

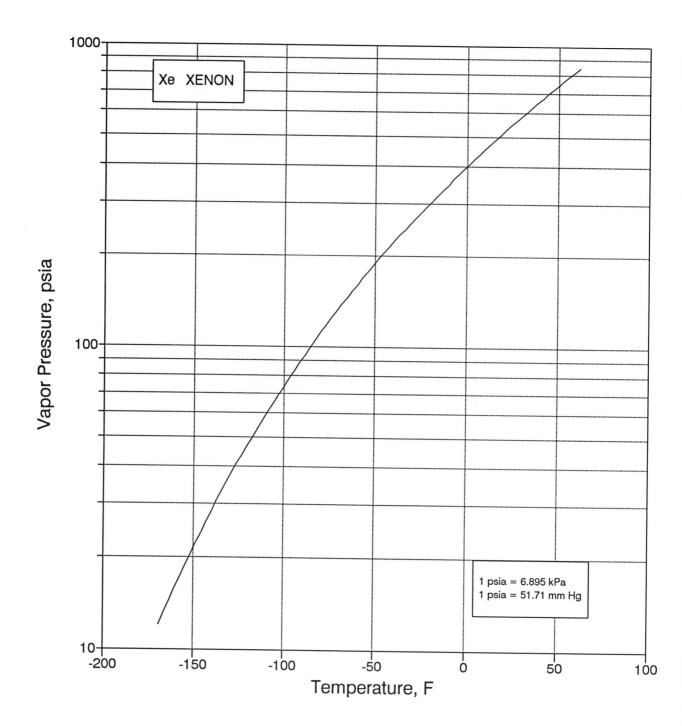

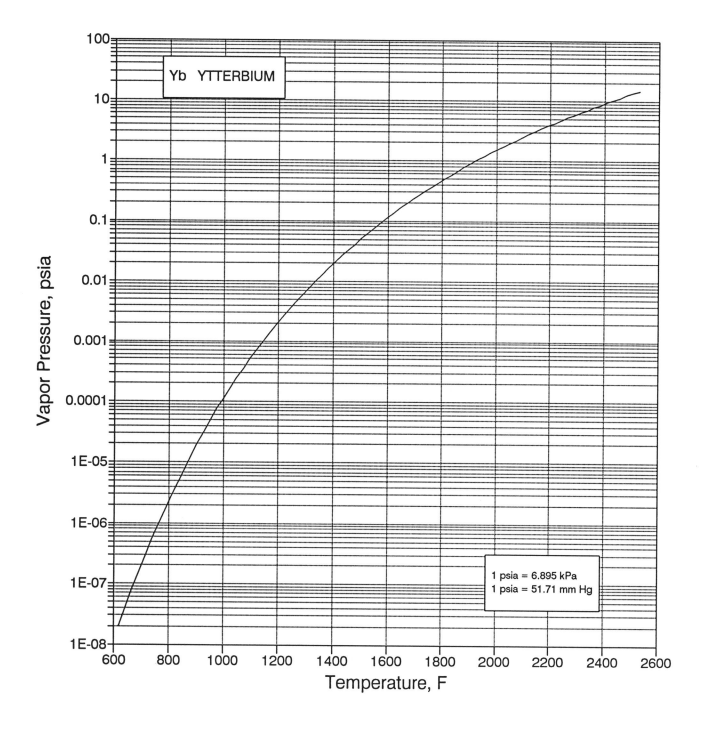

Yb YTTERBIUM

Vapor Pressure, psia

Temperature, F

1 psia = 6.895 kPa
1 psia = 51.71 mm Hg

333

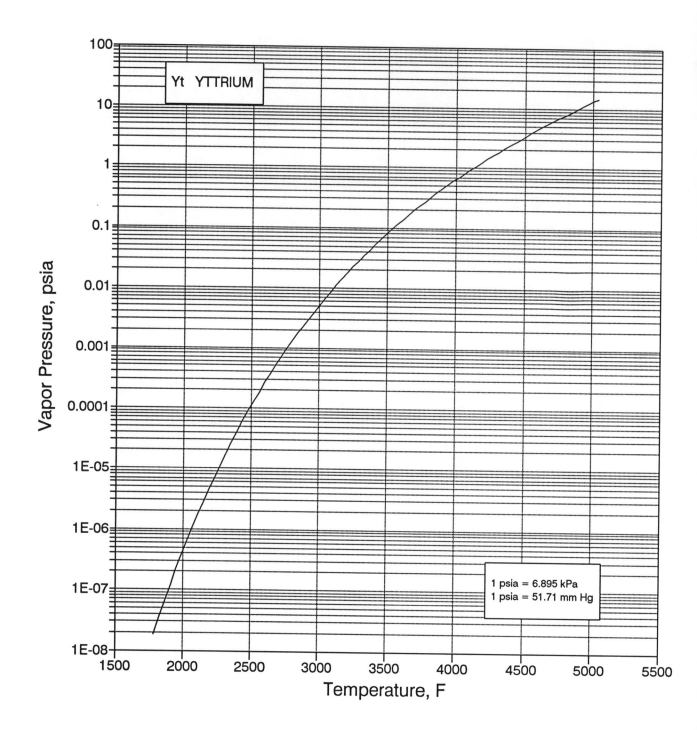

Yt  YTTRIUM

Vapor Pressure, psia

Temperature, F

1 psia = 6.895 kPa
1 psia = 51.71 mm Hg

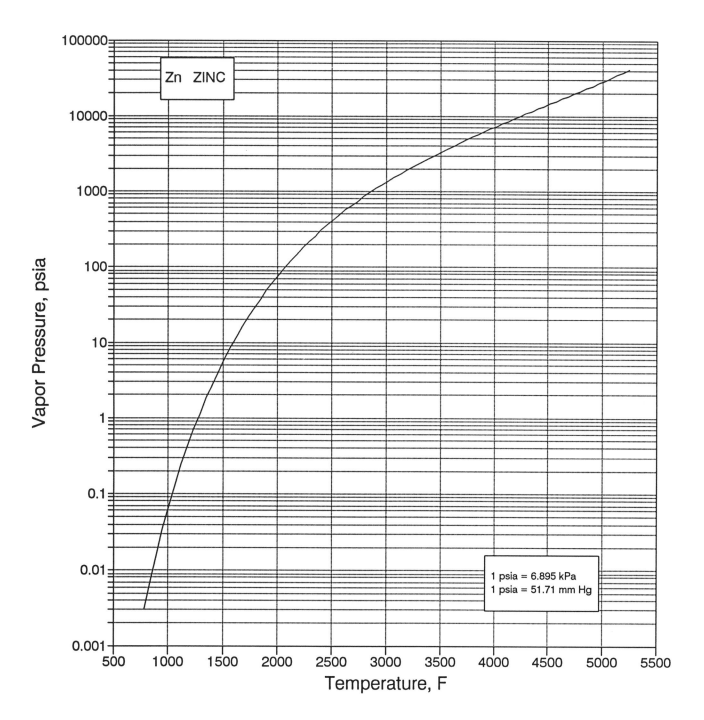

Zn ZINC

1 psia = 6.895 kPa
1 psia = 51.71 mm Hg

Vapor Pressure, psia

Temperature, F

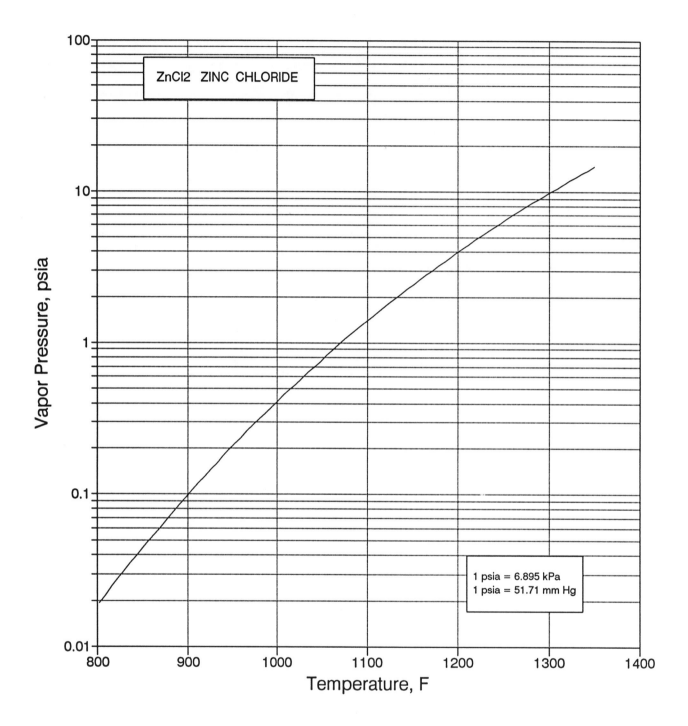

ZnCl2 ZINC CHLORIDE

Vapor Pressure, psia

Temperature, F

1 psia = 6.895 kPa
1 psia = 51.71 mm Hg

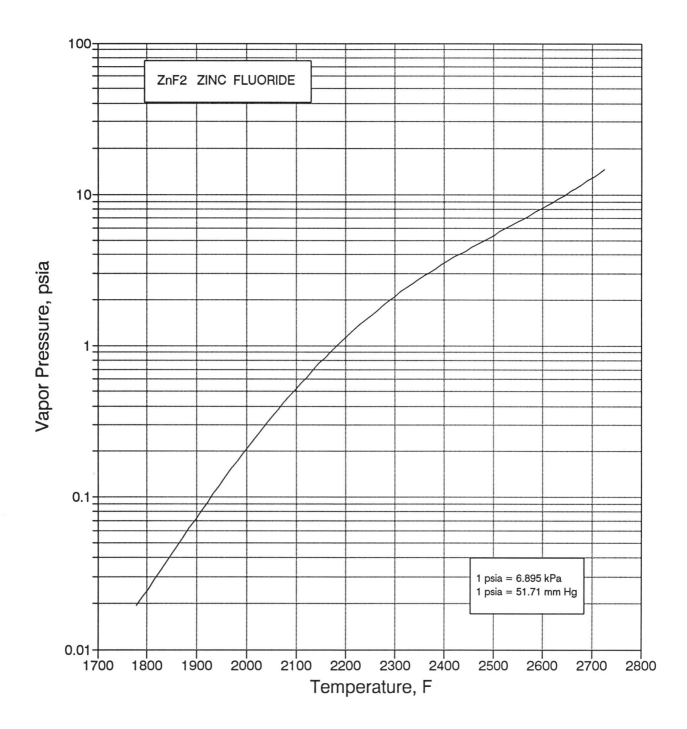

ZnF2 ZINC FLUORIDE

Vapor Pressure, psia

Temperature, F

1 psia = 6.895 kPa
1 psia = 51.71 mm Hg

337

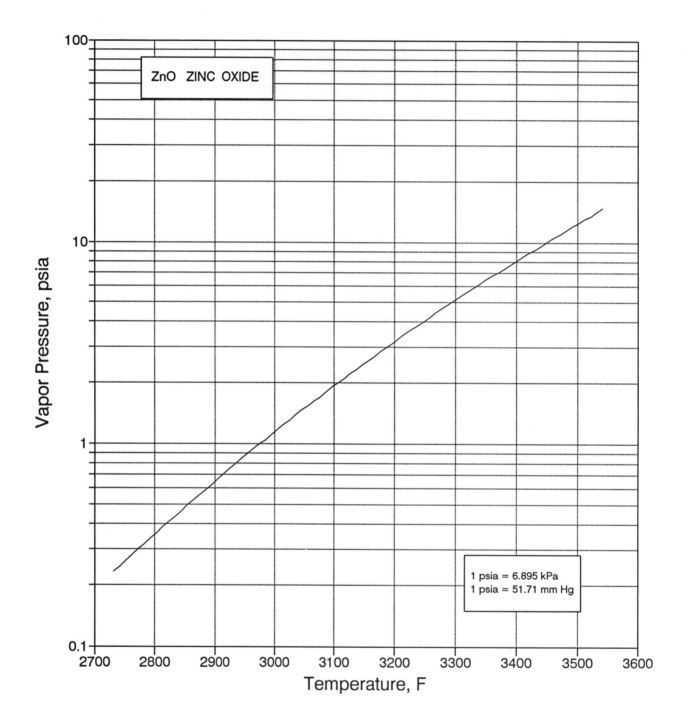

ZnO   ZINC OXIDE

1 psia = 6.895 kPa
1 psia = 51.71 mm Hg

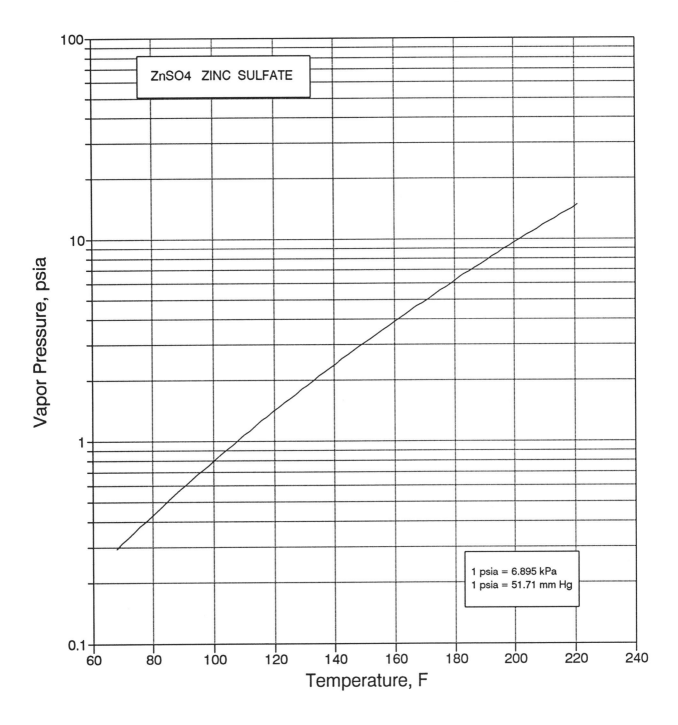

ZnSO4  ZINC SULFATE

Vapor Pressure, psia

Temperature, F

1 psia = 6.895 kPa
1 psia = 51.71 mm Hg

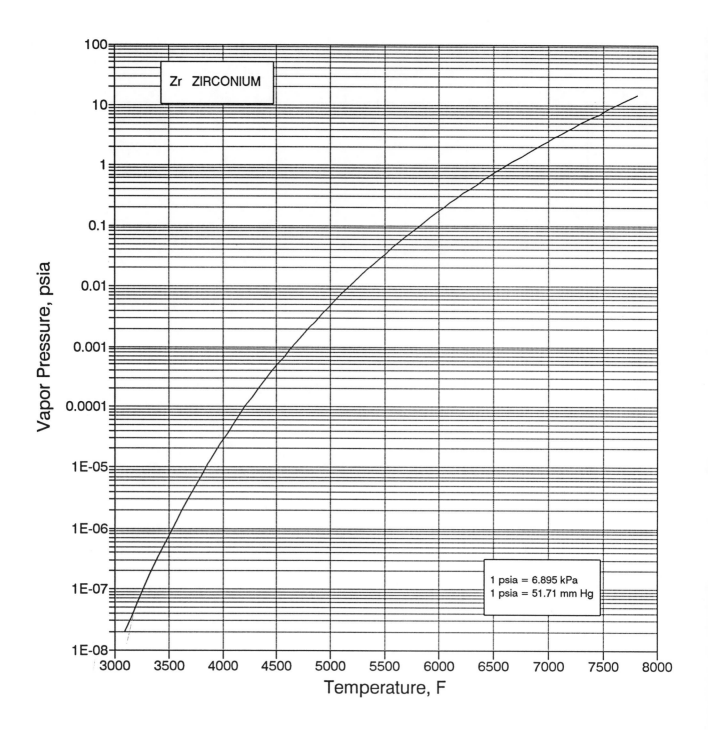

Zr  ZIRCONIUM

Vapor Pressure, psia

Temperature, F

1 psia = 6.895 kPa
1 psia = 51.71 mm Hg

340

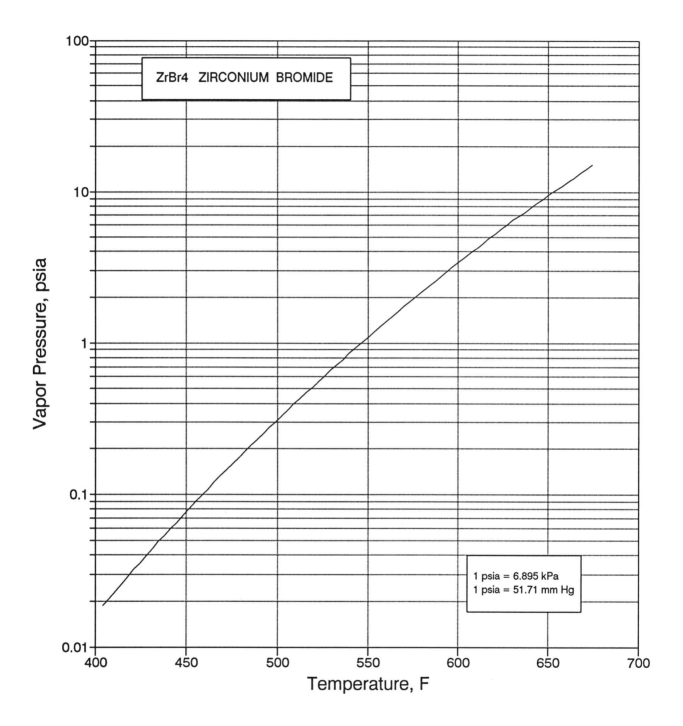

ZrBr4 ZIRCONIUM BROMIDE

1 psia = 6.895 kPa
1 psia = 51.71 mm Hg

Vapor Pressure, psia

Temperature, F

341

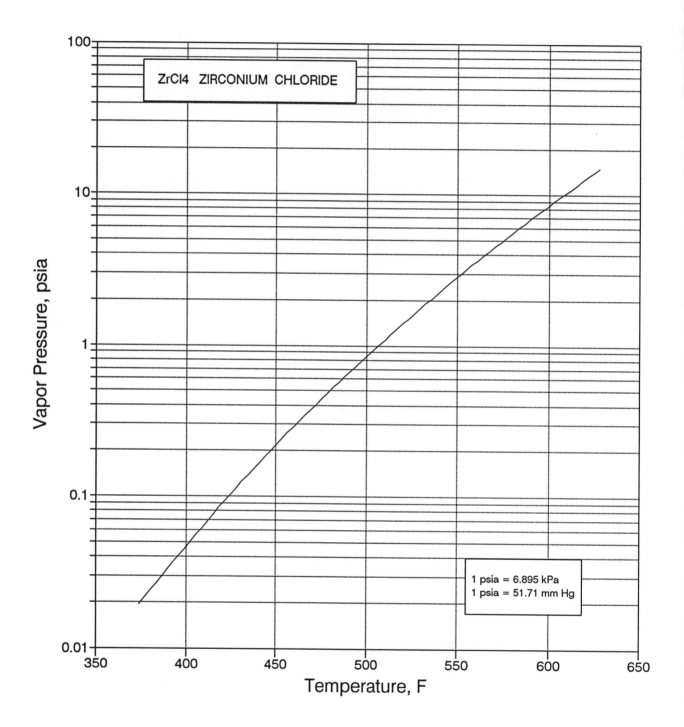

ZrCl4 ZIRCONIUM CHLORIDE

Vapor Pressure, psia

Temperature, F

1 psia = 6.895 kPa
1 psia = 51.71 mm Hg

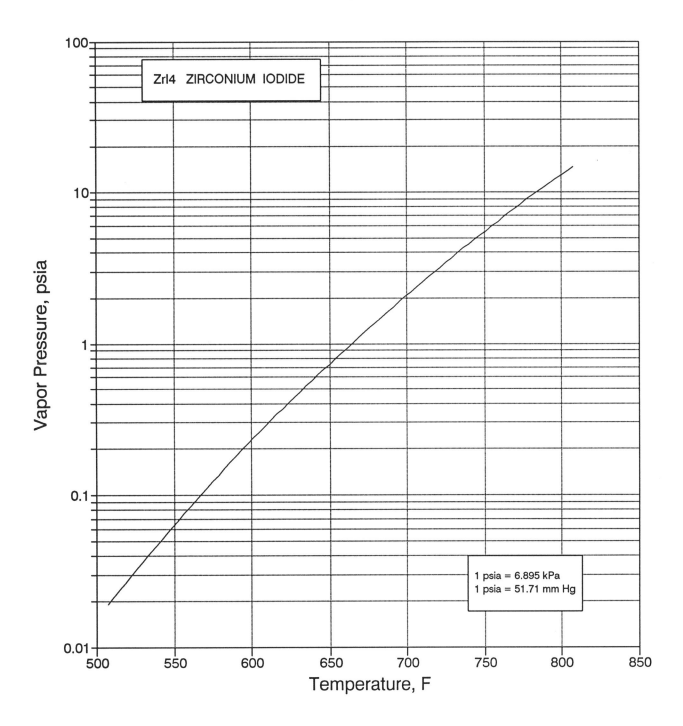

ZrI4  ZIRCONIUM  IODIDE

1 psia = 6.895 kPa
1 psia = 51.71 mm Hg

Vapor Pressure, psia

Temperature, F

343

# REFERENCES

1. SELECTED VALUES OF PROPERTIES OF HYDROCARBONS AND RELATED COMPOUNDS, Thermodynamics Research Center, TAMU, College Station, TX (1977, 1984).

2. SELECTED VALUES OF PROPERTIES OF CHEMICAL COMPOUNDS, Thermodynamics Research Center, TAMU, College Station, TX (1977, 1987).

3. Daubert, T. E. and R. P. Danner, DATA COMPILATION OF PROPERTIES OF PURE COMPOUNDS, Parts 1, 2, 3 and 4, Supplements 1 and 2, DIPPR Project, AIChE, New York, NY (1985-1992).

4. Nesmeyanov, A. N., VAPOR PRESSURE OF THE CHEMICAL ELEMENTS, Elsevier, New York, NY (1963).

5. Ambrose, D., VAPOUR-LIQUID CRITICAL PROPERTIES, National Physical Laboratory, Teddington, England, NPL Report Chem 107 (Feb., 1980).

6. Simmrock, K. H., R. Janowsky and A. Ohnsorge, CRITICAL DATA OF PURE SUBSTANCES, Vol. II, Parts 1 and 2, Dechema Chemistry Data Series, 6000 Frankfurt/Main, Germany (1986).

7. INTERNATIONAL CRITICAL TABLES, McGraw-Hill, New York, NY (1926).

8. Braker, W. and A. L. Mossman, MATHESON GAS DATA BOOK, 6th ed., Matheson Gas Products, Secaucaus, NJ (1980).

9. CRC HANDBOOK OF CHEMISTRY AND PHYSICS, 66th - 75th eds., CRC Press, Inc., Boca Raton, FL (1985-1994).

10. LANGE'S HANDBOOK OF CHEMISTRY, 13th and 14th eds., McGraw-Hill, New York, NY (1985, 1992).

11. PERRY'S CHEMICAL ENGINEERING HANDBOOK, 6th ed., McGraw-Hill, New York, NY (1984).

12. Kaye, G. W. C. and T. H. Laby, TABLES OF PHYSICAL AND CHEMICAL CONSTANTS, Longman Group Limited, London, England (1973).

13. Raznjevic, Kuzman, HANDBOOK OF THERMODYNAMIC TABLES AND CHARTS, Hemisphere Publishing Corporation, New York, NY (1976).

14. Vargaftik, N. B., TABLES ON THE THERMOPHYSICAL PROPERTIES OF LIQUIDS AND GASES, 2nd ed., English translation, Hemisphere Publishing Corporation, New York, NY (1975, 1983).

15. Lyman, W. J., W. F. Reehl and D. H. Rosenblatt, HANDBOOK OF CHEMICAL PROPERTY ESTIMATION METHODS, McGraw-Hill, New York, NY (1982).

16. Reid, R. C., J. M. Prausnitz and B. E. Poling, THE PROPERTIES OF GASES AND LIQUIDS, 3rd ed. (R. C. Reid and T. K. Sherwood), 4th ed., McGraw-Hill, New York, NY (1977, 1987).

17. Kirk, R. E. and D. F. Othmer, editors, ENCYCLOPEDIA OF CHEMICAL TECHNOLOGY, 3rd ed., Vols. 1-24, John Wiley and Sons, Inc., New York, NY (1978-1984).

18. CONDENSED CHEMICAL DICTIONARY, 10th (G. G. Hawley) and 11th eds. (N. I. Sax and R. J. Lewis, Jr.), Van Nostrand Reinhold Co., New York, NY (1981,1987).

19. Boublick, T., V. Fried and E. Hala, THE VAPOUR PRESSURES OF PURE SUBSTANCES, 1st and 2nd eds., Elsevier, New York, NY (1975, 1984).

20. Ohe, S., COMPUTER AIDED DATA BOOK OF VAPOR PRESSURE, Data Book Publishing Company, Tokyo, Japan (1976).

21. Hultgren, R., P. D. Desai, D. T. Hawkins, M. Gleiser, K. K. Kelley and D. D. Wagman, SELECTED VALUES OF THE THERMODYNAMIC PROPERTIES OF THE ELEMENTS, American Society for Metals, Metals Park, OH (1973).

22. Simmrock, K. H., R. Janowsky and A. Ohnsorge, CRITICAL DATA OF PURE SUBSTANCES, Vol. II, Parts 1 and 2, Dechema Chemistry Data Series, 6000 Frankfurt/Main, Germany (1986).

23. Yaws, C. L. and others, Solid State Technology, 16, No. 1, 39 (1973).

24. Yaws, C. L. and others, Solid State Technology, 17, No. 1, 47 (1974).

25. Yaws, C. L. and others, Solid State Technology, <u>17</u>, No. 11, 31 (1974).
26. Yaws, C. L. and others, Solid State Technology, <u>18</u>, No. 1, 35 (1975).
27. Yaws, C. L. and others, Solid State Technology, <u>21</u>, No. 1, 43 (1978).
28. Yaws, C. L. and others, Solid State Technology, <u>24</u>, No. 1, 87 (1981).
29. Yaws, C. L. and others, J. Ch. I. Ch. E., <u>12</u>, 33 (1981).
30. Yaws, C. L. and others, J. Ch. I. Ch. E., <u>14</u>, 205 (1983).
31. Yaws, C. L. and others, Ind. Eng. Chem. Process Des. Dev., <u>23</u>, 48 (1984).
32. Yaws, C. L., <u>PHYSICAL PROPERTIES</u>, McGraw-Hill, New York, NY (1977).
33. Yaws, C. L., <u>THERMODYNAMIC AND PHYSICAL PROPERTY DATA</u>, Gulf Publishing Co., Houston, TX (1992).
34. Yaws, C. L. and R. W. Gallant, <u>PHYSICAL PROPERTIES OF HYDROCARBONS</u>, Vols. 1 (2nd ed.), Vol 2, (3rd ed.) and Vol 3 (1st ed.), Gulf Publishing Co., Houston, TX (1992, 1993, 1993).

# Appendix A

# COMPOUND LIST BY FORMULA

| Formula | Name | Page |
|---|---|---|
| Ag | SILVER | 1 |
| AgCl | SILVER CHLORIDE | 2 |
| AgI | SILVER IODIDE | 3 |
| Al | ALUMINUM | 4 |
| AlB3H12 | ALUMINUM BOROHYDRIDE | 5 |
| AlBr3 | ALUMINUN BROMIDE | 6 |
| AlCl3 | ALUMINUM CHLORIDE | 7 |
| AlF3 | ALUMINUM FLUORIDE | 8 |
| AlI3 | ALUMINUM IODIDE | 9 |
| Al2O3 | ALUMINUM OXIDE | 10 |
| Al2S3O12 | ALUMINUM SULFATE | 11 |
| Ar | ARGON | 12 |
| As | ARSENIC | 13 |
| AsBr3 | ARSENIC TRIBROMIDE | 14 |
| AsCl3 | ARSENIC TRICHLORIDE | 15 |
| AsF3 | ARSENIC TRIFLUORIDE | 16 |
| AsF5 | ARSENIC PENTAFLUORIDE | 17 |
| AsH3 | ARSINE | 18 |
| AsI3 | ARSENIC TRIIODIDE | 19 |
| As2O3 | ARSENIC TRIOXIDE | 20 |
| At | ASTATINE | 21 |
| Au | GOLD | 22 |
| B | BORON | 23 |
| BBr3 | BORON TRIBROMIDE | 24 |
| BCl3 | BORON TRICHLORIDE | 25 |
| BF3 | BORON TRIFLUORIDE | 26 |
| BH2CO | BORINE CARBONYL | 27 |
| BH3O3 | BORIC ACID | 28 |
| B2D6 | DEUTERODIBORANE | 29 |
| B2H5Br | DIBORANE HYDROBROMIDE | 30 |
| B2H6 | DIBORANE | 31 |
| B3N3H6 | BORINE TRIAMINE | 32 |
| B4H10 | TETRABORANE | 33 |
| B5H9 | PENTABORANE | 34 |
| B5H11 | TETRAHYDROPENTABORANE | 35 |
| B10H14 | DECABORANE | 36 |
| Ba | BARIUM | 37 |
| Be | BERYLLIUM | 38 |
| BeB2H8 | BERYLLIUM BOROHYDRIDE | 39 |
| BeBr2 | BERYLLIUM BROMIDE | 40 |
| BeCl2 | BERYLLIUM CHLORIDE | 41 |
| BeF2 | BERYLLIUM FLUORIDE | 42 |
| BeI2 | BERYLLIUM IODIDE | 43 |
| Bi | BISMUTH | 44 |
| BiBr3 | BISMUTH TRIBROMIDE | 45 |
| BiCl3 | BISMUTH TRICHLORIDE | 46 |
| BrF5 | BROMINE PENTAFLUORIDE | 47 |
| Br2 | BROMINE | 48 |
| C | CARBON | 49 |
| CCl2O | PHOSGENE | 50 |
| CF2O | CARBONYL FLUORIDE | 51 |
| CH4N2O | UREA | 52 |
| CH4N2S | THIOUREA | 53 |
| CNBr | CYANOGEN BROMIDE | 54 |
| CNCl | CYANOGEN CHLORIDE | 55 |
| CNF | CYANOGEN FLUORIDE | 56 |
| CO | CARBON MONOXIDE | 57 |
| COS | CARBONYL SULFIDE | 58 |
| COSe | CARBON OXYSELENIDE | 59 |
| CO2 | CARBON DIOXIDE | 60 |
| CS2 | CARBON DISULFIDE | 61 |
| CSeS | CARBON SELENOSULFIDE | 62 |
| C2N2 | CYANOGEN | 63 |
| C3S2 | CARBON SUBSULFIDE | 64 |
| Ca | CALCIUM | 65 |
| CaF2 | CALCIUM FLUORIDE | 66 |
| CbF5 | COLUMBIUM FLUORIDE | 67 |
| Cd | CADMIUM | 68 |
| CdCl2 | CADMIUM CHLORIDE | 69 |
| CdF2 | CADMIUM FLUORIDE | 70 |
| CdI2 | CADMIUM IODIDE | 71 |
| CdO | CADMIUM OXIDE | 72 |
| ClF | CHLORINE MONOFLUORIDE | 73 |
| ClFO3 | PERCHLORYL FLUORIDE | 74 |
| ClF3 | CHLORINE TRIFLUORIDE | 75 |
| ClF5 | CHLORINE PENTAFLUORIDE | 76 |
| ClHO3S | CHLOROSULFONIC ACID | 77 |
| ClHO4 | PERCHLORIC ACID | 78 |
| ClO2 | CHLORINE DIOXIDE | 79 |
| Cl2 | CHLORINE | 80 |
| Cl2O | CHLORINE MONOXIDE | 81 |
| Cl2O7 | CHLORINE HEPTOXIDE | 82 |
| Co | COBALT | 83 |
| CoCl2 | COBALT CHLORIDE | 84 |
| CoNC3O4 | COBALT NITROSYL TRICARBONYL | 85 |
| Cr | CHROMIUM | 86 |
| CrC6O6 | CHROMIUM CARBONYL | 87 |
| CrO2Cl2 | CHROMIUM OXYCHLORIDE | 88 |
| Cs | CESIUM | 89 |
| CsBr | CESIUM BROMIDE | 90 |
| CsCl | CESIUM CHLORIDE | 91 |
| CsF | CESIUM FLUORIDE | 92 |
| CsI | CESIUM IODIDE | 93 |
| Cu | COPPER | 94 |
| CuBr | CUPROUS BROMIDE | 95 |
| CuCl | CUPROUS CHLORIDE | 96 |
| CuCl2 | CUPRIC CHLORIDE | 97 |
| CuI | COPPER IODIDE | 98 |
| DCN | DEUTERIUM CYANIDE | 99 |
| D2 | DEUTERIUM | 100 |
| D2O | DEUTERIUM OXIDE | 101 |
| Eu | EUROPIUM | 102 |
| F2 | FLUORINE | 103 |
| F2O | FLUORINE OXIDE | 104 |
| Fe | IRON | 105 |
| FeC5O5 | IRON PENTACARBONYL | 106 |
| FeCl2 | FERROUS CHLORIDE | 107 |
| FeCl3 | FERRIC CHLORIDE | 108 |
| Fr | FRANCIUM | 109 |
| Ga | GALLIUM | 110 |
| GaCl3 | GALLIUM TRICHLORIDE | 111 |
| Gd | GADOLINIUM | 112 |
| Ge | GERMANIUM | 113 |
| GeBr4 | GERMANIUM BROMIDE | 114 |
| GeCl4 | GERMANIUM CHLORIDE | 115 |
| GeHCl3 | TRICHLORO GERMANE | 116 |
| GeH4 | GERMANE | 117 |
| Ge2H6 | DIGERMANE | 118 |
| Ge3H8 | TRIGERMANE | 119 |
| HBr | HYDROGEN BROMIDE | 120 |
| HCN | HYDROGEN CYANIDE | 121 |
| HCl | HYDROGEN CHLORIDE | 122 |
| HF | HYDROGEN FLUORIDE | 123 |
| HI | HYDROGEN IODIDE | 124 |
| HNO3 | NITRIC ACID | 125 |
| H2 | HYDROGEN | 126 |
| H2O | WATER | 127 |
| H2O2 | HYDROGEN PEROXIDE | 128 |
| H2S | HYDROGEN SULFIDE | 129 |
| H2SO4 | SULFURIC ACID | 130 |
| H2S2 | HYDROGEN DISULFIDE | 131 |
| H2Se | HYDROGEN SELENIDE | 132 |
| H2Te | HYDROGEN TELLURIDE | 133 |
| H3NO3S | SULFAMIC ACID | 134 |
| He | HELIUM-3 | 135 |
| He | HELIUM-4 | 136 |
| Hf | HAFNIUM | 137 |
| Hg | MERCURY | 138 |

# Appendix B

# COMPOUND LIST BY NAME

351

# Appendix C

# COEFFICIENTS FOR VAPOR PRESSURE EQUATION[*]

Carl L. Yaws, Sachin Nijhawan and Li Bu
Lamar University, Beaumont, Texas

$$\log_{10} P = A + B/T + C \log_{10} T + D\,T + E\,T^2 \qquad (P - mm\ Hg,\ T - K)$$

| NO | FORMULA | NAME | A | B | C | D | E | TMIN | TMAX |
|---|---|---|---|---|---|---|---|---|---|
| 1 | Ag | SILVER | 23.6822 | -1.6026E+04 | -4.5239E+00 | 4.0517E-04 | 1.0895E-16 | 1234.00 | 6410.00 |
| 2 | AgCl | SILVER CHLORIDE | 35.4158 | -1.2320E+04 | -8.7445E+00 | 1.7514E-03 | -1.4905E-07 | 1185.15 | 1837.15 |
| 3 | AgI | SILVER IODIDE | 69.1111 | -1.3908E+04 | -1.9966E+01 | 4.3499E-03 | -3.9755E-07 | 1093.15 | 1779.15 |
| 4 | Al | ALUMINUM | 9.9884 | -1.3837E+04 | -3.4595E-01 | 1.1361E-11 | -1.2182E-15 | 933.00 | 2329.10 |
| 5 | AlB3H12 | ALUMINUM BOROHYDRIDE | 65.4884 | -2.8974E+03 | -2.4421E+01 | 2.8261E-02 | -1.3715E-05 | 220.95 | 319.05 |
| 6 | AlBr3 | ALUMINUN BROMIDE | 5597.9805 | -1.5335E+05 | -2.2913E+03 | 2.1944E+00 | -8.0708E-04 | 354.45 | 529.45 |
| 7 | AlCl3 | ALUMINUM CHLORIDE | 27.8613 | -6.7364E+03 | -3.8080E+00 | -1.3181E-05 | 5.2521E-09 | 373.15 | 465.75 |
| 8 | AlF3 | ALUMINUM FLUORIDE | 4703.8748 | -4.6135E+05 | -1.5238E+03 | 3.4062E-01 | -3.0079E-05 | 1511.15 | 1810.15 |
| 9 | AlI3 | ALUMINUM IODIDE | 585.1783 | -2.6198E+04 | -2.1938E+02 | 1.3820E-01 | -3.5008E-05 | 451.15 | 658.65 |
| 10 | Al2O3 | ALUMINUM OXIDE | 14.1611 | -2.8238E+04 | -7.3843E-01 | -3.7413E-07 | 2.2086E-11 | 2421.10 | 3253.10 |
| 11 | Al2S3O12 | ALUMINUM SULFATE | -464.7417 | 4.9953E+04 | 1.3895E+02 | 5.7398E-04 | -1.0158E-07 | 845.15 | 1043.20 |
| 12 | Ar | ARGON | 14.9138 | -4.5675E+02 | -3.5895E+00 | 3.5490E-08 | 2.1907E-05 | 83.78 | 150.86 |
| 13 | As | ARSENIC | 70.7356 | -9.6421E+03 | -2.2451E+01 | 1.3810E-02 | -3.8579E-06 | 420.00 | 885.00 |
| 14 | AsBr3 | ARSENIC TRIBROMIDE | 69.8773 | -4.4831E+03 | -2.4415E+01 | 1.8936E-02 | -6.1554E-06 | 314.95 | 493.15 |
| 15 | AsCl3 | ARSENIC TRICHLORIDE | 93.9840 | -4.4550E+03 | -3.5080E+01 | 3.3505E-02 | -1.3338E-05 | 261.75 | 403.55 |
| 16 | AsF3 | ARSENIC TRIFLUORIDE | 27.7465 | -2.4305E+03 | -7.8250E+00 | 7.8140E-03 | -3.3333E-06 | 270.65 | 329.45 |
| 17 | AsF5 | ARSENIC PENTAFLUORIDE | 1944.5157 | -3.0430E+04 | -9.1148E+02 | 1.8231E+00 | -1.4325E-03 | 155.25 | 220.35 |
| 18 | AsH3 | ARSINE | -10.3462 | -7.7134E+02 | 9.1086E+00 | -2.4654E-02 | 2.0673E-05 | 156.23 | 373.00 |
| 19 | AsI3 | ARSENIC TRIIODIDE | 96.0056 | -6.8120E+03 | -3.3239E+01 | 2.0112E-02 | -5.0888E-06 | 437.15 | 602.65 |
| 20 | As2O3 | ARSENIC TRIOXIDE | -2811.4735 | 7.9643E+04 | 1.1255E+03 | -9.0385E-01 | 2.6735E-04 | 485.65 | 730.35 |
| 21 | At | ASTATINE | -17.3579 | -4.5040E+03 | 1.3457E+01 | -2.2096E-02 | 9.7987E-06 | 279.00 | 607.00 |
| 22 | Au | GOLD | 126.1594 | -2.9160E+04 | -3.7975E+01 | 8.0922E-03 | -6.6670E-07 | 1226.00 | 3120.00 |
| 23 | B | BORON | 38.5877 | -3.6230E+04 | -7.5640E+00 | -1.1450E-04 | 5.1894E-08 | 1821.00 | 4133.00 |
| 24 | BBr3 | BORON TRIBROMIDE | 104.0215 | -4.2744E+03 | -3.9717E+01 | 3.9970E-02 | -1.6726E-05 | 273.15 | 361.05 |
| 25 | BCl3 | BORON TRICHLORIDE | 35.3538 | -2.3048E+03 | -1.0108E+01 | 5.5834E-10 | 5.1455E-06 | 166.15 | 451.95 |
| 26 | BF3 | BORON TRIFLUORIDE | 75.2534 | -2.4715E+03 | -2.6531E+01 | 8.5366E-08 | 4.2709E-05 | 144.78 | 260.90 |
| 27 | BH2CO | BORINE CARBONYL | 433.1860 | -6.8163E+03 | -2.0499E+02 | 4.5121E-01 | -3.7515E-04 | 133.95 | 209.15 |
| 28 | BH3O3 | BORIC ACID | -81.1257 | 2.1815E+03 | 2.9986E+01 | -2.2693E-05 | 1.0956E-08 | 293.15 | 401.15 |
| 29 | B2D6 | DEUTERODIBORANE | 120.8038 | -2.4092E+03 | -5.4329E+01 | 1.1943E-01 | -1.0820E-04 | 118.25 | 179.25 |
| 30 | B2H5Br | DIBORANE HYDROBROMIDE | 488.4916 | -8.8892E+03 | -2.2351E+02 | 4.1313E-01 | -2.9011E-04 | 179.85 | 289.15 |
| 31 | B2H6 | DIBORANE | 11.5597 | -9.4549E+02 | -1.3754E+00 | -2.3575E-03 | 2.6782E-06 | 107.65 | 289.80 |
| 32 | B3N3H6 | BORINE TRIAMINE | 58.7997 | -2.8667E+03 | -2.1466E+01 | 2.5002E-02 | -1.2241E-05 | 210.15 | 323.75 |
| 33 | B4H10 | TETRABORANE | 23.0730 | -1.7369E+03 | -6.6041E+00 | 8.4552E-03 | -4.6056E-06 | 182.25 | 289.25 |
| 34 | B5H9 | PENTABORANE | -163.4159 | 1.4743E+03 | 7.5896E+01 | -1.0578E-01 | 5.0905E-05 | 232.75 | 568.45 |
| 35 | B5H11 | TETRAHYDROPENTABORANE | 7.6458 | -1.8417E+03 | 2.8719E-01 | -3.0007E-04 | 1.3472E-07 | 222.95 | 340.15 |
| 36 | B10H14 | DECABORANE | 4813.9118 | -1.2837E+05 | -1.9845E+03 | 1.9935E+00 | -7.8068E-04 | 333.15 | 436.95 |
| 37 | Ba | BARIUM | -18.1369 | -8.1603E+03 | 1.0107E+01 | -5.9955E-03 | 9.8349E-07 | 638.00 | 1907.00 |
| 38 | Be | BERYLLIUM | -4.7459 | -1.6799E+04 | 5.4503E+00 | -2.6044E-03 | 2.8638E-07 | 1097.00 | 2744.00 |
| 39 | BeB2H8 | BERYLLIUM BOROHYDRIDE | -94.5732 | 2.2630E+02 | 4.1281E+01 | -2.7627E-02 | 8.9618E-06 | 274.15 | 363.15 |
| 40 | BeBr2 | BERYLLIUM BROMIDE | 2669.4633 | -9.8679E+04 | -1.0326E+03 | 7.2331E-01 | -1.9329E-04 | 562.15 | 747.15 |
| 41 | BeCl2 | BERYLLIUM CHLORIDE | -289.7317 | 1.0348E+04 | 1.0311E+02 | -6.6198E-02 | -9.9081E-06 | 564.15 | 760.15 |
| 42 | BeF2 | BERYLLIUM FLUORIDE | 224.7647 | -2.8762E+04 | -7.0897E+01 | 1.7991E-02 | -1.8955E-06 | 1145.55 | 1372.15 |
| 43 | BeI2 | BERYLLIUM IODIDE | 3417.9820 | -1.2770E+05 | -1.3156E+03 | 8.8819E-01 | -2.2863E-04 | 556.15 | 760.15 |
| 44 | Bi | BISMUTH | 1449.1712 | -8.3717E+04 | -5.1359E+02 | 2.0492E-01 | -2.9858E-05 | 569.00 | 1700.00 |
| 45 | BiBr3 | BISMUTH TRIBROMIDE | 216.7751 | -1.3734E+04 | -7.6683E+01 | 3.9665E-02 | -8.4421E-06 | 534.15 | 734.15 |
| 46 | BiCl3 | BISMUTH TRICHLORIDE | -6.1334 | -3.1936E+03 | 5.2620E+00 | -2.4866E-03 | 5.0977E-07 | 503.65 | 710.55 |
| 47 | BrF5 | BROMINE PENTAFLUORIDE | 51.1403 | -2.6286E+03 | -1.8290E+01 | 2.1840E-02 | -1.0991E-05 | 203.85 | 313.55 |
| 48 | Br2 | BROMINE | 23.7200 | -2.2840E+03 | -5.6145E+00 | 2.2602E-09 | 1.7888E-06 | 265.85 | 584.15 |
| 49 | C | CARBON | -2.0686 | -2.5987E+04 | 3.5504E+00 | -5.3440E-04 | 2.9310E-08 | 3259.10 | 4399.10 |
| 50 | CCl2O | PHOSGENE | 46.6551 | -2.4657E+03 | -1.5351E+01 | 9.2288E-03 | -4.9658E-14 | 145.37 | 455.00 |
| 51 | CF2O | CARBONYL FLUORIDE | -70.1866 | -3.8315E+02 | 4.1225E+01 | -1.2049E-01 | 1.1291E-04 | 161.89 | 297.00 |
| 52 | CH4N2O | UREA | 6.7305 | -4.4855E+03 | 1.5588E+00 | -1.7855E-03 | 8.4006E-07 | 340.15 | 368.05 |
| 53 | CH4N2S | THIOUREA | -35.4224 | -2.3772E+03 | 1.9025E+01 | -2.0819E-02 | 6.8609E-06 | 454.15 | 854.00 |
| 54 | CNBr | CYANOGEN BROMIDE | 2302.7452 | -4.9937E+04 | -9.9806E+02 | 1.3391E+00 | -7.0662E-04 | 273.01 | 313.09 |
| 55 | CNCl | CYANOGEN CHLORIDE | -9.0018 | -1.2842E+03 | 8.2222E+00 | -1.6105E-02 | 9.5986E-06 | 266.65 | 449.00 |

* A computer program, containing coefficients for vapor pressure for all compounds, is available for a nominal fee (Carl L. Yaws, Box 10053, Lamar University, Beaumont, TX 77710, phone/FAX 409-880-8787). The computer program is in ASCII which can be accessed by other software.

$$\log_{10} P = A + B/T + C \log_{10} T + D T + E T^2 \qquad (P - mm\ Hg,\ T - K)$$

| NO | FORMULA | NAME | A | B | C | D | E | TMIN | TMAX |
|---|---|---|---|---|---|---|---|---|---|
| 56 | CNF | CYANOGEN FLUORIDE | 179.8136 | -4.3677E+03 | -7.6561E+01 | 1.1639E-01 | -7.2513E-05 | 196.75 | 226.35 |
| 57 | CO | CARBON MONOXIDE | 51.8145 | -7.8824E+02 | -2.2734E+01 | 5.1225E-02 | 4.6603E-11 | 68.15 | 132.92 |
| 58 | COS | CARBONYL SULFIDE | 36.8556 | -1.7187E+03 | -1.2036E+01 | 8.9612E-03 | 2.0283E-13 | 134.30 | 378.80 |
| 59 | COSe | CARBON OXYSELENIDE | 525.5520 | -9.1072E+03 | -2.4172E+02 | 4.5341E-01 | -3.1974E-04 | 156.05 | 251.25 |
| 60 | CO2 | CARBON DIOXIDE | 35.0187 | -1.5119E+03 | -1.1335E+01 | 9.3383E-03 | 7.7626E-10 | 216.58 | 304.19 |
| 61 | CS2 | CARBON DISULFIDE | 25.1475 | -2.0439E+03 | -6.7794E+00 | 3.4828E-03 | 3.4373E-15 | 161.11 | 552.00 |
| 62 | CSeS | CARBON SELENOSULFIDE | -101.6898 | -4.5082E+02 | 5.5220E+01 | -1.0400E-01 | 6.9601E-05 | 225.85 | 358.75 |
| 63 | C2N2 | CYANOGEN | 4016.4115 | -5.9887E+04 | -1.9008E+03 | 3.9066E+00 | -3.0816E-03 | 177.35 | 252.15 |
| 64 | C3S2 | CARBON SUBSULFIDE | -243.5223 | 3.1498E+03 | 1.0795E+02 | -1.3131E-01 | 5.9581E-05 | 287.15 | 403.95 |
| 65 | Ca | CALCIUM | -75.9862 | -4.9573E+03 | 3.0625E+01 | -1.3574E-02 | 1.9938E-06 | 625.00 | 1762.00 |
| 66 | CaF2 | CALCIUM FLUORIDE | 49.0700 | -2.7912E+04 | -1.0510E+01 | 1.4288E-08 | -1.0745E-12 | 1691.00 | 2806.50 |
| 67 | CbF5 | COLUMBIUM FLUORIDE | -42.2353 | -1.1915E+03 | 2.0862E+01 | -2.2002E-02 | 8.8506E-06 | 359.45 | 498.15 |
| 68 | Cd | CADMIUM | -145.3985 | -8.3891E+02 | 6.0509E+01 | -4.3639E-02 | 1.1003E-05 | 393.00 | 1043.00 |
| 69 | CdCl2 | CADMIUM CHLORIDE | -4189.3798 | 2.2619E+05 | 1.4896E+03 | -5.9464E-01 | 9.0541E-05 | 891.15 | 1240.15 |
| 70 | CdF2 | CADMIUM FLUORIDE | -1738.0052 | 1.5268E+05 | 5.6319E+02 | -1.1301E-01 | 7.8517E-06 | 1385.15 | 2024.15 |
| 71 | CdI2 | CADMIUM IODIDE | -899.5410 | 3.5567E+04 | 3.3430E+02 | -1.6865E-01 | 3.2257E-05 | 689.15 | 1069.15 |
| 72 | CdO | CADMIUM OXIDE | 42.8498 | -1.5443E+04 | -1.0651E+01 | 2.0649E-03 | -1.7038E-07 | 1273.15 | 1832.15 |
| 73 | ClF | CHLORINE MONOFLUORIDE | 44.9140 | -1.6526E+03 | -1.6727E+01 | 3.3592E-02 | -2.8601E-05 | 129.75 | 172.65 |
| 74 | ClFO3 | PERCHLORYL FLUORIDE | 40.5028 | -1.8544E+03 | -1.3458E+01 | 9.9792E-03 | 3.8242E-13 | 125.41 | 368.40 |
| 75 | ClF3 | CHLORINE TRIFLUORIDE | 25.4578 | -1.6807E+03 | -7.7581E+00 | 9.8092E-03 | -5.2739E-06 | 192.75 | 284.65 |
| 76 | ClF5 | CHLORINE PENTAFLUORIDE | 131.4931 | -3.7872E+03 | -5.3872E+01 | 7.2145E-02 | -3.9712E-05 | 193.95 | 297.95 |
| 77 | ClHO3S | CHLOROSULFONIC ACID | -5.6040 | -2.7604E+03 | 8.4466E+00 | -2.2029E-02 | 1.1699E-05 | 193.15 | 700.00 |
| 78 | ClHO4 | PERCHLORIC ACID | 7.5448 | -2.0352E+03 | 1.0732E+00 | -7.2499E-03 | 4.2925E-06 | 171.95 | 631.00 |
| 79 | ClO2 | CHLORINE DIOXIDE | 121.9791 | -5.2893E+03 | -4.1838E+01 | -2.0950E-08 | 2.6848E-05 | 213.55 | 465.00 |
| 80 | Cl2 | CHLORINE | 28.8659 | -1.6745E+03 | -8.5216E+00 | 5.3792E-03 | -7.7867E-13 | 172.12 | 417.15 |
| 81 | Cl2O | CHLORINE MONOXIDE | 20.7506 | -1.6377E+03 | -5.5984E+00 | 7.4816E-03 | -2.4600E-06 | 174.65 | 275.35 |
| 82 | Cl2O7 | CHLORINE HEPTOXIDE | 5.5051 | -1.7898E+03 | 1.0930E+00 | -1.1071E-03 | 4.8227E-07 | 227.85 | 351.95 |
| 83 | Co | COBALT | 16.7750 | -1.8953E+04 | -1.7830E+00 | 2.8243E-05 | -6.2846E-08 | 1095.00 | 2528.00 |
| 84 | CoCl2 | COBALT CHLORIDE | 90.9544 | -1.2963E+04 | -2.7733E+01 | 7.3497E-03 | -8.1551E-07 | 1043.15 | 1323.15 |
| 85 | CoNC3O4 | COBALT NITROSYL TRICARBONYL | -2295.1356 | 4.3067E+04 | 1.0148E+03 | -1.4317E+00 | 7.7031E-04 | 271.85 | 353.15 |
| 86 | Cr | CHROMIUM | -80.3456 | -1.2221E+04 | 2.9746E+01 | -6.8400E-03 | 5.2454E-07 | 1229.00 | 2840.00 |
| 87 | CrC6O6 | CHROMIUM CARBONYL | -3135.3223 | 6.6813E+04 | 1.3443E+03 | -1.6151E+00 | 7.4196E-04 | 309.15 | 424.15 |
| 88 | CrO2Cl2 | CHROMIUM OXYCHLORIDE | 226.3675 | -7.6454E+03 | -9.0472E+01 | 9.3809E-02 | -3.9753E-05 | 254.75 | 390.25 |
| 89 | Cs | CESIUM | -2.9708 | -3.6286E+03 | 4.1641E+00 | -3.6632E-03 | 7.9575E-07 | 295.00 | 959.00 |
| 90 | CsBr | CESIUM BROMIDE | 1537.6463 | -1.0591E+05 | -5.2949E+02 | 1.7992E-01 | -2.3362E-05 | 1021.15 | 1573.15 |
| 91 | CsCl | CESIUM CHLORIDE | 438.8475 | -3.4903E+04 | -1.5008E+02 | 5.3499E-02 | -7.3394E-06 | 1017.15 | 1573.15 |
| 92 | CsF | CESIUM FLUORIDE | 1768.1436 | -1.1822E+05 | -6.1137E+02 | 2.1264E-01 | -2.8322E-05 | 985.15 | 1524.15 |
| 93 | CsI | CESIUM IODIDE | 276.4031 | -2.6150E+04 | -9.2366E+01 | 3.0791E-02 | -4.0411E-06 | 1011.15 | 1553.15 |
| 94 | Cu | COPPER | -82.6254 | -1.1231E+04 | 3.1336E+01 | -8.9074E-03 | 7.5663E-07 | 1130.00 | 3150.00 |
| 95 | CuBr | CUPROUS BROMIDE | 208.6672 | -2.0301E+04 | -6.8060E+01 | 1.9026E-02 | -2.1495E-06 | 845.15 | 1628.15 |
| 96 | CuCl | CUPROUS CHLORIDE | 27.2556 | -7.5654E+03 | -6.1858E+00 | 1.2376E-01 | -1.7508E-14 | 703.00 | 1763.10 |
| 97 | CuCl2 | CUPRIC CHLORIDE | -238.6809 | 1.7213E+04 | 7.5807E+01 | 5.5333E-05 | -6.1652E-09 | 582.85 | 794.15 |
| 98 | CuI | COPPER IODIDE | 855.8840 | -5.7746E+04 | -2.9506E+02 | 1.0189E-01 | -1.3492E-05 | 883.15 | 1609.15 |
| 99 | DCN | DEUTERIUM CYANIDE | 798.8279 | -1.7354E+04 | -3.4780E+02 | 4.9610E-01 | -2.8192E-04 | 204.25 | 299.35 |
| 100 | D2 | DEUTERIUM | 6.1037 | -6.7085E+01 | -5.7226E-01 | 1.6894E-02 | -1.7612E-11 | 18.73 | 38.35 |
| 101 | D2O | DEUTERIUM OXIDE | -12.8257 | -2.1886E+03 | 1.0645E+01 | -1.9029E-02 | 9.1375E-06 | 276.97 | 643.89 |
| 102 | Eu | EUROPIUM | -55.5456 | -4.6880E+03 | 2.3529E+01 | -1.0970E-02 | 1.6487E-06 | 640.00 | 1742.00 |
| 103 | F2 | FLUORINE | 27.1409 | -5.7201E+02 | -1.0015E+01 | 2.1078E-02 | 8.9567E-13 | 53.48 | 144.31 |
| 104 | F2O | FLUORINE OXIDE | -24.8186 | -1.9007E+02 | 1.5985E+01 | -4.1421E-02 | 4.8042E-05 | 77.05 | 128.55 |
| 105 | Fe | IRON | 11.5549 | -1.9538E+04 | -6.2549E-01 | -2.7182E-09 | 1.9086E-13 | 1808.15 | 3008.20 |
| 106 | FeC5O5 | IRON PENTACARBONYL | 317.6618 | -9.3157E+03 | -1.2965E+02 | 1.3467E-01 | -4.8175E-05 | 266.65 | 378.15 |
| 107 | FeCl2 | FERROUS CHLORIDE | -1350.9354 | 6.2772E+04 | 4.8764E+02 | -2.0598E-01 | 3.2434E-05 | 973.15 | 1299.15 |
| 108 | FeCl3 | FERRIC CHLORIDE | 64969.7167 | -1.9169E+06 | -2.6114E+04 | 2.2347E+01 | -7.3076E-03 | 467.15 | 592.15 |
| 109 | Fr | FRANCIUM | -38.3826 | -2.2043E+03 | 1.8322E+01 | -1.5441E-02 | 4.4042E-06 | 267.00 | 879.00 |
| 110 | Ga | GALLIUM | -4.4968 | -1.2924E+04 | 4.3403E+00 | -9.5226E-04 | 2.3443E-08 | 954.00 | 2517.00 |
| 111 | GaCl3 | GALLIUM TRICHLORIDE | 54.9805 | -4.8102E+03 | -1.6009E+01 | -8.4664E-09 | 3.9331E-06 | 350.90 | 694.00 |
| 112 | Gd | GADOLINIUM | 9.6612 | -1.0909E+04 | -4.7131E-01 | 1.2082E-03 | -3.9062E-07 | 728.00 | 1700.00 |
| 113 | Ge | GERMANIUM | 24.3265 | -1.9756E+04 | -4.9583E+00 | 9.5753E-04 | -8.0668E-08 | 1230.00 | 3125.00 |
| 114 | GeBr4 | GERMANIUM BROMIDE | 151.2941 | -6.5181E+03 | -5.7438E+01 | 4.8303E-02 | -1.6731E-05 | 316.45 | 462.15 |
| 115 | GeCl4 | GERMANIUM CHLORIDE | 207.3749 | -6.4745E+03 | -8.4232E+01 | 9.6332E-02 | -4.5015E-05 | 228.15 | 357.15 |
| 116 | GeHCl3 | TRICHLORO GERMANE | 175.3633 | -5.9201E+03 | -7.0301E+01 | 7.9543E-02 | -3.7043E-05 | 231.85 | 348.15 |
| 117 | GeH4 | GERMANE | 32.2743 | -1.3704E+03 | -9.9425E+00 | 5.1455E-10 | 1.6203E-05 | 107.26 | 308.00 |
| 118 | Ge2H6 | DIGERMANE | 45.6425 | -2.1567E+03 | -1.6528E+01 | 2.1047E-02 | -1.1265E-05 | 184.45 | 304.65 |
| 119 | Ge3H8 | TRIGERMANE | 36.7997 | -2.5324E+03 | -1.2064E+01 | 1.1954E-02 | -5.0050E-06 | 236.25 | 383.95 |
| 120 | HBr | HYDROGEN BROMIDE | 34.4939 | -1.6379E+03 | -1.0909E+01 | 7.5732E-03 | -2.5521E-12 | 185.15 | 363.15 |
| 121 | HCN | HYDROGEN CYANIDE | -57.0540 | -3.6256E+02 | 2.9415E+01 | -4.7528E-02 | 2.8406E-05 | 259.83 | 456.65 |

* A computer program, containing coefficients for vapor pressure for all compounds, is available for a nominal fee (Carl L. Yaws, Box 10053, Lamar University, Beaumont, TX 77710, phone/FAX 409-880-8787). The computer program is in ASCII which can be accessed by other software.

353

| NO | FORMULA | NAME | A | B | C | D | E | TMIN | TMAX |
|---|---|---|---|---|---|---|---|---|---|
| 122 | HCl | HYDROGEN CHLORIDE | 43.5455 | -1.6279E+03 | -1.5214E+01 | 1.3783E-02 | -1.4984E-11 | 158.97 | 324.65 |
| 123 | HF | HYDROGEN FLUORIDE | 23.7347 | -1.7996E+03 | -6.1764E+00 | -5.0046E-10 | 6.1500E-06 | 189.79 | 461.15 |
| 124 | HI | HYDROGEN IODIDE | 21.4282 | -1.4515E+03 | -5.5756E+00 | 3.3878E-03 | -8.9335E-12 | 222.38 | 423.85 |
| 125 | HNO3 | NITRIC ACID | 71.7653 | -4.3768E+03 | -2.2769E+01 | -4.5988E-07 | 1.1856E-05 | 231.55 | 376.10 |
| 126 | H2 | HYDROGEN | 3.4132 | -4.1316E+01 | 1.0947E+00 | -6.6896E-10 | 1.4589E-04 | 13.95 | 33.18 |
| 127 | H2O | WATER | 29.8605 | -3.1522E+03 | -7.3037E+00 | 2.4247E-09 | 1.8090E-06 | 273.16 | 647.13 |
| 128 | H2O2 | HYDROGEN PEROXIDE | 33.3222 | -3.7350E+03 | -8.3458E+00 | -1.2351E-10 | 1.6917E-06 | 272.74 | 730.15 |
| 129 | H2S | HYDROGEN SULFIDE | 18.6383 | -1.3446E+03 | -4.1034E+00 | 3.1815E-09 | 2.4664E-06 | 187.68 | 373.53 |
| 130 | H2SO4 | SULFURIC ACID | 2.0582 | -4.1924E+03 | 3.2578E+00 | -1.1224E-03 | 5.5371E-07 | 283.15 | 603.15 |
| 131 | H2S2 | HYDROGEN DISULFIDE | 22.3747 | -2.4043E+03 | -5.5539E+00 | 5.8402E-03 | -2.6241E-06 | 229.95 | 337.15 |
| 132 | H2Se | HYDROGEN SELENIDE | 573.9933 | -1.0568E+04 | -2.6071E+02 | 4.7258E-01 | -3.4316E-04 | 157.85 | 232.05 |
| 133 | H2Te | HYDROGEN TELLURIDE | 532.0079 | -1.0876E+04 | -2.3540E+02 | 3.7306E-01 | -2.3649E-04 | 176.75 | 271.15 |
| 134 | H3NO3S | SULFAMIC ACID | 0.1130 | -6.8449E+02 | 5.2613E-05 | -6.8934E-08 | 3.4587E-11 | 293.15 | 373.15 |
| 135 | He | HELIUM-3 | 2.1239 | -1.1769E+00 | 2.0493E+00 | -1.7652E-10 | 9.4980E-03 | 1.01 | 3.31 |
| 136 | He | HELIUM-4 | 2.8838 | -3.9043E+00 | 6.7240E-01 | 1.1913E-01 | -2.1010E-11 | 1.76 | 5.20 |
| 137 | Hf | HAFNIUM | -187.3128 | -4.7256E+03 | 5.9397E+01 | -7.1673E-03 | 2.6648E-07 | 2117.00 | 5960.00 |
| 138 | Hg | MERCURY | 11.3169 | -3.3515E+03 | -1.1296E+00 | -2.9698E-13 | 1.1699E-07 | 234.31 | 1735.00 |
| 139 | HgBr2 | MERCURIC BROMIDE | -6001.9471 | 1.6586E+05 | 2.4352E+03 | -2.1646E+00 | 7.2767E-04 | 409.65 | 592.15 |
| 140 | HgCl2 | MERCURIC CHLORIDE | 8091.0938 | -2.3113E+05 | -3.2811E+03 | 2.9716E+00 | -1.0294E-03 | 409.35 | 577.15 |
| 141 | HgI2 | MERCURIC IODIDE | -9817.2743 | 2.9088E+05 | 3.9308E+03 | -3.2560E+00 | 1.0229E-03 | 430.65 | 627.15 |
| 142 | IF7 | IODINE HEPTAFLUORIDE | 2.2894 | -1.4949E+03 | 2.7875E+00 | -3.5044E-03 | 1.9005E-06 | 186.15 | 277.15 |
| 143 | I2 | IODINE | 46.9335 | -5.1763E+03 | -1.1170E+01 | -1.0130E-02 | 7.7500E-06 | 242.00 | 819.15 |
| 144 | In | INDIUM | 112.3721 | -1.8894E+04 | -3.5777E+01 | 1.1333E-02 | -1.3444E-06 | 850.00 | 2323.00 |
| 145 | Ir | IRIDIUM | -197.4829 | -6.5755E+03 | 6.4387E+01 | -9.6603E-03 | 5.0122E-07 | 1944.00 | 4450.00 |
| 146 | K | POTASSIUM | 10.8410 | -4.6934E+03 | -1.1916E+00 | 1.5875E-04 | 1.7454E-17 | 336.35 | 2223.00 |
| 147 | KBr | POTASSIUM BROMIDE | 86.7862 | -1.5677E+04 | -2.5680E+01 | 5.9097E-03 | -5.6987E-07 | 1068.15 | 1656.15 |
| 148 | KCl | POTASSIUM CHLORIDE | -8.0224 | -9.3722E+03 | 6.4641E+00 | -3.1639E-03 | 3.2745E-07 | 1044.00 | 3470.00 |
| 149 | KF | POTASSIUM FLUORIDE | 86.1553 | -1.7030E+04 | -2.5138E+01 | 5.3564E-03 | -4.7879E-07 | 1158.15 | 1775.15 |
| 150 | KI | POTASSIUM IODIDE | 88.6708 | -1.5026E+04 | -2.6513E+01 | 6.3780E-03 | -6.4206E-07 | 1018.15 | 1597.15 |
| 151 | KOH | POTASSIUM HYDROXIDE | 20.9787 | -9.5262E+03 | -3.8001E+00 | -2.9030E-10 | 1.2312E-08 | 679.00 | 1600.00 |
| 152 | Kr | KRYPTON | -12.6883 | -3.1111E+02 | 1.0610E+01 | -3.8518E-02 | 5.0870E-05 | 115.78 | 209.35 |
| 153 | La | LANTHANUM | 190.2107 | -4.1131E+04 | -5.7186E+01 | 9.9276E-03 | -6.4353E-07 | 1441.50 | 3643.00 |
| 154 | Li | LITHIUM | 12.1182 | -8.4301E+03 | -1.3510E+00 | 2.2909E-04 | -6.2641E-18 | 453.69 | 4085.00 |
| 155 | LiBr | LITHIUM BROMIDE | 65.3934 | -1.3245E+04 | -1.8782E+01 | 4.4647E-03 | -4.4694E-07 | 1021.15 | 1583.15 |
| 156 | LiCl | LITHIUM CHLORIDE | 1659.2051 | -1.1514E+05 | -5.7029E+02 | 1.9064E-01 | -2.4324E-05 | 1056.15 | 1655.15 |
| 157 | LiF | LITHIUM FLUORIDE | 69.3367 | -1.8147E+04 | -1.9178E+01 | 3.6011E-03 | -2.8544E-07 | 1320.15 | 1954.15 |
| 158 | LiI | LITHIUM IODIDE | 72.2997 | -1.4450E+04 | -2.0834E+01 | 5.2464E-03 | -5.5773E-07 | 996.15 | 1444.15 |
| 159 | Lu | LUTETIUM | -47.6746 | -1.1929E+04 | 1.8989E+01 | -4.5961E-03 | 3.5367E-07 | 1057.00 | 2535.00 |
| 160 | Mg | MAGNESIUM | -72.6513 | -4.2014E+03 | 3.0264E+01 | -1.5558E-02 | 2.6429E-06 | 517.00 | 1376.00 |
| 161 | MgCl2 | MAGNESIUM CHLORIDE | 1239.2680 | -9.0236E+04 | -4.2304E+02 | 1.3654E-01 | -1.6884E-05 | 1051.15 | 1691.15 |
| 162 | MgO | MAGNESIUM OXIDE | -41.2727 | -1.4025E+04 | 1.5392E+01 | -2.3614E-03 | 1.1265E-07 | 3105.00 | 5950.00 |
| 163 | Mn | MANGANESE | -123.9176 | -5.9845E+03 | 4.6074E+01 | -1.4659E-02 | 1.5195E-06 | 924.00 | 2392.00 |
| 164 | MnCl2 | MANGANESE CHLORIDE | 1275.0362 | -8.8381E+04 | -4.3842E+02 | 1.4858E-01 | -1.9347E-05 | 1009.15 | 1463.15 |
| 165 | Mo | MOLYBDENUM | 74.9735 | -4.1955E+04 | -2.0072E+01 | 3.2166E-03 | -2.2507E-07 | 1677.00 | 5100.00 |
| 166 | MoF6 | MOLYBDENUM FLUORIDE | 300.7229 | -7.9655E+03 | -1.2644E+02 | 1.6582E-01 | -8.8423E-05 | 207.65 | 309.15 |
| 167 | MoO3 | MOLYBDENUM OXIDE | 3015.8551 | -2.1859E+05 | -1.0259E+03 | 3.1920E-01 | -3.8938E-05 | 1007.15 | 1424.15 |
| 168 | NCl3 | NITROGEN TRICHLORIDE | 88.8600 | -4.2008E+03 | -3.1565E+01 | 1.8305E-02 | -6.0409E-11 | 246.15 | 367.15 |
| 169 | ND3 | HEAVY AMMONIA | 125.5395 | -3.5614E+03 | -5.1766E+01 | 7.4954E-02 | -4.4958E-05 | 199.15 | 239.75 |
| 170 | NF3 | NITROGEN TRIFLUORIDE | 8.4514 | -7.4300E+02 | 3.9105E-01 | -1.1295E-02 | 1.7758E-05 | 66.36 | 233.85 |
| 171 | NH3 | AMMONIA | 37.1575 | -2.0277E+03 | -1.1601E+01 | 7.4625E-03 | -9.5811E-12 | 195.41 | 405.65 |
| 172 | NH3O | HYDROXYLAMINE | 2.4590 | -4.0142E+03 | 6.3297E+00 | -1.8547E-02 | 1.1266E-05 | 306.25 | 383.00 |
| 173 | NH4Br | AMMONIUM BROMIDE | 23.7261 | -5.2014E+03 | -5.1721E+00 | 2.6919E-03 | -5.9955E-07 | 471.45 | 669.15 |
| 174 | NH4Cl | AMMONIUM CHLORIDE | -893.5430 | 1.9484E+04 | 3.6716E+02 | -3.2056E-01 | 1.0034E-04 | 553.15 | 793.15 |
| 175 | NH4I | AMMONIUM IODIDE | 0.4097 | -4.4032E+03 | 3.5216E+00 | -1.7562E-03 | 3.7842E-07 | 484.05 | 678.05 |
| 176 | NH5O | AMMONIUM HYDROXIDE | 42.3381 | -2.3577E+03 | -1.3597E+01 | 8.5486E-03 | 4.7298E-11 | 203.15 | 353.15 |
| 177 | NH5S | AMMONIUM HYDROGENSULFIDE | 178.9587 | -6.0657E+03 | -7.2004E+01 | 8.7809E-02 | -4.4365E-05 | 222.05 | 306.45 |
| 178 | NO | NITRIC OXIDE | 61.2046 | -1.5365E+03 | -2.3621E+01 | 2.9377E-02 | -1.3066E-09 | 109.50 | 180.15 |
| 179 | NOCl | NITROSYL CHLORIDE | 24.1469 | -1.8469E+03 | -6.0513E+00 | 9.2881E-09 | 4.5319E-06 | 213.55 | 440.65 |
| 180 | NOF | NITROSYL FLUORIDE | 478.8331 | -7.7441E+03 | -2.2562E+02 | 4.9213E-01 | -4.2279E-04 | 141.15 | 217.15 |
| 181 | NO2 | NITROGEN DIOXIDE | 32.1203 | -2.2563E+03 | -9.7702E+00 | 8.6560E-03 | -5.1036E-11 | 261.95 | 431.35 |
| 182 | N2 | NITROGEN | 23.8572 | -4.7668E+02 | -8.6689E+00 | 2.0128E-02 | -2.4139E-11 | 63.15 | 126.10 |
| 183 | N2F4 | TETRAFLUOROHYDRAZINE | 16.4473 | -9.6387E+02 | -4.3123E+00 | 5.9876E-03 | 2.2680E-13 | 111.65 | 309.35 |
| 184 | N2H4 | HYDRAZINE | 31.2541 | -3.1466E+03 | -8.2200E+00 | 2.6734E-03 | 4.6004E-13 | 274.68 | 653.15 |
| 185 | N2H4C | AMMONIUM CYANIDE | 179.6608 | -6.1294E+03 | -7.2215E+01 | 8.0090E-02 | -4.4545E-05 | 222.55 | 304.85 |
| 186 | N2H6CO2 | AMMONIUM CARBAMATE | 93.9638 | -4.8160E+03 | -3.4502E+01 | 3.6858E-02 | -1.6578E-05 | 247.05 | 331.45 |
| 187 | N2O | NITROUS OXIDE | 61.5168 | -2.1016E+03 | -2.2337E+01 | 1.8232E-02 | -1.1348E-10 | 182.30 | 309.57 |

* A computer program, containing coefficients for vapor pressure for all compounds, is available for a nominal fee (Carl L. Yaws, Box 10053, Lamar University, Beaumont, TX 77710, phone/FAX 409-880-8787). The computer program is in ASCII which can be accessed by other software.

$$\log_{10} P = A + B/T + C \log_{10} T + D T + E T^2 \qquad (P - mm\ Hg,\ T - K)$$

| NO | FORMULA | NAME | A | B | C | D | E | TMIN | TMAX |
|---|---|---|---|---|---|---|---|---|---|
| 188 | N2O3 | NITROGEN TRIOXIDE | -2.5932 | -1.8582E+03 | 7.4627E+00 | -2.7402E-02 | 2.0570E-05 | 170.00 | 425.00 |
| 189 | N2O4 | NITROGEN TETRAOXIDE | -197.7926 | 7.7599E+02 | 9.9702E+01 | -2.0780E-01 | 1.4907E-04 | 261.90 | 320.65 |
| 190 | N2O5 | NITROGEN PENTOXIDE | -270.4814 | 4.6851E+02 | 1.3740E+02 | -2.9833E-01 | 2.3064E-04 | 236.35 | 305.55 |
| 191 | Na | SODIUM | 7.2828 | -5.4249E+03 | 1.5358E-01 | -2.0200E-04 | 3.7007E-08 | 370.97 | 2573.00 |
| 192 | NaBr | SODIUM BROMIDE | 18.4164 | -1.0501E+04 | -2.8637E+00 | 8.9249E-10 | -1.1019E-13 | 1020.00 | 1720.00 |
| 193 | NaCN | SODIUM CYANIDE | -2.2303 | -8.2019E+03 | 3.8990E+00 | -2.3458E-03 | 3.9325E-07 | 836.85 | 1769.10 |
| 194 | NaCl | SODIUM CHLORIDE | 22.4317 | -1.1358E+04 | -4.2035E+00 | 3.4674E-04 | -3.9472E-12 | 1073.90 | 1738.10 |
| 195 | NaF | SODIUM FLUORIDE | 11.6744 | -1.1663E+04 | -9.6084E-01 | 1.2957E-04 | -1.1232E-14 | 1269.00 | 2060.00 |
| 196 | NaI | SODIUM IODIDE | 69.6407 | -1.4137E+04 | -2.0041E+01 | 4.7338E-03 | -4.7106E-07 | 1040.15 | 1577.15 |
| 197 | NaOH | SODIUM HYDROXIDE | -48.2774 | -1.9340E+03 | 1.7000E+01 | 2.9640E-11 | -8.7510E-07 | 596.00 | 1830.00 |
| 198 | Na2SO4 | SODIUM SULFATE | -2.2687 | -1.5051E+04 | 3.5005E+00 | -1.2712E-03 | 1.7716E-07 | 1173.10 | 1223.10 |
| 199 | Nb | NIOBIUM | -64.3485 | -2.9438E+04 | 2.3622E+01 | -3.9155E-03 | 2.0660E-07 | 2250.00 | 5115.00 |
| 200 | Nd | NEODYMIUM | 130.3312 | -2.6812E+04 | -3.9789E+01 | 8.4714E-03 | -6.7800E-07 | 1144.00 | 3384.00 |
| 201 | Ne | NEON | 0.0175 | -5.9863E+01 | 3.6341E+00 | -8.8025E-09 | -1.8115E-04 | 24.56 | 44.40 |
| 202 | Ni | NICKEL | -57.4301 | -1.3533E+04 | 2.3611E+01 | -7.6670E-03 | 7.8143E-07 | 1061.00 | 2415.00 |
| 203 | NiC4O4 | NICKEL CARBONYL | 24.7530 | -1.9573E+03 | -7.0802E+00 | 7.4594E-03 | -3.3536E-06 | 250.15 | 315.65 |
| 204 | NiF2 | NICKEL FLUORIDE | 634.6103 | -7.2704E+04 | -2.0233E+02 | 4.4627E-02 | -3.9862E-06 | 1350.15 | 1556.15 |
| 205 | Np | NEPTUNIUM | 88.3172 | -3.1522E+04 | -2.5083E+01 | 4.1165E-03 | -2.8679E-07 | 1617.99 | 2073.99 |
| 206 | O2 | OXYGEN | 20.6695 | -5.2697E+02 | -6.7062E+00 | 1.2926E-02 | -9.8832E-13 | 54.35 | 154.58 |
| 207 | O3 | OZONE | 38.6910 | -1.4144E+03 | -1.2543E+01 | -1.3045E-10 | 2.4393E-05 | 80.15 | 261.00 |
| 208 | Os | OSMIUM | -252.0332 | -6.3868E+03 | 8.1758E+01 | -1.2735E-02 | 7.0592E-07 | 2234.00 | 4880.00 |
| 209 | OsOF5 | OSMIUM OXIDE PENTAFLUORIDE | 18125.3278 | -4.6581E+05 | -7.4493E+03 | 6.8812E+00 | -2.0309E-03 | 304.80 | 330.90 |
| 210 | OsO4 | OSMIUM TETROXIDE - YELLOW | -6976.1921 | 1.3882E+05 | 3.0478E+03 | -4.0659E+00 | 2.0494E-03 | 276.35 | 403.15 |
| 211 | OsO4 | OSMIUM TETROXIDE - WHITE | -1362.6815 | 2.3640E+04 | 6.0656E+02 | -8.6236E-01 | 4.5669E-04 | 267.55 | 403.15 |
| 212 | P | PHOSPHORUS - WHITE | 37.5747 | -4.5200E+03 | -9.8304E+00 | -1.4385E-06 | 1.4357E-06 | 404.15 | 590.15 |
| 213 | PBr3 | PHOSPHORUS TRIBROMIDE | 75.1763 | -4.1298E+03 | -2.7120E+01 | 2.3495E-02 | -8.5043E-06 | 280.95 | 448.45 |
| 214 | PCl2F3 | PHOSPHORUS DICHLORIDE TRIFLUORIDE | 41.7838 | -2.0432E+03 | -1.4841E+01 | 2.0345E-02 | -1.1817E-05 | 193.35 | 250.35 |
| 215 | PCl3 | PHOSPHORUS TRICHLORIDE | 56.7046 | -3.2295E+03 | -1.8915E+01 | 1.0097E-02 | -7.5546E-13 | 181.15 | 374.15 |
| 216 | PCl5 | PHOSPHORUS PENTACHLORIDE | 30.0396 | -2.6579E+03 | -7.4964E+00 | -2.9515E-02 | 1.0957E-05 | 433.15 | 465.00 |
| 217 | PH3 | PHOSPHINE | 17.6034 | -1.0512E+03 | -4.0706E+00 | -1.5186E-09 | 5.1920E-06 | 139.37 | 324.75 |
| 218 | PH4Br | PHOSPHONIUM BROMIDE | 79.2715 | -4.0989E+03 | -2.8848E+01 | 3.2807E-02 | -1.5741E-05 | 229.45 | 311.45 |
| 219 | PH4Cl | PHOSPHONIUM CHLORIDE | 255.9276 | -6.5388E+03 | -1.0929E+02 | 1.6761E-01 | -1.0572E-04 | 182.15 | 246.15 |
| 220 | PH4I | PHOSPHONIUM IODIDE | 90.3833 | -4.6958E+03 | -3.3063E+01 | 3.5026E-02 | -1.5625E-05 | 247.95 | 335.45 |
| 221 | POCl3 | PHOSPHORUS OXYCHLORIDE | 89.5904 | -4.4038E+03 | -3.1847E+01 | 1.8570E-02 | 1.2780E-08 | 274.33 | 378.65 |
| 222 | PSBr3 | PHOSPHORUS THIOBROMIDE | 100.8066 | -6.1441E+03 | -3.5921E+01 | 2.9018E-02 | -9.8408E-06 | 323.15 | 448.15 |
| 223 | PSCl3 | PHOSPHORUS THIOCHLORIDE | 30.9267 | -2.9337E+03 | -8.3323E+00 | 2.4967E-03 | -3.2735E-11 | 236.95 | 398.15 |
| 224 | P4O6 | PHOSPHORUS TRIOXIDE | -9.1727 | -1.7181E+03 | 6.7269E+00 | -5.0441E-03 | 1.6463E-06 | 312.85 | 446.25 |
| 225 | P4O10 | PHOSPHORUS PENTOXIDE | -55.9316 | -2.8529E+03 | 2.7900E+01 | -2.9138E-02 | 9.4669E-06 | 693.15 | 758.15 |
| 226 | P4S10 | PHOSPHORUS PENTASULFIDE | 18.3195 | -5.1772E+03 | -3.3346E+00 | 1.0110E-03 | -6.7993E-14 | 561.15 | 1291.00 |
| 227 | Pb | LEAD | -17.6204 | -8.5777E+03 | 9.2106E+00 | -3.9318E-03 | 5.4789E-07 | 708.00 | 2024.00 |
| 228 | PbBr2 | LEAD BROMIDE | -392.9133 | 9.8716E+03 | 1.4928E+02 | -7.7340E-02 | 1.4423E-05 | 786.15 | 1187.15 |
| 229 | PbCl2 | LEAD CHLORIDE | 178.2156 | -1.8711E+04 | -5.7943E+01 | 1.8346E-02 | -2.4020E-06 | 820.15 | 1227.15 |
| 230 | PbF2 | LEAD FLUORIDE | 797.7308 | -6.5142E+04 | -2.6816E+02 | 8.0837E-02 | -9.4270E-06 | 1134.15 | 1566.15 |
| 231 | PbI2 | LEAD IODIDE | 1790.0196 | -9.5547E+04 | -6.4509E+02 | 2.9522E-01 | -5.2262E-05 | 752.15 | 1145.15 |
| 232 | PbO | LEAD OXIDE | -357.2600 | 1.4278E+04 | 1.2467E+02 | -3.7266E-02 | 4.2118E-06 | 1216.15 | 1745.15 |
| 233 | PbS | LEAD SULFIDE | -2243.0170 | 1.5079E+05 | 7.6189E+02 | -2.1739E-01 | 2.2835E-05 | 1125.15 | 1554.15 |
| 234 | Pd | PALLADIUM | 90.8138 | -2.9630E+04 | -2.5363E+01 | 3.7450E-03 | -2.0394E-07 | 1336.00 | 3385.00 |
| 235 | Po | POLONIUM | 220.5827 | -1.5775E+04 | -7.8208E+01 | 3.8994E-02 | -7.4047E-06 | 448.00 | 1235.00 |
| 236 | Pt | PLATINUM | 87.6383 | -4.0548E+04 | -2.2701E+01 | 1.8600E-03 | -1.5887E-08 | 1744.00 | 3980.00 |
| 237 | Ra | RADIUM | 31.4507 | -1.0221E+04 | -7.3958E+00 | 4.1178E-04 | 1.3090E-07 | 593.00 | 1809.00 |
| 238 | Rb | RUBIDIUM | 13.7111 | -4.4634E+03 | -2.1307E+00 | -7.2787E-05 | 1.8422E-07 | 310.00 | 978.00 |
| 239 | RbBr | RUBIDIUM BROMIDE | 78.4584 | -1.4856E+04 | -2.2965E+01 | 5.3426E-03 | -5.2198E-07 | 1054.15 | 1625.15 |
| 240 | RbCl | RUBIDIUM CHLORIDE | 95.4464 | -1.6358E+04 | -2.8542E+01 | 6.6137E-03 | -6.4090E-07 | 1065.15 | 1654.15 |
| 241 | RbF | RUBIDIUM FLUORIDE | 4858.3696 | -3.9369E+05 | -1.6193E+03 | 4.3432E-01 | -4.5433E-05 | 1194.15 | 1681.15 |
| 242 | RbI | RUBIDIUM IODIDE | 87.3359 | -1.5106E+04 | -2.6017E+01 | 6.2775E-03 | -6.3467E-07 | 1021.15 | 1577.15 |
| 243 | Re | RHENIUM | -31.5392 | -3.2254E+04 | 1.2215E+01 | -1.2695E-03 | 3.7363E-08 | 2480.00 | 5915.00 |
| 244 | Re2O7 | RHENIUM HEPTOXIDE | -50918.9674 | 1.6274E+06 | 2.0103E+04 | -1.5444E+01 | 4.5207E-03 | 485.65 | 635.55 |
| 245 | Rh | RHODIUM | -83.3270 | -2.3619E+04 | 3.2126E+01 | -8.8651E-03 | 7.4827E-07 | 1735.00 | 3940.00 |
| 246 | Rn | RADON | 168.4046 | -3.3600E+03 | -7.5618E+01 | 1.4923E-01 | -1.2016E-04 | 128.95 | 211.75 |
| 247 | Ru | RUTHENIUM | 3.1324 | -3.5567E+04 | 3.0770E+00 | -1.1098E-03 | 6.9449E-08 | 2051.00 | 4500.00 |
| 248 | RuF5 | RUTHENIUM PENTAFLUORIDE | 253.0895 | -1.0396E+04 | -9.8031E+01 | 8.5375E-02 | -3.0601E-05 | 322.75 | 429.95 |
| 249 | S | SULFUR | 86.7925 | -7.8894E+03 | -2.7433E+01 | 7.5706E-03 | 5.7656E-15 | 388.36 | 1313.00 |
| 250 | SF4 | SULFUR TETRAFLUORIDE | 340.1740 | -6.8405E+03 | -1.5222E+02 | 2.6997E-01 | -1.9417E-04 | 160.85 | 223.95 |
| 251 | SF6 | SULFUR HEXAFLUORIDE | 10.5389 | -1.0352E+03 | -1.1341E+00 | -1.8565E-07 | 1.1504E-10 | 223.15 | 318.69 |
| 252 | SOBr2 | THIONYL BROMIDE | 1.5135 | -1.9716E+03 | 2.6530E+00 | -2.2814E-03 | 8.4571E-07 | 266.45 | 412.65 |
| 253 | SOCl2 | THIONYL CHLORIDE | 66.4546 | -3.3385E+03 | -2.3318E+01 | 1.5153E-02 | -8.4247E-12 | 172.00 | 372.15 |

* A computer program, containing coefficients for vapor pressure for all compounds, is available for a nominal fee (Carl L. Yaws, Box 10053, Lamar University, Beaumont, TX 77710, phone/FAX 409-880-8787). The computer program is in ASCII which can be accessed by other software.

$$\log_{10} P = A + B/T + C \log_{10} T + D T + E T^2 \qquad (P - mm\ Hg,\ T - K)$$

| NO | FORMULA | NAME | A | B | C | D | E | TMIN | TMAX |
|----|---------|------|---|---|---|---|---|------|------|
| 254 | SOF2 | SULFUROUS OXYFLUORIDE | 238.2631 | -5.2181E+03 | -1.0431E+02 | 1.7358E-01 | -1.1781E-04 | 174.42 | 229.05 |
| 255 | SO2 | SULFUR DIOXIDE | 19.7418 | -1.8132E+03 | -4.1458E+00 | -4.4284E-09 | 8.4918E-07 | 197.67 | 430.75 |
| 256 | SO2Cl2 | SULFURYL CHLORIDE | 5.9028 | -2.0407E+03 | 2.2796E+00 | -1.0358E-02 | 6.0161E-06 | 222.00 | 545.00 |
| 257 | SO3 | SULFUR TRIOXIDE | 114.0529 | -6.4619E+03 | -3.6784E+01 | -1.7530E-07 | 1.1919E-05 | 289.95 | 490.85 |
| 258 | S2Cl2 | SULFUR MONOCHLORIDE | 162.6274 | -6.2944E+03 | -6.3327E+01 | 6.1401E-02 | -2.4506E-05 | 265.75 | 411.15 |
| 259 | Sb | ANTIMONY | -14.9322 | -1.3754E+04 | 1.5738E+01 | -2.3835E-02 | 5.1947E-06 | 617.00 | 1898.00 |
| 260 | SbBr3 | ANTIMONY TRIBROMIDE | -68.4797 | -6.2413E+02 | 3.0034E+01 | -2.0378E-02 | 4.6884E-06 | 367.05 | 548.15 |
| 261 | SbCl3 | ANTIMONY TRICHLORIDE | 68.0484 | -5.1047E+03 | -2.1724E+01 | 7.4742E-03 | 5.7270E-13 | 346.55 | 794.00 |
| 262 | SbCl5 | ANTIMONY PENTACHLORIDE | 73.6650 | -4.3530E+03 | -2.6389E+01 | 2.3832E-02 | -9.0652E-06 | 295.85 | 387.25 |
| 263 | SbH3 | STIBINE | 65.7241 | -2.2895E+03 | -2.5586E+01 | 3.4766E-02 | -1.7809E-05 | 177.87 | 440.35 |
| 264 | SbI3 | ANTIMONY TRIIODIDE | -1834.2309 | 5.3081E+04 | 7.3328E+02 | -5.9001E-01 | 1.8014E-04 | 436.75 | 674.15 |
| 265 | Sb2O3 | ANTIMONY TRIOXIDE | 3557.6905 | -2.2934E+05 | -1.2298E+03 | 4.1528E-01 | -5.3009E-05 | 847.15 | 1698.15 |
| 266 | Sc | SCANDIUM | -410.3222 | 1.5184E+04 | 1.4020E+02 | -3.6544E-02 | 3.4523E-06 | 1110.00 | 2700.00 |
| 267 | Se | SELENIUM | 994.5705 | -4.3994E+04 | -3.7357E+02 | 2.2452E-01 | -5.1145E-05 | 397.00 | 930.00 |
| 268 | SeCl4 | SELENIUM TETRACHLORIDE | 62.7934 | -5.6601E+03 | -2.0127E+01 | 1.5021E-02 | -4.7573E-06 | 347.15 | 464.65 |
| 269 | SeF6 | SELENIUM HEXAFLUORIDE | -13.7376 | -9.6701E+02 | 1.0083E+01 | -1.4935E-02 | 9.6551E-06 | 154.55 | 227.35 |
| 270 | SeOCl2 | SELENIUM OXYCHLORIDE | -31.1449 | -1.5960E+03 | 1.5901E+01 | -1.1690E-02 | 3.7904E-06 | 307.95 | 441.15 |
| 271 | SeO2 | SELENIUM DIOXIDE | -26.7091 | -2.9884E+03 | 1.3896E+01 | -7.6193E-03 | 1.8305E-06 | 430.15 | 590.15 |
| 272 | Si | SILICON | 315.0687 | -7.1384E+04 | -8.9680E+01 | 8.3445E-03 | -2.5806E-09 | 1997.10 | 2560.10 |
| 273 | SiBrCl2F | BROMODICHLOROFLUOROSILANE | 221.2325 | -5.5951E+03 | -9.3501E+01 | 1.2431E-01 | -6.0950E-05 | 186.65 | 474.30 |
| 274 | SiBrF3 | TRIFLUOROBROMOSILANE | -316.1829 | 5.0196E+03 | 1.4386E+02 | -2.1246E-01 | 1.1958E-04 | 203.35 | 354.90 |
| 275 | SiBr2ClF | DIBROMOCHLOROFLUOROSILANE | 114.7880 | -3.9867E+03 | -4.5478E+01 | 5.1723E-02 | -2.2118E-05 | 207.95 | 515.92 |
| 276 | SiClF3 | TRIFLUOROCHLOROSILANE | 102.6712 | -2.5416E+03 | -4.3347E+01 | 7.1921E-02 | -4.5231E-05 | 129.15 | 308.83 |
| 277 | SiCl2F2 | DICHLORODIFLUOROSILANE | 65.5754 | -2.0450E+03 | -2.6643E+01 | 4.5134E-02 | -2.8299E-05 | 148.45 | 367.35 |
| 278 | SiCl3F | TRICHLOROFLUOROSILANE | 106.0275 | -3.4770E+03 | -4.2166E+01 | 5.0626E-02 | -2.3053E-05 | 180.55 | 434.85 |
| 279 | SiCl4 | SILICON TETRACHLORIDE | 28.4503 | -2.3911E+03 | -7.3965E+00 | -9.3193E-10 | 2.7569E-06 | 204.30 | 507.00 |
| 280 | SiF4 | SILICON TETRAFLUORIDE | 250.9551 | -6.9843E+03 | -9.4613E+01 | -5.3053E-05 | 1.3121E-04 | 186.35 | 259.00 |
| 281 | SiHBr3 | TRIBROMOSILANE | 29.8264 | -2.6956E+03 | -8.0772E+00 | 2.0334E-03 | 1.0063E-06 | 242.65 | 617.50 |
| 282 | SiHCl3 | TRICHLOROSILANE | 8.8008 | -1.6896E+03 | 5.0043E-01 | -6.9024E-03 | 5.1631E-06 | 144.95 | 479.00 |
| 283 | SiHF3 | TRIFLUOROSILANE | -122.7759 | 5.7853E+01 | 6.9914E+01 | -2.1980E-01 | 2.2343E-04 | 121.15 | 276.65 |
| 284 | SiH2Br2 | DIBROMOSILANE | 63.8405 | -2.9308E+03 | -2.3568E+01 | 2.4715E-02 | -9.7683E-06 | 212.25 | 559.24 |
| 285 | SiH2Cl2 | DICHLOROSILANE | 30.1827 | -2.0844E+03 | -8.2717E+00 | -3.7469E-10 | 4.5636E-06 | 151.15 | 449.00 |
| 286 | SiH2F2 | DIFLUOROSILANE | 23.8427 | -1.5675E+03 | -4.4921E+00 | -1.9742E-02 | 3.1144E-05 | 126.45 | 310.75 |
| 287 | SiH2I2 | DIIODOSILANE | -132.4973 | 2.1632E+03 | 5.6398E+01 | -4.9010E-02 | 1.6020E-05 | 276.95 | 683.81 |
| 288 | SiH3Br | MONOBROMOSILANE | 48.9247 | -2.1416E+03 | -1.7831E+01 | 2.1780E-02 | -1.0087E-05 | 187.45 | 455.15 |
| 289 | SiH3Cl | MONOCHLOROSILANE | 97.0716 | -3.0578E+03 | -3.8232E+01 | 4.4033E-02 | -1.8687E-05 | 155.35 | 396.79 |
| 290 | SiH3F | MONOFLUOROSILANE | 262.3858 | -5.0922E+03 | -1.1579E+02 | 1.8678E-01 | -1.1081E-04 | 120.15 | 285.87 |
| 291 | SiH3I | IODOSILANE | 81.5431 | -3.2034E+03 | -3.0771E+01 | 3.1175E-02 | -1.1973E-05 | 220.15 | 524.59 |
| 292 | SiH4 | SILANE | 49.8037 | -1.3946E+03 | -1.8981E+01 | 2.2497E-02 | 3.0530E-13 | 88.48 | 269.70 |
| 293 | SiO2 | SILICON DIOXIDE | -378.5210 | 6.5473E+03 | 1.3150E+02 | -3.5774E-02 | 3.4220E-06 | 1883.00 | 2503.20 |
| 294 | Si2Cl6 | HEXACHLORODISILANE | 28.0093 | -2.9842E+03 | -7.7427E+00 | 6.7458E-03 | -2.5044E-06 | 277.15 | 412.15 |
| 295 | Si2F6 | HEXAFLUORODISILANE | -3.5918 | -1.9401E+03 | 6.6355E+00 | -8.5356E-03 | 4.7603E-06 | 192.15 | 254.25 |
| 296 | Si2H5Cl | DISILANYL CHLORIDE | 50.2854 | -2.4890E+03 | -1.8081E+01 | 2.1374E-02 | -1.0671E-05 | 226.95 | 291.15 |
| 297 | Si2H6 | DISILANE | 19.4083 | -1.5000E+03 | -4.5432E+00 | -2.0804E-10 | 3.3390E-06 | 143.85 | 432.00 |
| 298 | Si2OCl3F3 | TRICHLOROTRIFLUORODISILOXANE | -39.7219 | -2.7322E+02 | 1.9305E+01 | -1.7428E-02 | 7.2191E-06 | 235.15 | 316.35 |
| 299 | Si2OCl6 | HEXACHLORODISILOXANE | 173.0145 | -6.6562E+03 | -6.7483E+01 | 6.5445E-02 | -2.6121E-05 | 268.15 | 408.75 |
| 300 | Si2OH6 | DISILOXANE | 23.4097 | -1.5324E+03 | -6.9541E+00 | 1.0056E-02 | -6.1820E-06 | 160.65 | 257.75 |
| 301 | Si3Cl8 | OCTACHLOROTRISILANE | 0.9548 | -2.4388E+03 | 8.9120E+00 | -2.1105E-03 | 6.5985E-07 | 319.45 | 484.55 |
| 302 | Si3H8 | TRISILANE | 61.7030 | -2.7796E+03 | -2.2974E+01 | 2.7178E-02 | -1.3469E-05 | 204.25 | 326.25 |
| 303 | Si3H9N | TRISILAZANE | 77.8169 | -3.1701E+03 | -2.9814E+01 | 3.5861E-02 | -1.8007E-05 | 204.45 | 321.85 |
| 304 | Si4H10 | TETRASILANE | 72.5811 | -3.7126E+03 | -2.6507E+01 | 2.6806E-02 | -1.1367E-05 | 245.45 | 373.15 |
| 305 | Sm | SAMARIUM | -63.3814 | -6.9759E+03 | 2.5576E+01 | -9.5945E-03 | 1.2128E-06 | 733.00 | 1874.00 |
| 306 | Sn | TIN | -11.8452 | -1.3744E+04 | 6.4004E+00 | -9.7861E-04 | -4.2795E-10 | 1096.00 | 2995.00 |
| 307 | SnBr4 | STANNIC BROMIDE | 62.0925 | -4.1212E+03 | -2.1351E+01 | 1.6385E-02 | -5.2861E-06 | 331.45 | 477.85 |
| 308 | SnCl2 | STANNOUS CHLORIDE | 430.9683 | -2.2551E+04 | -1.5874E+02 | 9.2347E-02 | -2.1208E-05 | 589.15 | 896.15 |
| 309 | SnCl4 | STANNIC CHLORIDE | 146.4762 | -5.5847E+03 | -5.7244E+01 | 5.8634E-02 | -2.4782E-05 | 250.45 | 386.15 |
| 310 | SnH4 | STANNIC HYDRIDE | 7.8482 | -9.7474E+02 | -2.7435E-01 | 4.6026E-04 | -3.3081E-07 | 133.15 | 220.85 |
| 311 | SnI4 | STANNIC IODIDE | -5.9850 | -2.4424E+03 | 5.1095E+00 | -2.7912E-03 | 6.6165E-07 | 429.15 | 621.15 |
| 312 | Sr | STRONTIUM | -111.2297 | -2.8962E+03 | 4.3785E+01 | -2.0560E-02 | 3.2941E-06 | 582.00 | 1630.00 |
| 313 | SrO | STRONTIUM OXIDE | 2848.9447 | -4.0303E+05 | -8.7252E+02 | 1.3051E-01 | -7.7456E-06 | 2341.15 | 2683.15 |
| 314 | Ta | TANTALUM | 90.4608 | -5.3107E+04 | -2.4076E+01 | 2.9553E-03 | -1.3908E-07 | 2511.00 | 5565.00 |
| 315 | Tc | TECHNETIUM | -240.5191 | -5.7928E+03 | 7.8794E+01 | -1.3112E-02 | 7.4635E-07 | 1660.00 | 5000.00 |
| 316 | Te | TELLURIUM | -130.9948 | -5.2810E+03 | 5.7873E+01 | -1.4686E-02 | 1.0781E-05 | 497.00 | 1285.00 |
| 317 | TeCl4 | TELLURIUM TETRACHLORIDE | 225.5681 | -1.3194E+04 | -8.0899E+01 | 4.5316E-02 | -1.0441E-05 | 506.15 | 665.15 |
| 318 | TeF6 | TELLURIUM HEXAFLUORIDE | 103.1152 | -3.1620E+03 | -4.2245E+01 | 6.7520E-02 | -4.5015E-05 | 161.85 | 234.55 |
| 319 | Ti | TITANIUM | -194.8742 | -8.2733E+03 | 6.8261E+01 | -1.7329E-02 | 1.5517E-06 | 1508.00 | 3442.00 |

* A computer program, containing coefficients for vapor pressure for all compounds, is available for a nominal fee (Carl L. Yaws, Box 10053, Lamar University, Beaumont, TX 77710, phone/FAX 409-880-8787). The computer program is in ASCII which can be accessed by other software.

$$\log_{10} P = A + B/T + C \log_{10} T + D\,T + E\,T^2 \qquad (P - \text{mm Hg}, \ T - K)$$

| NO | FORMULA | NAME | A | B | C | D | E | TMIN | TMAX |
|----|---------|------|-----|-----|-----|-----|-----|------|------|
| 320 | TiCl4 | TITANIUM TETRACHLORIDE | 65.9073 | -4.0187E+03 | -2.2002E+01 | 1.0422E-02 | -8.7715E-13 | 249.05 | 638.00 |
| 321 | Tl | THALLIUM | 149.2845 | -1.7175E+04 | -4.9494E+01 | 1.8084E-02 | -2.5186E-06 | 636.00 | 1745.00 |
| 322 | TlBr | THALLOUS BROMIDE | 65.7795 | -9.3154E+03 | -1.9913E+01 | 6.6371E-03 | -9.3147E-07 | 763.15 | 1092.15 |
| 323 | TlI | THALLOUS IODIDE | 107.4718 | -1.2047E+04 | -3.4449E+01 | 1.2081E-02 | -1.7669E-06 | 713.15 | 1096.15 |
| 324 | Tm | THULIUM | 581.9203 | -4.0634E+04 | -2.0847E+02 | 9.8996E-02 | -1.8204E-05 | 661.00 | 1237.00 |
| 325 | U | URANIUM | 5.1916 | -2.3655E+04 | 1.4051E+00 | -6.7084E-04 | 6.4472E-08 | 1600.00 | 4135.00 |
| 326 | UF6 | URANIUM FLUORIDE | 141.7284 | -5.4522E+03 | -5.5519E+01 | 6.2879E-02 | -2.9623E-05 | 234.35 | 328.85 |
| 327 | V | VANADIUM | 52.0677 | -3.1989E+04 | -1.2620E+01 | 1.6179E-03 | -1.0505E-07 | 1604.00 | 3665.00 |
| 328 | VCl4 | VANADIUM TETRACHLORIDE | 12.7215 | -2.2338E+03 | -1.7693E+00 | -9.8525E-11 | 3.4861E-07 | 247.45 | 697.00 |
| 329 | VOCl3 | VANADIUM OXYTRICHLORIDE | 31.8135 | -2.8585E+03 | -8.5879E+00 | 3.9375E-09 | 3.4924E-06 | 193.65 | 400.00 |
| 330 | W | TUNGSTEN | -19.5111 | -3.8683E+04 | 8.6887E+00 | -7.1355E-04 | 2.1195E-08 | 2667.00 | 5645.00 |
| 331 | WF6 | TUNGSTEN FLUORIDE | 354.8192 | -8.8753E+03 | -1.5107E+02 | 2.0875E-01 | -1.1715E-04 | 201.55 | 290.45 |
| 332 | Xe | XENON | 15.6530 | -8.1035E+02 | -3.9013E+00 | 4.7985E-03 | -1.7020E-11 | 161.36 | 289.74 |
| 333 | Yb | YTTERBIUM | -61.8092 | -5.3349E+03 | 2.5566E+01 | -1.1528E-02 | 1.7104E-06 | 599.00 | 1660.00 |
| 334 | Yt | YTTRIUM | -198.7054 | -2.2991E+03 | 6.8766E+01 | -1.6442E-02 | 1.3848E-06 | 1246.00 | 3055.00 |
| 335 | Zn | ZINC | -20.3143 | -4.6362E+03 | 1.0073E+01 | -3.8085E-03 | 4.8860E-07 | 692.70 | 3170.00 |
| 336 | ZnCl2 | ZINC CHLORIDE | 256.5500 | -1.7845E+04 | -9.1583E+01 | 4.9680E-02 | -1.0787E-05 | 701.15 | 1005.15 |
| 337 | ZnF2 | ZINC FLUORIDE | -17799.4171 | 1.2561E+06 | 6.0444E+03 | -1.7915E+00 | 2.0155E-04 | 1243.15 | 1770.15 |
| 338 | ZnO | ZINC OXIDE | 10.0724 | -1.5790E+04 | -3.0065E-02 | 6.5659E-06 | -5.4927E-10 | 1773.10 | 2223.10 |
| 339 | ZnSO4 | ZINC SULFATE | 8.7415 | -2.2158E+03 | -5.1178E-04 | 6.6577E-07 | -3.3159E-10 | 293.15 | 378.15 |
| 340 | Zr | ZIRCONIUM | 95.1134 | -4.3264E+04 | -2.5683E+01 | 2.9830E-03 | -1.1696E-07 | 1975.00 | 4598.00 |
| 341 | ZrBr4 | ZIRCONIUM BROMIDE | 35.8920 | -6.9028E+03 | -8.7767E+00 | 4.6821E-03 | -1.0688E-06 | 480.15 | 630.15 |
| 342 | ZrCl4 | ZIRCONIUM CHLORIDE | 80.7853 | -8.5643E+03 | -2.5618E+01 | 1.4577E-02 | -3.5146E-06 | 463.15 | 604.15 |
| 343 | ZrI4 | ZIRCONIUM IODIDE | 95.3927 | -1.0449E+04 | -3.0444E+01 | 1.5025E-02 | -3.1340E-06 | 537.15 | 704.15 |

* A computer program, containing coefficients for vapor pressure for all compounds, is available for a nominal fee (Carl L. Yaws, Box 10053, Lamar University, Beaumont, TX 77710, phone/FAX 409-880-8787). The computer program is in ASCII which can be accessed by other software.

# Appendix D

# CRITICAL PROPERTIES AND ACENTRIC FACTOR*

Carl L. Yaws, Sachin Nijhawan and Li Bu
Lamar University, Beaumont, Texas

| NO | FORMULA | NAME | MW g/mol | $T_F$ K | $T_B$ K | $T_C$ K | $P_C$ bar | $V_C$ cm$^3$/mol | $\rho_C$ g/cm$^3$ | $Z_C$ | $\omega$ | SOURCE |
|----|---------|------|------|------|------|------|------|------|------|------|------|------|
| 1 | Ag | SILVER | 107.868 | 1234.00 | 2485.00 | 7480.00 | 5066.0 | 58.20 | 1.8534 | 0.474 | 0.150 | 1,6 |
| 2 | AgCl | SILVER CHLORIDE | 143.321 | 728.15 | 1837.15 | --- | --- | --- | --- | --- | --- | 2 |
| 3 | AgI | SILVER IODIDE | 234.773 | 825.15 | 1779.15 | --- | --- | --- | --- | --- | --- | 2 |
| 4 | Al | ALUMINUM | 26.982 | 933.00 | 2329.15 | 7151.00 | 5458.0 | 39.00 | 0.6918 | 0.358 | --- | 1 |
| 5 | AlB3H12 | ALUMINUM BOROHYDRIDE | 71.510 | 209.15 | 319.05 | --- | --- | --- | --- | --- | --- | 2 |
| 6 | AlBr3 | ALUMINUN BROMIDE | 266.694 | 390.15 | 529.45 | 763.00 | --- | 124.20 | 2.1473 | --- | --- | 3,6 |
| 7 | AlCl3 | ALUMINUM CHLORIDE | 133.340 | 465.70 | 453.15 | 629.00 | 26.35 | 261.45 | 0.5100 | 0.132 | --- | 2 |
| 8 | AlF3 | ALUMINUM FLUORIDE | 83.977 | 1313.15 | 1810.15 | --- | --- | --- | --- | --- | --- | 2 |
| 9 | AlI3 | ALUMINUM IODIDE | 407.695 | 464.15 | 658.65 | --- | --- | --- | --- | --- | --- | 2,10 |
| 10 | Al2O3 | ALUMINUM OXIDE | 101.961 | 2325.00 | 3253.15 | 5335.00 | --- | --- | --- | --- | --- | 1 |
| 11 | Al2S3O12 | ALUMINUM SULFATE | 342.154 | 1043.20 | --- | --- | --- | --- | --- | --- | --- | 1 |
| 12 | Ar | ARGON | 39.948 | 83.80 | 87.28 | 150.86 | 48.98 | 74.59 | 0.5356 | 0.291 | 0.000 | 1 |
| 13 | As | ARSENIC | 74.922 | 1090.15 | 885.00 | 1673.15 | --- | --- | --- | --- | --- | 4,5 |
| 14 | AsBr3 | ARSENIC TRIBROMIDE | 314.634 | 306.15 | 493.15 | --- | --- | --- | --- | --- | --- | 2,10 |
| 15 | AsCl3 | ARSENIC TRICHLORIDE | 181.280 | 255.15 | 403.55 | --- | --- | --- | --- | --- | --- | 2 |
| 16 | AsF3 | ARSENIC TRIFLUORIDE | 131.917 | 267.25 | 329.45 | --- | --- | --- | --- | --- | --- | 2 |
| 17 | AsF5 | ARSENIC PENTAFLUORIDE | 169.914 | 193.35 | 220.35 | --- | --- | --- | --- | --- | --- | 2 |
| 18 | AsH3 | ARSINE | 77.945 | 156.28 | 210.67 | 373.00 | 65.50 | 97.80 | 0.7970 | 0.207 | 0.006 | 1 |
| 19 | AsI3 | ARSENIC TRIIODIDE | 455.635 | 419.15 | 676.15 | --- | --- | --- | --- | --- | --- | 2,5 |
| 20 | As2O3 | ARSENIC TRIOXIDE | 197.841 | 585.95 | 730.35 | --- | --- | --- | --- | --- | --- | 3 |
| 21 | At | ASTATINE | 210.000 | 575.15 | 607.00 | --- | --- | --- | --- | --- | --- | 4,5 |
| 22 | Au | GOLD | 196.967 | 1337.33 | 3120.00 | 4398.00 | --- | 50.30 | 3.9158 | --- | --- | 4,5,6 |
| 23 | B | BORON | 10.811 | 2348.15 | 4133.00 | --- | --- | --- | --- | --- | --- | 4,5 |
| 24 | BBr3 | BORON TRIBROMIDE | 250.523 | 228.15 | 364.85 | --- | --- | --- | --- | --- | --- | 2 |
| 25 | BCl3 | BORON TRICHLORIDE | 117.169 | 166.15 | 285.65 | 451.95 | 38.71 | 148.34 | 0.7899 | 0.153 | 0.151 | 1 |
| 26 | BF3 | BORON TRIFLUORIDE | 67.806 | 146.05 | 173.35 | 260.90 | 49.85 | 123.61 | 0.5485 | 0.284 | 0.430 | 1 |
| 27 | BH2CO | BORINE CARBONYL | 40.837 | 136.15 | 209.15 | --- | --- | --- | --- | --- | --- | 3 |
| 28 | BH3O3 | BORIC ACID | 61.833 | 458.15 | --- | --- | --- | --- | --- | --- | --- | 1 |
| 29 | B2D6 | DEUTERODIBORANE | 33.718 | --- | 179.87 | --- | --- | --- | --- | --- | --- | 2 |
| 30 | B2H5Br | DIBORANE HYDROBROMIDE | 106.566 | 168.95 | 289.45 | --- | --- | --- | --- | --- | --- | 3 |
| 31 | B2H6 | DIBORANE | 27.670 | 107.65 | 180.65 | 289.80 | 40.53 | 173.10 | 0.1598 | 0.291 | 0.125 | 1 |
| 32 | B3N3H6 | BORINE TRIAMINE | 80.501 | 214.95 | 323.75 | --- | --- | --- | --- | --- | --- | 2 |
| 33 | B4H10 | TETRABORANE | 53.323 | 153.25 | 289.25 | --- | --- | --- | --- | --- | --- | 2 |
| 34 | B5H9 | PENTABORANE | 63.126 | 226.35 | 331.55 | 568.45 | 46.41 | 285.1 | 0.2214 | 0.280 | --- | 2,8 |
| 35 | B5H11 | TETRAHYDROPENTABORANE | 65.142 | --- | 340.15 | --- | --- | --- | --- | --- | --- | 2 |
| 36 | B10H14 | DECABORANE | 122.221 | 372.75 | 486.15 | --- | --- | --- | --- | --- | --- | 2,10 |
| 37 | Ba | BARIUM | 137.327 | 1000.15 | 1907.00 | --- | --- | --- | --- | --- | --- | 2,4,5 |
| 38 | Be | BERYLLIUM | 9.012 | 1560.15 | 2744.00 | --- | --- | --- | --- | --- | --- | 4,5 |
| 39 | BeB2H8 | BERYLLIUM BOROHYDRIDE | 38.698 | 396.15 | 363.15 | --- | --- | --- | --- | --- | --- | 2 |
| 40 | BeBr2 | BERYLLIUM BROMIDE | 168.820 | 763.15 | 747.15 | --- | --- | --- | --- | --- | --- | 3 |
| 41 | BeCl2 | BERYLLIUM CHLORIDE | 79.918 | 678.15 | 760.15 | --- | --- | --- | --- | --- | --- | 3 |
| 42 | BeF2 | BERYLLIUM FLUORIDE | 47.009 | 1073.15 | --- | --- | --- | --- | --- | --- | --- | 2,10 |
| 43 | BeI2 | BERYLLIUM IODIDE | 262.821 | 761.15 | 761.15 | --- | --- | --- | --- | --- | --- | 3 |
| 44 | Bi | BISMUTH | 208.980 | 544.15 | 1698.15 | 4620.00 | --- | 79.40 | 2.6320 | --- | --- | 3,6 |
| 45 | BiBr3 | BISMUTH TRIBROMIDE | 448.692 | 491.15 | 734.15 | 1220.00 | --- | 302.00 | 1.4857 | --- | --- | 2,6 |
| 46 | BiCl3 | BISMUTH TRICHLORIDE | 315.338 | 503.15 | 714.15 | 1178.00 | --- | 261.70 | 1.2050 | --- | --- | 2,6 |
| 47 | BrF5 | BROMINE PENTAFLUORIDE | 174.896 | 211.75 | 313.55 | 470.00 | --- | --- | --- | --- | --- | 2,6 |
| 48 | Br2 | BROMINE | 159.808 | 265.90 | 331.90 | 584.15 | 103.35 | 135.00 | 1.1838 | 0.287 | 0.119 | 1 |
| 49 | C | CARBON | 12.011 | 4247.00 | 4203.00 | 6810.00 | 2230.0 | 18.80 | 0.6389 | 0.074 | 1.566 | 1 |
| 50 | CCl2O | PHOSGENE | 98.916 | 145.37 | 280.71 | 455.00 | 56.74 | 190.22 | 0.5200 | 0.285 | 0.201 | 1 |
| 51 | CF2O | CARBONYL FLUORIDE | 66.007 | 161.89 | 188.58 | 297.00 | 57.60 | 141.00 | 0.4681 | 0.329 | 0.283 | 1 |
| 52 | CH4N2O | UREA | 60.056 | 405.85 | 465.00 | 705.00 | 90.50 | 218.00 | 0.2755 | 0.337 | --- | 1 |
| 53 | CH4N2S | THIOUREA | 76.122 | 454.15 | 536.00 | 854.00 | 82.30 | 248.00 | 0.3069 | 0.287 | 0.359 | 1 |
| 54 | CNBr | CYANOGEN BROMIDE | 105.922 | 331.15 | 334.65 | --- | --- | --- | --- | --- | --- | 1 |
| 55 | CNCl | CYANOGEN CHLORIDE | 61.470 | 266.65 | 286.00 | 449.00 | 59.90 | 163.00 | 0.3771 | 0.262 | 0.320 | 1 |
| 56 | CNF | CYANOGEN FLUORIDE | 45.016 | --- | 227.17 | --- | --- | --- | --- | --- | --- | 2 |
| 57 | CO | CARBON MONOXIDE | 28.010 | 68.15 | 81.70 | 132.92 | 34.99 | 93.10 | 0.3009 | 0.295 | 0.066 | 1 |
| 58 | COS | CARBONYL SULFIDE | 60.076 | 134.35 | 223.00 | 378.80 | 63.49 | 135.10 | 0.4447 | 0.272 | 0.097 | 1 |

* A computer program, containing data for all compounds, is available for a nominal fee (Carl L. Yaws, Box 10053, Lamar University, Beaumont, TX 77710, phone/FAX 409-880-8787). The computer program is in ASCII which can be accessed by other software.

| NO | FORMULA | NAME | MW g/mol | $T_F$ K | $T_B$ K | $T_C$ K | $P_C$ bar | $V_C$ cm³/mol | $\rho_C$ g/cm³ | $Z_C$ | $\omega$ | SOURCE |
|----|---------|------|----------|---------|---------|---------|-----------|---------------|----------------|-------|----------|--------|
| 59 | COSe | CARBON OXYSELENIDE | 106.970 | --- | 251.25 | --- | --- | --- | --- | --- | --- | 3 |
| 60 | CO2 | CARBON DIOXIDE | 44.010 | 216.58 | 194.70 | 304.19 | 73.82 | 94.00 | 0.4682 | 0.274 | 0.228 | 1 |
| 61 | CS2 | CARBON DISULFIDE | 76.143 | 161.58 | 319.37 | 552.00 | 79.03 | 160.00 | 0.4759 | 0.276 | 0.108 | 1 |
| 62 | CSeS | CARBON SELENOSULFIDE | 123.037 | 197.95 | 358.75 | --- | --- | --- | --- | --- | --- | 3 |
| 63 | C2N2 | CYANOGEN | 52.035 | 238.75 | 252.15 | 399.90 | 60.60 | --- | --- | --- | --- | 3,6 |
| 64 | C3S2 | CARBON SUBSULFIDE | 100.165 | 273.55 | --- | --- | --- | --- | --- | --- | --- | 3 |
| 65 | Ca | CALCIUM | 40.078 | 1115.15 | 1762.00 | --- | --- | --- | --- | --- | --- | 4,5 |
| 66 | CaF2 | CALCIUM FLUORIDE | 78.075 | 1691.00 | 2806.50 | --- | --- | --- | --- | --- | --- | 1 |
| 67 | CbF5 | COLUMBIUM FLUORIDE | 187.898 | 348.65 | 498.15 | --- | --- | --- | --- | --- | --- | 3 |
| 68 | Cd | CADMIUM | 112.411 | 594.05 | 1043.15 | 2291.00 | --- | 37.90 | 2.9660 | --- | --- | 3,6 |
| 69 | CdCl2 | CADMIUM CHLORIDE | 183.316 | 841.15 | 1240.15 | --- | --- | --- | --- | --- | --- | 3 |
| 70 | CdF2 | CADMIUM FLUORIDE | 150.408 | 793.15 | 2024.15 | --- | --- | --- | --- | --- | --- | 3 |
| 71 | CdI2 | CADMIUM IODIDE | 366.220 | 658.15 | 1069.15 | --- | --- | --- | --- | --- | --- | 3 |
| 72 | CdO | CADMIUM OXIDE | 128.410 | --- | 1832.15 | --- | --- | --- | --- | --- | --- | 2 |
| 73 | ClF | CHLORINE MONOFLUORIDE | 54.451 | 128.15 | 172.65 | --- | --- | --- | --- | --- | --- | 2 |
| 74 | ClFO3 | PERCHLORYL FLUORIDE | 102.449 | 125.41 | 226.49 | 368.40 | 53.70 | 161.00 | 0.6363 | 0.282 | 0.173 | 1 |
| 75 | ClF3 | CHLORINE TRIFLUORIDE | 92.448 | 190.15 | 284.65 | --- | --- | --- | --- | --- | --- | 2 |
| 76 | ClF5 | CHLORINE PENTAFLUORIDE | 130.445 | --- | 260.05 | 415.90 | 52.60 | 230.40 | 0.5662 | 0.350 | --- | 2,6 |
| 77 | ClHO3S | CHLOROSULFONIC ACID | 116.525 | 193.15 | 427.00 | 700.00 | 85.00 | 195.00 | 0.5976 | 0.285 | 0.301 | 1 |
| 78 | ClHO4 | PERCHLORIC ACID | 100.458 | 171.95 | 385.00 | 631.00 | 38.60 | 168.00 | 0.5980 | 0.124 | 0.050 | 1 |
| 79 | ClO2 | CHLORINE DIOXIDE | 67.452 | 213.55 | 284.05 | 465.00 | 86.13 | --- | --- | --- | 0.356 | 1 |
| 80 | Cl2 | CHLORINE | 70.905 | 172.12 | 239.12 | 417.15 | 77.11 | 123.75 | 0.5730 | 0.275 | 0.069 | 1 |
| 81 | Cl2O | CHLORINE MONOXIDE | 86.905 | 157.15 | 275.35 | --- | --- | --- | --- | --- | --- | 2 |
| 82 | Cl2O7 | CHLORINE HEPTOXIDE | 182.901 | 182.15 | 351.95 | --- | --- | --- | --- | --- | --- | 2 |
| 83 | Co | COBALT | 58.933 | 1768.15 | 2528.00 | --- | --- | --- | --- | --- | --- | 4,5 |
| 84 | CoCl2 | COBALT CHLORIDE | 129.839 | 1008.15 | 1323.15 | --- | --- | --- | --- | --- | --- | 2 |
| 85 | CoNC3O4 | COBALT NITROSYL TRICARBONYL | 172.971 | 262.15 | 353.15 | --- | --- | --- | --- | --- | --- | 3 |
| 86 | Cr | CHROMIUM | 51.996 | 2180.15 | 2840.00 | --- | --- | --- | --- | --- | --- | 2,4,5 |
| 87 | CrC6O6 | CHROMIUM CARBONYL | 220.059 | 423.65 | 424.15 | --- | --- | --- | --- | --- | --- | 3,10 |
| 88 | CrO2Cl2 | CHROMIUM OXYCHLORIDE | 154.900 | 176.65 | 390.25 | --- | --- | --- | --- | --- | --- | 2,10 |
| 89 | Cs | CESIUM | 132.905 | 301.65 | 963.15 | 2048.10 | 116.50 | 316.40 | 0.4201 | 0.216 | --- | 2,6 |
| 90 | CsBr | CESIUM BROMIDE | 212.809 | 909.15 | 1573.15 | --- | --- | --- | --- | --- | --- | 3 |
| 91 | CsCl | CESIUM CHLORIDE | 168.358 | 919.15 | 1573.15 | --- | --- | --- | --- | --- | --- | 3 |
| 92 | CsF | CESIUM FLUORIDE | 151.904 | 956.15 | 1524.15 | --- | --- | --- | --- | --- | --- | 3 |
| 93 | CsI | CESIUM IODIDE | 259.810 | 894.15 | 1553.15 | --- | --- | --- | --- | --- | --- | 3 |
| 94 | Cu | COPPER | 63.546 | 1357.77 | 3150.00 | 5123.00 | --- | 61.00 | 1.0417 | --- | --- | 4,5,6 |
| 95 | CuBr | CUPROUS BROMIDE | 143.450 | 777.15 | 1628.15 | --- | --- | --- | --- | --- | --- | 2 |
| 96 | CuCl | CUPROUS CHLORIDE | 98.999 | 703.00 | 1763.15 | 2435.00 | --- | --- | --- | --- | --- | 1 |
| 97 | CuCl2 | CUPRIC CHLORIDE | 134.451 | 906.15 | 1266.15 | 2010.00 | --- | --- | --- | --- | --- | 1 |
| 98 | CuI | COPPER IODIDE | 190.450 | 878.15 | 1609.15 | --- | --- | --- | --- | --- | --- | 3 |
| 99 | DCN | DEUTERIUM CYANIDE | 28.034 | 261.15 | 299.35 | --- | --- | --- | --- | --- | --- | 2 |
| 100 | D2 | DEUTERIUM | 4.032 | 18.73 | 23.65 | 38.35 | 16.64 | 60.26 | 0.0669 | 0.314 | -0.145 | 1 |
| 101 | D2O | DEUTERIUM OXIDE | 20.031 | 276.95 | 374.55 | 643.89 | 219.41 | 56.30 | 0.3558 | 0.231 | 0.368 | 1 |
| 102 | Eu | EUROPIUM | 151.965 | 1095.15 | 1742.00 | 5150.00 | --- | --- | --- | --- | --- | 4,5,6 |
| 103 | F2 | FLUORINE | 37.997 | 53.53 | 84.95 | 144.31 | 52.15 | 66.20 | 0.5740 | 0.288 | 0.059 | 1 |
| 104 | F2O | FLUORINE OXIDE | 53.996 | 49.25 | 128.55 | 215.10 | 49.50 | 97.60 | 0.5532 | 0.270 | --- | 2,6 |
| 105 | Fe | IRON | 55.847 | 1808.15 | 3000.00 | 9340.00 | 10150 | 28.00 | 1.9945 | 0.366 | -0.298 | 1 |
| 106 | FeC5O5 | IRON PENTACARBONYL | 195.899 | 252.15 | 378.15 | --- | --- | --- | --- | --- | --- | 3 |
| 107 | FeCl2 | FERROUS CHLORIDE | 126.752 | 945.15 | 1299.15 | --- | --- | --- | --- | --- | --- | 3,10 |
| 108 | FeCl3 | FERRIC CHLORIDE | 162.205 | 577.15 | 592.15 | --- | --- | --- | --- | --- | --- | 3 |
| 109 | Fr | FRANCIUM | 223.000 | 300.15 | 879.00 | --- | --- | --- | --- | --- | --- | 4,5 |
| 110 | Ga | GALLIUM | 69.723 | 302.91 | 2517.00 | 7620.00 | --- | 75.30 | 0.9259 | --- | --- | 4,5,6 |
| 111 | GaCl3 | GALLIUM TRICHLORIDE | 176.081 | 350.90 | 474.15 | 694.00 | 38.20 | 263.00 | 0.6695 | 0.174 | 0.458 | 1 |
| 112 | Gd | GADOLINIUM | 157.250 | 1587.15 | 1770.00 | --- | --- | --- | --- | --- | --- | 4,5 |
| 113 | Ge | GERMANIUM | 72.610 | 1211.40 | 3125.00 | 8400.00 | --- | --- | --- | --- | --- | 4,5,6 |
| 114 | GeBr4 | GERMANIUM BROMIDE | 392.226 | 299.25 | 462.15 | --- | --- | --- | --- | --- | --- | 2 |
| 115 | GeCl4 | GERMANIUM CHLORIDE | 214.421 | 223.65 | 357.15 | --- | --- | --- | --- | --- | --- | 2 |
| 116 | GeHCl3 | TRICHLORO GERMANE | 179.976 | 202.05 | 348.15 | --- | --- | --- | --- | --- | --- | 2 |
| 117 | GeH4 | GERMANE | 76.642 | 107.26 | 185.00 | 308.00 | 55.50 | 140.00 | 0.5474 | 0.303 | 0.151 | 1 |
| 118 | Ge2H6 | DIGERMANE | 151.268 | 164.15 | 304.65 | --- | --- | --- | --- | --- | --- | 2 |
| 119 | Ge3H8 | TRIGERMANE | 225.894 | 167.55 | 383.95 | --- | --- | --- | --- | --- | --- | 2 |
| 120 | HBr | HYDROGEN BROMIDE | 80.912 | 186.34 | 206.45 | 363.15 | 85.52 | 100.26 | 0.8070 | 0.284 | 0.069 | 1 |
| 121 | HCN | HYDROGEN CYANIDE | 27.026 | 259.91 | 298.85 | 456.65 | 53.91 | 138.59 | 0.1950 | 0.197 | 0.410 | 1 |
| 122 | HCl | HYDROGEN CHLORIDE | 36.461 | 158.97 | 188.15 | 324.65 | 83.09 | 81.02 | 0.4500 | 0.249 | 0.132 | 1 |
| 123 | HF | HYDROGEN FLUORIDE | 20.006 | 189.79 | 292.67 | 461.15 | 64.85 | 69.00 | 0.2899 | 0.117 | 0.383 | 1 |
| 124 | HI | HYDROGEN IODIDE | 127.912 | 222.38 | 237.55 | 423.85 | 83.10 | 121.94 | 1.0490 | 0.288 | 0.038 | 1 |
| 125 | HNO3 | NITRIC ACID | 63.013 | 231.55 | 356.15 | 520.00 | 68.90 | 145.00 | 0.4346 | 0.231 | 0.714 | 1 |
| 126 | H2 | HYDROGEN | 2.016 | 13.95 | 20.39 | 33.18 | 13.13 | 64.15 | 0.0314 | 0.305 | -0.215 | 1 |
| 127 | H2O | WATER | 18.015 | 273.15 | 373.15 | 647.13 | 220.55 | 55.95 | 0.3220 | 0.229 | 0.345 | 1 |

* A computer program, containing data for all compounds, is available for a nominal fee (Carl L. Yaws, Box 10053, Lamar University, Beaumont, TX 77710, phone/FAX 409-880-8787). The computer program is in ASCII which can be accessed by other software.

| NO | FORMULA | NAME | MW g/mol | $T_F$ K | $T_B$ K | $T_C$ K | $P_C$ bar | $V_C$ cm$^3$/mol | $\rho_C$ g/cm$^3$ | $Z_C$ | $\omega$ | SOURCE |
|----|---------|------|----------|---------|---------|---------|-----------|------------------|-------------------|-------|----------|--------|
| 128 | H2O2 | HYDROGEN PEROXIDE | 34.015 | 272.72 | 423.35 | 730.15 | 216.84 | 77.70 | 0.4378 | 0.278 | 0.360 | 1 |
| 129 | H2S | HYDROGEN SULFIDE | 34.082 | 187.68 | 212.80 | 373.53 | 89.63 | 98.49 | 0.3460 | 0.284 | 0.083 | 1 |
| 130 | H2SO4 | SULFURIC ACID | 98.079 | 283.46 | 610.00 | 925.00 | 64.00 | 177.03 | 0.5540 | 0.147 | --- | 1 |
| 131 | H2S2 | HYDROGEN DISULFIDE | 66.148 | 183.45 | 337.15 | --- | --- | --- | --- | --- | --- | 2 |
| 132 | H2Se | HYDROGEN SELENIDE | 80.976 | 209.15 | 232.05 | 411.10 | 90.30 | --- | --- | --- | --- | 2,6 |
| 133 | H2Te | HYDROGEN TELLURIDE | 129.616 | 224.15 | 271.15 | --- | --- | --- | --- | --- | --- | 2 |
| 134 | H3NO3S | SULFAMIC ACID | 97.095 | 478.00 | --- | --- | --- | 225.00 | 0.4315 | --- | --- | 1 |
| 135 | He | HELIUM-3 | 3.016 | 1.01 | 3.20 | 3.31 | 1.17 | 72.50 | 0.0416 | 0.308 | -0.472 | 1 |
| 136 | He | HELIUM-4 | 4.003 | 1.76 | 4.22 | 5.20 | 2.28 | 57.30 | 0.0699 | 0.302 | -0.390 | 1 |
| 137 | Hf | HAFNIUM | 178.490 | 2506.15 | 5960.00 | --- | --- | --- | --- | --- | --- | 4,5 |
| 138 | Hg | MERCURY | 200.590 | 234.29 | 629.73 | 1735.00 | 1608.0 | 56.35 | 3.5597 | 0.628 | -0.164 | 1 |
| 139 | HgBr2 | MERCURIC BROMIDE | 360.398 | 510.15 | 592.15 | --- | --- | --- | --- | --- | --- | 3 |
| 140 | HgCl2 | MERCURIC CHLORIDE | 271.495 | 550.15 | 577.15 | --- | --- | --- | --- | --- | --- | 3 |
| 141 | HgI2 | MERCURIC IODIDE | 454.399 | 532.15 | 627.15 | 1078.10 | 100.00 | --- | --- | --- | --- | 3,6 |
| 142 | IF7 | IODINE HEPTAFLUORIDE | 259.893 | 278.65 | 277.15 | --- | --- | --- | --- | --- | --- | 1 |
| 143 | I2 | IODINE | 253.809 | 386.75 | 458.39 | 819.15 | 116.54 | 155.00 | 1.6375 | 0.265 | 0.117 | 1 |
| 144 | In | INDIUM | 114.818 | 429.75 | 2323.00 | 6730.00 | 2432.0 | 82.60 | 1.3900 | 0.359 | --- | 4,5,6 |
| 145 | Ir | IRIDIUM | 192.220 | 2719.15 | 4450.00 | --- | --- | --- | --- | --- | --- | 4,5 |
| 146 | K | POTASSIUM | 39.098 | 336.35 | 1037.00 | 2223.00 | 162.12 | 209.00 | 0.1871 | 0.183 | -0.183 | 1 |
| 147 | KBr | POTASSIUM BROMIDE | 119.002 | 1003.15 | 1656.15 | --- | --- | --- | --- | --- | --- | 2 |
| 148 | KCl | POTASSIUM CHLORIDE | 74.551 | 1044.00 | 1688.87 | 3470.00 | 180.00 | 625.00 | 0.1193 | 0.390 | -0.121 | 1 |
| 149 | KF | POTASSIUM FLUORIDE | 58.097 | 1153.15 | 1775.15 | --- | --- | --- | --- | --- | --- | 2 |
| 150 | KI | POTASSIUM IODIDE | 166.003 | 996.15 | 1597.15 | --- | --- | --- | --- | --- | --- | 2 |
| 151 | KOH | POTASSIUM HYDROXIDE | 56.106 | 679.00 | 1600.00 | --- | --- | --- | --- | --- | --- | 1 |
| 152 | Kr | KRYPTON | 83.800 | 115.78 | 119.80 | 209.35 | 55.02 | 91.20 | 0.9189 | 0.288 | 0.000 | 1 |
| 153 | La | LANTHANUM | 138.906 | 1193.15 | 3643.00 | 9511.00 | 5460.0 | 36.50 | 3.8056 | 0.252 | --- | 4,5,6 |
| 154 | Li | LITHIUM | 6.941 | 453.69 | 1597.00 | 4085.00 | 1722.5 | 47.00 | 0.1477 | 0.238 | -0.045 | 1 |
| 155 | LiBr | LITHIUM BROMIDE | 86.845 | 820.15 | 1583.15 | --- | --- | --- | --- | --- | --- | 2 |
| 156 | LiCl | LITHIUM CHLORIDE | 42.394 | 887.15 | 1655.15 | --- | --- | --- | --- | --- | --- | 3 |
| 157 | LiF | LITHIUM FLUORIDE | 25.939 | 1143.15 | 1954.15 | --- | --- | --- | --- | --- | --- | 2 |
| 158 | LiI | LITHIUM IODIDE | 133.845 | 719.15 | 1444.15 | --- | --- | --- | --- | --- | --- | 2 |
| 159 | Lu | LUTECIUM | 174.967 | 1936.15 | 2535.00 | --- | --- | --- | --- | --- | --- | 4,5 |
| 160 | Mg | MAGNESIUM | 24.305 | 923.15 | 1376.00 | --- | --- | --- | --- | --- | --- | 4,5 |
| 161 | MgCl2 | MAGNESIUM CHLORIDE | 95.210 | 985.15 | 1691.15 | --- | --- | --- | --- | --- | --- | 3 |
| 162 | MgO | MAGNESIUM OXIDE | 40.304 | 3105.00 | 3873.20 | 5950.00 | 33.91 | 209.50 | 0.1924 | 0.014 | 0.214 | 1 |
| 163 | Mn | MANGANESE | 54.938 | 1519.15 | 2392.00 | --- | --- | --- | --- | --- | --- | 4,5 |
| 164 | MnCl2 | MANGANESE CHLORIDE | 125.843 | 923.15 | 1463.15 | --- | --- | --- | --- | --- | --- | 3 |
| 165 | Mo | MOLYBDENUM | 95.940 | 2895.15 | 5081.15 | 9620.00 | --- | 38.30 | 2.5050 | --- | --- | 3,6 |
| 166 | MoF6 | MOLYBDENUM FLUORIDE | 209.930 | 290.15 | 309.15 | --- | --- | --- | --- | --- | --- | 2 |
| 167 | MoO3 | MOLYBDENUM OXIDE | 143.938 | 1068.15 | 1424.15 | --- | --- | --- | --- | --- | --- | 2 |
| 168 | NCl3 | NITROGEN TRICHLORIDE | 120.365 | 246.15 | 344.15 | 564.00 | 104.00 | --- | --- | --- | --- | 1 |
| 169 | ND3 | HEAVY AMMONIA | 20.055 | 199.15 | 239.75 | --- | --- | --- | --- | --- | --- | 2 |
| 170 | NF3 | NITROGEN TRIFLUORIDE | 71.002 | 66.36 | 144.09 | 233.85 | 45.30 | 118.75 | 0.5979 | 0.277 | 0.126 | 1 |
| 171 | NH3 | AMMONIA | 17.031 | 195.41 | 239.72 | 405.65 | 112.78 | 72.47 | 0.2350 | 0.242 | 0.252 | 1 |
| 172 | NH3O | HYDROXYLAMINE | 33.030 | 306.25 | 383.00 | 574.00 | 137.00 | --- | --- | --- | 0.694 | 1 |
| 173 | NH4Br | AMMONIUM BROMIDE | 97.943 | --- | 669.15 | --- | --- | --- | --- | --- | --- | 2 |
| 174 | NH4Cl | AMMONIUM CHLORIDE | 53.491 | 793.20 | 612.00 | 882.00 | 16.40 | --- | --- | --- | --- | 1 |
| 175 | NH4I | AMMONIUM IODIDE | 144.943 | --- | 678.05 | --- | --- | --- | --- | --- | --- | 2 |
| 176 | NH5O | AMMONIUM HYDROXIDE | 35.046 | 194.15 | --- | --- | --- | --- | --- | --- | --- | 1 |
| 177 | NH5S | AMMONIUM HYDROGENSULFIDE | 51.112 | --- | 306.45 | --- | --- | --- | --- | --- | --- | 2 |
| 178 | NO | NITRIC OXIDE | 30.006 | 112.15 | 121.38 | 180.15 | 64.85 | 57.70 | 0.5200 | 0.250 | 0.585 | 1 |
| 179 | NOCl | NITROSYL CHLORIDE | 65.459 | 213.55 | 267.77 | 440.65 | 91.19 | 139.30 | 0.4699 | 0.347 | 0.307 | 1 |
| 180 | NOF | NITROSYL FLUORIDE | 49.005 | 139.15 | 217.15 | --- | --- | --- | --- | --- | --- | 3 |
| 181 | NO2 | NITROGEN DIOXIDE | 46.006 | 261.95 | 294.00 | 431.35 | 101.33 | 82.49 | 0.5577 | 0.233 | 0.849 | 1 |
| 182 | N2 | NITROGEN | 28.013 | 63.15 | 77.35 | 126.10 | 33.94 | 90.10 | 0.3109 | 0.292 | 0.040 | 1 |
| 183 | N2F4 | TETRAFLUOROHYDRAZINE | 104.007 | 111.65 | 198.95 | 309.35 | 37.10 | 213.00 | 0.4883 | 0.307 | 0.223 | 1 |
| 184 | N2H4 | HYDRAZINE | 32.045 | 274.69 | 386.65 | 653.15 | 146.92 | 158.00 | 0.2028 | 0.427 | 0.314 | 1 |
| 185 | N2H4C | AMMONIUM CYANIDE | 44.056 | 309.15 | 304.85 | --- | --- | --- | --- | --- | --- | 2 |
| 186 | N2H6CO2 | AMMONIUM CARBAMATE | 78.071 | --- | 331.45 | --- | --- | --- | --- | --- | --- | 2 |
| 187 | N2O | NITROUS OXIDE | 44.013 | 182.33 | 184.67 | 309.57 | 72.45 | 97.37 | 0.4520 | 0.274 | 0.142 | 1 |
| 188 | N2O3 | NITROGEN TRIOXIDE | 76.012 | 170.00 | 275.15 | 425.00 | 69.90 | 195.00 | 0.3898 | 0.386 | 0.431 | 1 |
| 189 | N2O4 | NITROGEN TETRAOXIDE | 92.011 | 261.90 | 302.22 | 431.15 | 101.33 | 82.49 | 1.1154 | 0.233 | 1.007 | 1 |
| 190 | N2O5 | NITROGEN PENTOXIDE | 108.010 | --- | --- | --- | --- | --- | --- | --- | --- | 1 |
| 191 | Na | SODIUM | 22.990 | 370.98 | 1156.00 | 2573.00 | 354.64 | 116.00 | 0.1982 | 0.192 | -0.102 | 1 |
| 192 | NaBr | SODIUM BROMIDE | 102.894 | 1020.00 | 1663.82 | 4287.00 | 192.52 | 398.00 | 0.2585 | 0.215 | -0.800 | 1 |
| 193 | NaCN | SODIUM CYANIDE | 49.008 | 836.85 | 1769.15 | 2900.00 | --- | --- | --- | --- | --- | 1 |
| 194 | NaCl | SODIUM CHLORIDE | 58.442 | 1073.95 | 1738.15 | 3400.00 | 355.00 | 266.00 | 0.2197 | 0.334 | 0.134 | 1 |
| 195 | NaF | SODIUM FLUORIDE | 41.988 | 1269.00 | 1982.72 | 5530.00 | 531.96 | 185.00 | 0.2270 | 0.214 | -1.115 | 1 |
| 196 | NaI | SODIUM IODIDE | 149.894 | 924.15 | 1577.15 | --- | --- | --- | --- | --- | --- | 2 |

* A computer program, containing data for all compounds, is available for a nominal fee (Carl L. Yaws, Box 10053, Lamar University, Beaumont, TX 77710, phone/FAX 409-880-8787). The computer program is in ASCII which can be accessed by other software.

| NO | FORMULA | NAME | MW g/mol | $T_F$ K | $T_B$ K | $T_C$ K | $P_C$ bar | $V_C$ cm³/mol | $\rho_C$ g/cm³ | $Z_C$ | $\omega$ | SOURCE |
|----|---------|------|----------|---------|---------|---------|-----------|---------------|----------------|--------|----------|--------|
| 197 | NaOH | SODIUM HYDROXIDE | 39.997 | 596.00 | 1663.15 | 2820.00 | 253.31 | 200.00 | 0.2000 | 0.216 | --- | 1 |
| 198 | Na2SO4 | SODIUM SULFATE | 142.043 | 1157.00 | --- | --- | --- | --- | --- | --- | --- | 1 |
| 199 | Nb | NIOBIUM | 92.906 | 2750.15 | 5115.00 | --- | --- | --- | --- | --- | --- | 4,5 |
| 200 | Nd | NEODYMIUM | 144.240 | 1289.15 | 3384.00 | --- | --- | --- | --- | --- | --- | 4,5 |
| 201 | Ne | NEON | 20.180 | 24.55 | 27.09 | 44.40 | 26.53 | 41.70 | 0.4839 | 0.300 | -0.041 | 1 |
| 202 | Ni | NICKEL | 58.693 | 1728.15 | 2415.00 | --- | --- | --- | --- | --- | --- | 4,5 |
| 203 | NiC4O4 | NICKEL CARBONYL | 170.735 | 248.15 | 315.65 | --- | --- | --- | --- | --- | --- | 2 |
| 204 | NiF2 | NICKEL FLUORIDE | 96.690 | --- | --- | --- | --- | --- | --- | --- | --- | 2 |
| 205 | Np | NEPTUNIUM | 237.000 | 913.15 | --- | --- | --- | --- | --- | --- | --- | 2,10 |
| 206 | O2 | OXYGEN | 31.999 | 54.36 | 90.17 | 154.58 | 50.43 | 73.40 | 0.4360 | 0.288 | 0.022 | 1 |
| 207 | O3 | OZONE | 47.998 | 80.15 | 161.85 | 261.00 | 55.73 | 89.00 | 0.5393 | 0.229 | 0.227 | 1 |
| 208 | Os | OSMIUM | 190.230 | 3306.15 | 4880.00 | --- | --- | --- | --- | --- | --- | 4,5 |
| 209 | OsOF5 | OSMIUM OXIDE PENTAFLUORIDE | 301.221 | --- | --- | --- | --- | --- | --- | --- | --- | 2 |
| 210 | OsO4 | OSMIUM TETROXIDE - YELLOW | 254.228 | 329.15 | 403.15 | --- | --- | --- | --- | --- | --- | 3 |
| 211 | OsO4 | OSMIUM TETROXIDE - WHITE | 254.228 | 315.15 | 403.15 | --- | --- | --- | --- | --- | --- | 3 |
| 212 | P | PHOSPHORUS - WHITE | 30.974 | 870.00 | 553.45 | 993.75 | 83.29 | --- | --- | --- | --- | 1 |
| 213 | PBr3 | PHOSPHORUS TRIBROMIDE | 270.686 | 233.15 | 448.45 | --- | --- | --- | --- | --- | --- | 2 |
| 214 | PCl2F3 | PHOSPHORUS DICHLORIDE TRIFLUORIDE | 158.874 | --- | --- | --- | --- | --- | --- | --- | --- | 2 |
| 215 | PCl3 | PHOSPHORUS TRICHLORIDE | 137.332 | 181.15 | 349.25 | 563.15 | 56.70 | 260.00 | 0.5282 | 0.315 | 0.234 | 1 |
| 216 | PCl5 | PHOSPHORUS PENTACHLORIDE | 208.237 | 433.15 | 433.00 | 646.15 | --- | --- | --- | --- | --- | 1 |
| 217 | PH3 | PHOSPHINE | 33.998 | 139.37 | 185.41 | 324.75 | 65.36 | 113.32 | 0.3000 | 0.274 | 0.036 | 1 |
| 218 | PH4Br | PHOSPHONIUM BROMIDE | 114.910 | --- | 311.45 | --- | --- | --- | --- | --- | --- | 2 |
| 219 | PH4Cl | PHOSPHONIUM CHLORIDE | 70.458 | 244.65 | 246.15 | --- | --- | --- | --- | --- | --- | 2 |
| 220 | PH4I | PHOSPHONIUM IODIDE | 161.910 | 291.65 | 335.45 | --- | --- | --- | --- | --- | --- | 2,10 |
| 221 | POCl3 | PHOSPHORUS OXYCHLORIDE | 153.331 | 274.33 | 378.65 | 602.15 | --- | --- | --- | --- | --- | 1 |
| 222 | PSBr3 | PHOSPHORUS THIOBROMIDE | 302.752 | 311.15 | 448.15 | --- | --- | --- | --- | --- | --- | 2 |
| 223 | PSCl3 | PHOSPHORUS THIOCHLORIDE | 169.398 | 236.95 | 398.15 | --- | --- | --- | --- | --- | --- | 1 |
| 224 | P4O6 | PHOSPHORUS TRIOXIDE | 219.891 | 295.65 | 446.25 | --- | --- | --- | --- | --- | --- | 2 |
| 225 | P4O10 | PHOSPHORUS PENTOXIDE | 283.889 | 693.15 | --- | --- | --- | --- | --- | --- | --- | 1 |
| 226 | P4S10 | PHOSPHORUS PENTASULFIDE | 444.555 | 561.15 | 787.15 | 1291.00 | 232.00 | --- | --- | --- | 0.594 | 1 |
| 227 | Pb | LEAD | 207.200 | 600.61 | 2024.00 | 5400.00 | 861.30 | 93.20 | 2.2232 | 0.179 | --- | 4,5,6 |
| 228 | PbBr2 | LEAD BROMIDE | 367.008 | 646.15 | 1187.15 | --- | --- | --- | --- | --- | --- | 3 |
| 229 | PbCl2 | LEAD CHLORIDE | 278.105 | 774.15 | 1227.15 | --- | --- | --- | --- | --- | --- | 2 |
| 230 | PbF2 | LEAD FLUORIDE | 245.197 | 1128.15 | 1566.15 | --- | --- | --- | --- | --- | --- | 3 |
| 231 | PbI2 | LEAD IODIDE | 461.009 | 675.15 | 1145.15 | --- | --- | --- | --- | --- | --- | 3 |
| 232 | PbO | LEAD OXIDE | 223.199 | 1163.15 | 1745.15 | --- | --- | --- | --- | --- | --- | 3 |
| 233 | PbS | LEAD SULFIDE | 239.266 | 1387.15 | 1554.15 | --- | --- | --- | --- | --- | --- | 3 |
| 234 | Pd | PALLADIUM | 106.420 | 1828.05 | 3385.00 | --- | --- | --- | --- | --- | --- | 4,5 |
| 235 | Po | POLONIUM | 209.000 | 527.15 | 1235.00 | --- | --- | --- | --- | --- | --- | 4,5 |
| 236 | Pt | PLATINUM | 195.080 | 2041.55 | 3980.00 | 6983.00 | --- | 759.10 | 0.2570 | --- | --- | 4,5,6 |
| 237 | Ra | RADIUM | 226.000 | 973.15 | 1809.00 | --- | --- | --- | --- | --- | --- | 4,5 |
| 238 | Rb | RUBIDIUM | 85.468 | 312.46 | 978.00 | 2111.10 | 134.00 | 247.00 | 0.3460 | 0.189 | --- | 4,5,6 |
| 239 | RbBr | RUBIDIUM BROMIDE | 165.372 | 955.15 | 1625.15 | --- | --- | --- | --- | --- | --- | 2 |
| 240 | RbCl | RUBIDIUM CHLORIDE | 120.921 | 988.15 | 1654.15 | --- | --- | --- | --- | --- | --- | 2 |
| 241 | RbF | RUBIDIUM FLUORIDE | 104.466 | 1033.15 | 1681.15 | --- | --- | --- | --- | --- | --- | 2 |
| 242 | RbI | RUBIDIUM IODIDE | 212.372 | 915.15 | 1577.15 | --- | --- | --- | --- | --- | --- | 2 |
| 243 | Re | RHENIUM | 186.207 | 3459.15 | 5915.00 | --- | --- | 32.10 | 5.8008 | --- | --- | 4,5,6 |
| 244 | Re2O7 | RHENIUM HEPTOXIDE | 484.410 | 569.15 | 635.55 | --- | --- | --- | --- | --- | --- | 3 |
| 245 | Rh | RHODIUM | 102.906 | 2237.15 | 3940.00 | --- | --- | --- | --- | --- | --- | 4,5 |
| 246 | Rn | RADON | 222.000 | 202.15 | 211.35 | 377.40 | 63.00 | 140.00 | 1.5857 | 0.281 | --- | 2,6 |
| 247 | Ru | RUTHENIUM | 101.070 | 2607.15 | 4500.00 | --- | --- | --- | --- | --- | --- | 4,5 |
| 248 | RuF5 | RUTHENIUM PENTAFLUORIDE | 196.062 | --- | --- | --- | --- | --- | --- | --- | --- | 2 |
| 249 | S | SULFUR | 32.066 | 388.36 | 717.82 | 1313.00 | 182.08 | 158.00 | 0.2029 | 0.264 | 0.262 | 1 |
| 250 | SF4 | SULFUR TETRAFLUORIDE | 108.060 | 149.15 | 233.15 | --- | --- | --- | --- | --- | --- | 2,10 |
| 251 | SF6 | SULFUR HEXAFLUORIDE | 146.056 | 222.45 | 209.25 | 318.69 | 37.60 | 198.52 | 0.7357 | 0.282 | 0.215 | 1 |
| 252 | SOBr2 | THIONYL BROMIDE | 207.873 | 220.95 | 412.65 | --- | --- | --- | --- | --- | --- | 2 |
| 253 | SOCl2 | THIONYL CHLORIDE | 118.971 | 172.00 | 348.75 | 567.00 | --- | 203.00 | 0.5861 | --- | --- | 1 |
| 254 | SOF2 | SULFUROUS OXYFLUORIDE | 86.062 | --- | 228.90 | --- | --- | --- | --- | --- | --- | 1 |
| 255 | SO2 | SULFUR DIOXIDE | 64.065 | 200.00 | 263.13 | 430.75 | 78.84 | 122.00 | 0.5251 | 0.269 | 0.245 | 1 |
| 256 | SO2Cl2 | SULFURYL CHLORIDE | 134.970 | 222.00 | 342.55 | 545.00 | 46.10 | 224.00 | 0.6025 | 0.228 | 0.176 | 1 |
| 257 | SO3 | SULFUR TRIOXIDE | 80.064 | 289.95 | 317.90 | 490.85 | 82.07 | 127.08 | 0.6300 | 0.256 | 0.422 | 1 |
| 258 | S2Cl2 | SULFUR MONOCHLORIDE | 135.037 | 193.15 | 411.15 | --- | --- | --- | --- | --- | --- | 2 |
| 259 | Sb | ANTIMONY | 121.757 | 903.78 | 1898.00 | 5070.00 | --- | --- | --- | --- | --- | 4,5,6 |
| 260 | SbBr3 | ANTIMONY TRIBROMIDE | 361.469 | 369.75 | 548.15 | --- | --- | --- | --- | --- | --- | 1 |
| 261 | SbCl3 | ANTIMONY TRICHLORIDE | 228.115 | 346.55 | 493.40 | 794.00 | 48.20 | 270.00 | 0.8449 | 0.197 | 0.171 | 1 |
| 262 | SbCl5 | ANTIMONY PENTACHLORIDE | 299.021 | 275.95 | 413.15 | --- | --- | --- | --- | --- | --- | 2,11 |
| 263 | SbH3 | STIBINE | 124.781 | 185.15 | 255.15 | 440.35 | 73.06 | 157.2 | 0.7938 | 0.314 | --- | 7 |
| 264 | SbI3 | ANTIMONY TRIIODIDE | 502.470 | 440.15 | 674.15 | --- | --- | --- | --- | --- | --- | 3 |
| 265 | Sb2O3 | ANTIMONY TRIOXIDE | 291.512 | 929.15 | 1698.15 | --- | --- | --- | --- | --- | --- | 3 |

* A computer program, containing data for all compounds, is available for a nominal fee (Carl L. Yaws, Box 10053, Lamar University, Beaumont, TX 77710, phone/FAX 409-880-8787). The computer program is in ASCII which can be accessed by other software.

| NO | FORMULA | NAME | MW g/mol | $T_F$ K | $T_B$ K | $T_C$ K | $P_C$ bar | $V_C$ cm$^3$/mol | $\rho_C$ g/cm$^3$ | $Z_C$ | $\omega$ | SOURCE |
|----|---------|------|----------|---------|---------|---------|-----------|------------------|-------------------|-------|----------|--------|
| 266 | Sc | SCANDIUM | 44.956 | 1814.15 | 2700.00 | --- | --- | --- | --- | --- | --- | 4,5 |
| 267 | Se | SELENIUM | 78.960 | 494.15 | 930.00 | 1766.00 | 380.00 | 62.30 | 1.2674 | 0.161 | --- | 4,5,6 |
| 268 | SeCl4 | SELENIUM TETRACHLORIDE | 220.771 | --- | 464.65 | --- | --- | --- | --- | --- | --- | 2 |
| 269 | SeF6 | SELENIUM HEXAFLUORIDE | 192.950 | 238.45 | 227.35 | --- | --- | --- | --- | --- | --- | 2 |
| 270 | SeOCl2 | SELENIUM OXYCHLORIDE | 165.865 | 281.65 | 441.15 | --- | --- | --- | --- | --- | --- | 2 |
| 271 | SeO2 | SELENIUM DIOXIDE | 110.959 | 613.15 | 590.15 | --- | --- | --- | --- | --- | --- | 2 |
| 272 | Si | SILICON | 28.086 | 1685.00 | 3513.80 | 5159.00 | 537.00 | 233.00 | 0.1205 | 0.292 | --- | 1 |
| 273 | SiBrCl2F | BROMODICHLOROFLUOROSILANE | 197.893 | 160.85 | 308.55 | --- | --- | --- | --- | --- | --- | 3 |
| 274 | SiBrF3 | TRIFLUOROBROMOSILANE | 164.985 | 202.65 | 231.45 | --- | --- | --- | --- | --- | --- | 2 |
| 275 | SiBr2ClF | DIBROMOCHLOROFLUOROSILANE | 242.345 | 173.85 | 332.65 | --- | --- | --- | --- | --- | --- | 2 |
| 276 | SiClF3 | TRIFLUOROCHLOROSILANE | 120.533 | 131.15 | 203.15 | --- | --- | --- | --- | --- | --- | 2 |
| 277 | SiCl2F2 | DICHLORODIFLUOROSILANE | 136.988 | 133.45 | 241.35 | --- | --- | --- | --- | --- | --- | 2 |
| 278 | SiCl3F | TRICHLOROFLUOROSILANE | 153.442 | 152.35 | 285.35 | --- | --- | --- | --- | --- | --- | 2 |
| 279 | SiCl4 | SILICON TETRACHLORIDE | 169.896 | 204.30 | 330.00 | 507.00 | 35.93 | 326.00 | 0.5212 | 0.278 | 0.232 | 1 |
| 280 | SiF4 | SILICON TETRAFLUORIDE | 104.079 | 186.35 | 178.35 | 259.00 | 37.19 | 165.00 | 0.6308 | 0.285 | 0.385 | 1 |
| 281 | SiHBr3 | TRIBROMOSILANE | 268.805 | 199.65 | 384.95 | --- | --- | --- | --- | --- | --- | 2 |
| 282 | SiHCl3 | TRICHLOROSILANE | 135.452 | 144.95 | 305.00 | 479.00 | 41.70 | 268.00 | 0.5054 | 0.281 | 0.203 | 1 |
| 283 | SiHF3 | TRIFLUOROSILANE | 86.089 | 141.75 | 178.15 | --- | --- | --- | --- | --- | --- | 2 |
| 284 | SiH2Br2 | DIBROMOSILANE | 189.909 | 202.95 | 343.65 | --- | --- | --- | --- | --- | --- | 2 |
| 285 | SiH2Cl2 | DICHLOROSILANE | 101.007 | 151.15 | 281.45 | 449.00 | 44.30 | 228.00 | 0.4430 | 0.271 | 0.177 | 1 |
| 286 | SiH2F2 | DIFLUOROSILANE | 68.098 | --- | 195.35 | --- | --- | --- | --- | --- | --- | 2 |
| 287 | SiH2I2 | DIIODOSILANE | 283.910 | 272.15 | 422.65 | --- | --- | --- | --- | --- | --- | 2 |
| 288 | SiH3Br | MONOBROMOSILANE | 111.013 | 179.25 | 275.55 | --- | --- | --- | --- | --- | --- | 2 |
| 289 | SiH3Cl | MONOCHLOROSILANE | 66.562 | 155.05 | 242.75 | 396.65 | 48.43 | 174.00 | 0.3822 | 0.256 | 0.136 | 3,9 |
| 290 | SiH3F | MONOFLUOROSILANE | 50.108 | --- | 175.15 | --- | --- | --- | --- | --- | --- | 2 |
| 291 | SiH3I | IODOSILANE | 158.014 | 216.15 | 318.55 | --- | --- | --- | --- | --- | --- | 2 |
| 292 | SiH4 | SILANE | 32.117 | 88.15 | 161.00 | 269.70 | 48.43 | 132.70 | 0.2420 | 0.287 | 0.097 | 1 |
| 293 | SiO2 | SILICON DIOXIDE | 60.084 | 1883.00 | 2503.20 | --- | --- | --- | --- | --- | --- | 1 |
| 294 | Si2Cl6 | HEXACHLORODISILANE | 268.887 | 271.95 | 412.15 | --- | --- | --- | --- | --- | --- | 2 |
| 295 | Si2F6 | HEXAFLUORODISILANE | 170.161 | 254.55 | 254.25 | --- | --- | --- | --- | --- | --- | 2 |
| 296 | Si2H5Cl | DISILANYL CHLORIDE | 96.663 | --- | 314.70 | --- | --- | --- | --- | --- | --- | 2 |
| 297 | Si2H6 | DISILANE | 62.219 | 140.65 | 259.00 | 432.00 | 51.30 | 198.00 | 0.3142 | 0.283 | 0.102 | 1 |
| 298 | Si2OCl3F3 | TRICHLOROTRIFLUORODISILOXANE | 235.524 | --- | 315.89 | --- | --- | --- | --- | --- | --- | 2 |
| 299 | Si2OCl6 | HEXACHLORODISILOXANE | 284.887 | 239.95 | 408.75 | --- | --- | --- | --- | --- | --- | 2 |
| 300 | Si2OH6 | DISILOXANE | 78.218 | 128.95 | 257.75 | --- | --- | --- | --- | --- | --- | 2 |
| 301 | Si3Cl8 | OCTACHLOROTRISILANE | 367.878 | --- | 484.55 | --- | --- | --- | --- | --- | --- | 2 |
| 302 | Si3H8 | TRISILANE | 92.320 | 155.95 | 326.25 | --- | --- | --- | --- | --- | --- | 2 |
| 303 | Si3H9N | TRISILAZANE | 107.335 | 167.45 | 321.85 | --- | --- | --- | --- | --- | --- | 2 |
| 304 | Si4H10 | TETRASILANE | 122.421 | 179.55 | 373.15 | --- | --- | --- | --- | --- | --- | 2 |
| 305 | Sm | SAMARIUM | 150.360 | 1345.15 | 1874.00 | --- | --- | --- | --- | --- | --- | 4,5 |
| 306 | Sn | TIN | 118.710 | 505.08 | 2995.00 | 7400.00 | --- | 115.10 | 1.0314 | --- | --- | 4,5,6 |
| 307 | SnBr4 | STANNIC BROMIDE | 438.326 | 304.15 | 477.85 | --- | --- | --- | --- | --- | --- | 2 |
| 308 | SnCl2 | STANNOUS CHLORIDE | 189.615 | 519.95 | 896.15 | --- | --- | --- | --- | --- | --- | 3 |
| 309 | SnCl4 | STANNIC CHLORIDE | 260.521 | 242.95 | 386.15 | --- | --- | --- | --- | --- | --- | 2 |
| 310 | SnH4 | STANNIC HYDRIDE | 122.742 | 123.25 | 220.85 | --- | --- | --- | --- | --- | --- | 2 |
| 311 | SnI4 | STANNIC IODIDE | 626.328 | 417.65 | 621.15 | --- | --- | --- | --- | --- | --- | 2 |
| 312 | Sr | STRONTIUM | 87.620 | 1050.15 | 1630.00 | --- | --- | --- | --- | --- | --- | 4,5 |
| 313 | SrO | STRONTIUM OXIDE | 103.619 | 2703.15 | --- | --- | --- | --- | --- | --- | --- | 2 |
| 314 | Ta | TANTALUM | 180.948 | 3290.15 | 5565.00 | --- | --- | --- | --- | --- | --- | 4,5 |
| 315 | Tc | TECNNETIUM | 98.000 | 2430.15 | 5000.00 | --- | --- | --- | --- | --- | --- | 4,5 |
| 316 | Te | TELLURIUM | 127.600 | 722.66 | 1285.00 | 4840.00 | --- | --- | --- | --- | --- | 4,5,6 |
| 317 | TeCl4 | TELLURIUM TETRACHLORIDE | 269.411 | 497.15 | 665.15 | --- | --- | --- | --- | --- | --- | 2 |
| 318 | TeF6 | TELLURIUM HEXAFLUORIDE | 241.590 | 235.35 | 234.55 | --- | --- | --- | --- | --- | --- | 2 |
| 319 | Ti | TITANIUM | 47.880 | 1941.15 | 3442.00 | 6400.00 | --- | --- | --- | --- | --- | 4,5,6 |
| 320 | TiCl4 | TITANIUM TETRACHLORIDE | 189.691 | 249.05 | 409.00 | 638.00 | 46.61 | 340.00 | 0.5579 | 0.299 | 0.284 | 1 |
| 321 | Tl | THALLIUM | 204.383 | 577.15 | 1745.00 | --- | --- | --- | --- | --- | --- | 4,5 |
| 322 | TlBr | THALLOUS BROMIDE | 284.287 | 733.15 | 1092.15 | --- | --- | --- | --- | --- | --- | 2 |
| 323 | TlI | THALLOUS IODIDE | 331.288 | 713.15 | 1096.15 | --- | --- | --- | --- | --- | --- | 2 |
| 324 | Tm | THULIUM | 168.934 | 1818.15 | 2219.15 | --- | --- | --- | --- | --- | --- | 4,5 |
| 325 | U | URANIUM | 238.029 | 1408.15 | 4135.00 | --- | --- | --- | --- | --- | --- | 4,5 |
| 326 | UF6 | URANIUM FLUORIDE | 352.019 | 342.15 | 328.85 | --- | --- | --- | --- | --- | --- | 2 |
| 327 | V | VANADIUM | 50.942 | 2183.15 | 3665.00 | --- | --- | --- | --- | --- | --- | 4,5 |
| 328 | VCl4 | VANADIUM TETRACHLORIDE | 192.752 | 247.45 | 425.00 | 697.00 | 60.30 | 268.00 | 0.7192 | 0.279 | 0.186 | 1 |
| 329 | VOCl3 | VANADIUM OXYTRICHLORIDE | 173.299 | 193.65 | 400.00 | 636.00 | --- | 290.00 | 0.5976 | --- | --- | 1 |
| 330 | W | TUNGSTEN | 183.840 | 3695.15 | 5645.00 | 14756.0 | --- | 33.90 | 5.4230 | --- | --- | 4,5,6 |
| 331 | WF6 | TUNGSTEN FLUORIDE | 297.830 | 272.65 | 290.45 | --- | --- | --- | --- | --- | --- | 2 |
| 332 | Xe | XENON | 131.290 | 161.36 | 165.03 | 289.74 | 58.40 | 118.00 | 1.1126 | 0.286 | 0.000 | 1 |
| 333 | Yb | YTTERBIUM | 173.040 | 1097.15 | 1660.00 | --- | --- | --- | --- | --- | --- | 4,5 |
| 334 | Yt | YTTRIUM | 88.906 | 1799.15 | 3055.00 | --- | --- | --- | --- | --- | --- | 4,5 |

* A computer program, containing data for all compounds, is available for a nominal fee (Carl L. Yaws, Box 10053, Lamar University, Beaumont, TX 77710, phone/FAX 409-880-8787). The computer program is in ASCII which can be accessed by other software.

| NO | FORMULA | NAME | MW g/mol | $T_F$ K | $T_B$ K | $T_C$ K | $P_C$ bar | $V_C$ cm³/mol | $\rho_C$ g/cm³ | $Z_C$ | $\omega$ | SOURCE |
|----|---------|------|----------|---------|---------|---------|-----------|---------------|----------------|-------|----------|--------|
| 335 | Zn | ZINC | 65.390 | 692.70 | 1181.15 | 3170.00 | 2904.0 | 33.00 | 1.9815 | 0.364 | 0.078 | 1 |
| 336 | ZnCl2 | ZINC CHLORIDE | 136.295 | 638.15 | 1005.15 | --- | --- | --- | --- | --- | --- | 3 |
| 337 | ZnF2 | ZINC FLUORIDE | 103.387 | 1145.15 | 1770.15 | --- | --- | --- | --- | --- | --- | 3 |
| 338 | ZnO | ZINC OXIDE | 81.389 | 2248.20 | --- | --- | --- | --- | --- | --- | --- | 1 |
| 339 | ZnSO4 | ZINC SULFATE | 161.454 | 953.00 | --- | --- | --- | --- | --- | --- | --- | 1 |
| 340 | Zr | ZIRCONIUM | 91.224 | 2128.15 | 4598.00 | 8802.00 | --- | --- | --- | --- | --- | 4,5,6 |
| 341 | ZrBr4 | ZIRCONIUM BROMIDE | 410.840 | 723.15 | 630.15 | --- | --- | --- | --- | --- | --- | 2 |
| 342 | ZrCl4 | ZIRCONIUM CHLORIDE | 233.035 | 710.15 | 604.15 | --- | --- | --- | --- | --- | --- | 2 |
| 343 | ZrI4 | ZIRCONIUM IODIDE | 598.842 | 772.15 | 704.15 | --- | --- | --- | --- | --- | --- | 2 |

* A computer program, containing data for all compounds, is available for a nominal fee (Carl L. Yaws, Box 10053, Lamar University, Beaumont, TX 77710, phone/FAX 409-880-8787). The computer program is in ASCII which can be accessed by other software.

NOTE:

1. Sources for the property data are:

    1. Daubert, T. E. and R. P. Danner, DATA COMPILATION OF PROPERTIES OF PURE COMPOUNDS, Parts 1, 2, 3 and 4, Supplements 1 and 2, DIPPR Project, AIChE, New York, NY (1985-1992).
    2. Ohe, S., COMPUTER AIDED DATA BOOK OF VAPOR PRESSURE, Data Book Publishing Company, Tokyo, Japan (1976).
    3. PERRY'S CHEMICAL ENGINEERING HANDBOOK, 6th ed., McGraw-Hill, New York, NY (1984).
    4. Nesmeyanov, A. N., VAPOR PRESSURE OF THE CHEMICAL ELEMENTS, Elsevier, New York, NY (1963).
    5. CRC HANDBOOK OF CHEMISTRY AND PHYSICS, 66th - 75th eds., CRC Press, Inc., Boca Raton, FL (1985-1994).
    6. Simmrock, K. H., R. Janowsky and A. Ohnsorge, CRITICAL DATA OF PURE SUBSTANCES, Vol. II, Parts 1 and 2, Dechema Chemistry Data Series, 6000 Frankfurt/Main, Germany (1986).
    7. Yaws, C. L. and others, Solid State Technology, 17, No. 1, 47 (1974).
    8. Yaws, C. L. and others, Solid State Technology, 18, No. 1, 35 (1975).
    9. Yaws, C. L. and others, Solid State Technology, 16, No. 1, 39 (1973).
    10. CONDENSED CHEMICAL DICTIONARY, 10th (G. G. Hawley) and 11th eds. (N. I. Sax and R. J. Lewis, Jr.), Van Nostrand Reinhold Co., New York, NY (1981,1987).
    11. LANGE'S HANDBOOK OF CHEMISTRY, 13th and 14th eds., McGraw-Hill, New York, NY (1985, 1992).

2. Very limited experimental data for critical constants and acentric factor are available for inorganic compounds and elements which are solids at room temperature. Thus, the estimates for these substances should be considered rough approximations in the absence of experimental data.